Ancient West Mexicos

UNIVERSITY PRESS OF FLORIDA

Florida A&M University, Tallahassee
Florida Atlantic University, Boca Raton
Florida Gulf Coast University, Ft. Myers
Florida International University, Miami
Florida State University, Tallahassee
New College of Florida, Sarasota
University of Central Florida, Orlando
University of Florida, Gainesville
University of North Florida, Jacksonville
University of South Florida, Tampa
University of West Florida, Pensacola

ANCIENT WEST MEXICOS

TIME, SPACE, AND DIVERSITY

EDITED BY

Joshua D. Englehardt, Verenice Y. Heredia Espinoza, and Christopher S. Beekman

University Press of Florida

Gainesville · Tallahassee · Tampa · Boca Raton

Pensacola · Orlando · Miami · Jacksonville · Ft. Myers · Sarasota

Published in the United States of America.

25 24 23 22 21 20   6 5 4 3 2 1

Library of Congress Cataloging-in-Publication Data
Names: Englehardt, Joshua, editor. | Heredia Espinoza, Verenice Y., editor. | Beekman, Christopher, editor.
Title: Ancient West Mexicos : Time, Space, and Diversity / edited by Joshua D. Englehardt, Verenice Y. Heredia Espinoza, Christopher S. Beekman.
Description: Gainesville : University Press of Florida, [2020] | Includes bibliographical references and index.
Identifiers: LCCN 2019040157 (print) | LCCN 2019040158 (ebook) | ISBN 9780813066349 (hardback) | ISBN 9780813057453 (pdf)
Subjects: LCSH: Indians of Mexico—Antiquities. | Time perception—Mexico—History. | Space perception—Mexico—History. | Indian art—Mexico—History. | Mexico—Antiquities.
Classification: LCC F1219 .A5973 2019 (print) | LCC F1219 (ebook) | DDC 972/.01—dc23
LC record available at https://lccn.loc.gov/2019040157
LC ebook record available at https://lccn.loc.gov/2019040158

The University Press of Florida is the scholarly publishing agency for the State University System of Florida, comprising Florida A&M University, Florida Atlantic University, Florida Gulf Coast University, Florida International University, Florida State University, New College of Florida, University of Central Florida, University of Florida, University of North Florida, University of South Florida, and University of West Florida.

University Press of Florida
2046 NE Waldo Road
Suite 2100
Gainesville, FL 32609
http://upress.ufl.edu

Contents

Figures

Tables

Introduction

Ancient West Mexicos

CHRISTOPHER S. BEEKMAN, JOSHUA D. ENGLEHARDT, AND VERENICE Y. HEREDIA ESPINOZA

West Mexico in its broadest sense extends from the valley of Toluca westward to include most of the Neo-Volcanic Axis, plus the Sierra Madre Occidental and adjoining coast from the northern end of Sinaloa down to Michoacán (Figure I.1). The region is fully the equal in size to either the Maya region or to non-Maya Mesoamerica, but hosts only a fraction of the archaeologists that have been drawn to those areas. The lacunae in our understanding are vast, and a single well-placed field project can make a major contribution. This volume brings together researchers who are carrying out those field projects to address longstanding gaps in our knowledge of the varied ancient societies of West Mexico. Indeed, "the West" was not a single, monolithic entity; rather, the region was home to a multiplicity of complex societies and cultures that spanned the breadth of Mesoamerican prehistory.

Contributors to this volume present current data and interpretations spanning the Early Formative to Late Postclassic periods, and stretching from Durango to Colima to Michoacán. In doing so, contributors significantly advance our understanding of temporal, spatial, and cultural diversity in this expansive region. Simultaneously, they demonstrate that various societies in the West engaged in the same dynamic cultural processes evident in other Mesoamerican subregions, albeit in distinct contexts. Consequently, this collection presents the full richness of ancient West Mexico on its own terms, rather than considering the region's pre-Hispanic cultures as singular entities somehow unconnected with either the rest of Mesoamerica or each other, or, alternatively, as unrecognized or "lost" pockets of complex civilization. Thus, the volume not only seeks to contribute to our

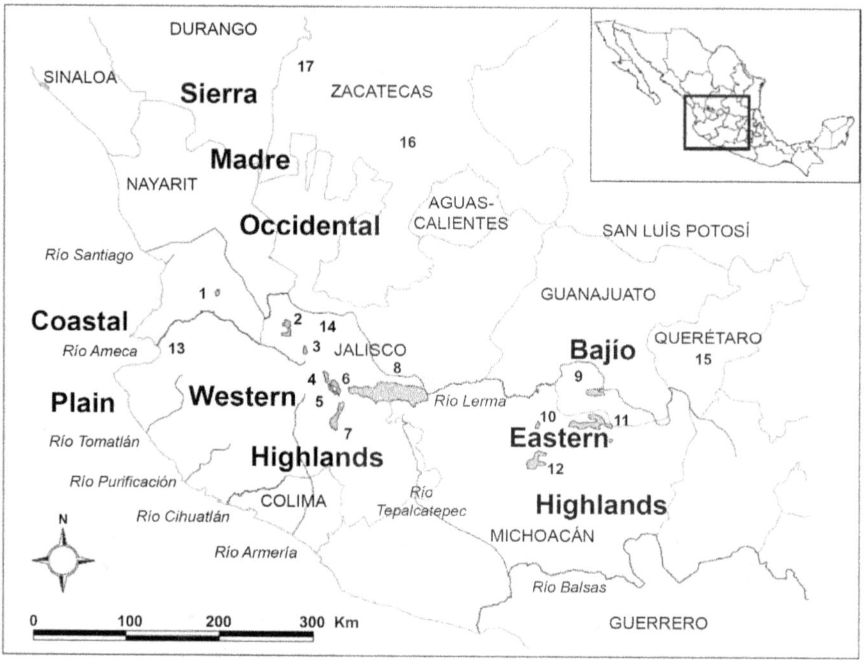

Figure I.1. Map of West Mexico showing important rivers and lakes, modern Mexican states, and cultural subregions. Lake basins: 1, San Pedro; 2, Magdalena; 3, La Vega; 4, Atotonilco; 5, Zacoalco; 6, San Marcos; 7, Sayula; 8, Chapala; 9, San Nicolas; 10, Zacapu; 11, Cuitzeo; 12, Pátzcuaro Basin; 13, Banderas; 14, Tequila; 15, San Juan; 16, Malpaso; and 17, Suchil (after Beekman 2010: 42, fig. 1).

knowledge of the archaeological past in West Mexico, but also to underline the West's long and complex relationship with other parts of Mesoamerica. Finally, this collection strives to highlight the potential value of data from the West in elucidating archaeological explanations of similar dynamics in better-known contexts in other Mesoamerican subregions, and beyond.

A Brief and Necessarily Selective History of Research

There are an increasing number of historiographic analyses of archaeological research in West Mexico (Beekman 2010; Cardona Machado 2016; Gorenstein and Foster 2000; Michelet 1995; Pickering and Beekman 2016; Williams 2004). They have exposed a series of interwoven tropes that have emerged over the years and contributed to the "Othering" of West Mexico. The region has been characterized as non-Mesoamerican, affiliated with South American cultures, its linguistic landscape has been exoticized, complex cultures have been reduced to a "cult of the dead," evidence of religious beliefs have been ignored in favor of an all-pervasive shamanism, and most recently the region has been given a new normative regional identity based on the exotic Teuchitlán architectural tradition. All of these entrenched themes promote the impression that West Mexico is something so different from what "traditional" Mesoamericanists study that it can be safely ignored. These problematic tropes will be addressed in detail at a future date, but as they permeate the history of West Mexican archaeology, they must be treated briefly here.

The Spanish conquest of West Mexico followed that of the Aztec empire, beginning shortly after the fall of Tenochtitlán. The comparative lack of resistance by the Tarascan Empire of Michoacán allowed the Spanish conquistadors to push quickly to Guerrero, Colima, and the Pacific coast. Francisco Cortés's conquests of far west Mexico in the early 1520s were followed by Nuño Beltrán de Guzmán's brutal murder of the Tarascan ruler and bloody establishment of Nueva Galicia in Jalisco and Nayarit in the 1530s (Altman 2010; Warren 1985). The imposition of political control in some cases took considerably longer. While Michoacán moved more peacefully into the Spanish colonial system, the far western area engaged in more overt resistance. The Mixtón Rebellion of the 1540s was an alliance of nomadic Zacatecos, settled Caxcanes, and older polities from central Jalisco that briefly threatened Spanish control in Nueva Galicia before being put down by Viceroy Mendoza (Weigand and García de Weigand 1996). Native people endured slave raiding, forced labor in the mines of Zacatecas, and the imposition of colonies of Tlaxcallans into their midst, and many fled into the mountains of Nayarit and Jalisco. Two decades after the conquest of the last Maya kingdom, Spanish troops finally conquered the Cora kingdom of eastern Nayarit in 1722 (McCarty and Matson 1975; see particularly Malvido Miranda 2000). Their accounts provide useful information on native religion and political systems that have still not been fully explored by archaeologists.

Manuel Orozco y Berra (1864) appears to have been the first to divide West Mexico into a Purhépecha-speaking Michoacán and a Southern Uto-Aztecan (SUA)–speaking West. Purhépecha remains an isolated language without clear relationships to other language families (Campbell et al. 1986; Foster 1969). This is no doubt due to the replacement of variants during the Tarascan imperial expansion in the final centuries before the conquest. Attempts to link Purhépecha to Zuñi, Quechua, Maya, or Chibcha (Greenberg 1987; Swadesh 1956, 1966) were poorly supported but repeatedly referenced, and have been selectively cited to support claims for their South American origins (Anawalt 1992; Corona Núñez and Estrada 1994; Furst 1967; Long 1967). Languages for the far western highlands and the Sierra Madre Occidental have been more difficult to recover, complicated by the importation of central Mexicans in the sixteenth century and the use of Nahuatl as the official language for the teaching of Catholic doctrine beginning circa 1550 (Santoscoy 1902: 317, 323). The oft-cited linguistic fragmentation for the far western states is based more on the multiplicity of names rather than linguistic evidence (Anguiano 1992; Yáñez Rosales 2001). Many ethnolinguistic terms are derived from the name of the town where they were spoken, and every language that has been studied for its linguistic affiliations has been assigned to one or another branch of SUA. This becomes more straightforward as one moves up the Sierra Madre Occidental, which Wilcox (1986) called the Tepiman Corridor, the route of closely related SUA languages that connected Mesoamerica with the American Southwest. Detailed studies of the ethnohistoric evidence have long suggested that SUA preceded Purhépecha on the western and southern fringes of the empire (Albiez-Wieck 2011), while Nahuatl or its variants were recorded for Colima (Sauer 1948: 63–64) and down the coast of Guerrero in the twentieth century (Guerrero and del Castillo 1948; Weitlaner 1948), with an isolate at Pochutla, Oaxaca (Boas 1917). Pochutec is believed to have separated from Nahuatl before Nahuatl and Pipil separated (Campbell and Langacker 1978), likely dating its separation to the Epiclassic or earlier.

The proposed separation between SUA and Purhépecha speakers became more pronounced as archaeological data began to emerge. Postclassic remains from Ixcuintla, Nayarit, were the first archaeological images to be published (Retes 1845), and they were quickly related to the Nahuatl-speaking peoples of central Mexico by reference to the well-known migration accounts that put the origins of the Aztec and others in the northwest (Ekholm 1942; Toro 1925). Chavero (1887: 460–463) went so far as to propose Mexcaltitan, Nayarit, as the original Aztlan based on these archaeological and ethnohistoric data. Apart from a brief period in the early twentieth century when the Formative period ceramic effigies from Jalisco and elsewhere were given a Purhépecha association (starting with Lumholtz 1902, II: 313), the figures have usually been assigned a broadly Nahua affiliation. Today, the gaps in linguistic data for the far western highlands continue to bedevil studies of its deeper history.

By the mid- to late nineteenth century, archaeological studies had begun to focus in on the most prominent remains in each subregion. Work in

Michoacán had begun to focus on the Tarascan Empire through early excavations at Tzintzuntzan and elsewhere (León 1888). The interest was so pronounced that Late Formative–Early Classic ceramics and associated remains were being conflated with Late Postclassic Tarascan artifacts (for example, Freddolino 1973 and many other early twentieth-century sources), seriously muddling all attempts to construct a continuous archaeological sequence until the French Centre d'Études Mexicaines et Centraméricaines (CEMCA) projects of the 1980s and 1990s (for example, Carot 2001). Along the Sierra Madre Occidental and its adjacent valleys, the mountaintop fortress of La Quemada and the ceremonial center of Alta Vista attracted a great deal of attention from an early date (Batres 1903; Berghes 1996; Hrdlička 1903; Medina González and García Uranga 2010), primarily for their apparent isolation. Both were interpreted as centers imposed upon the local populations by the central Mexican metropoli of Teotihuacan or Tula, whether as economic colonies (Weigand 1968, 1978), astronomical centers (Aveni et al. 1982), or bulwarks against the marauding Chichimecs of the barbarous north (Kelley 1971). In the far western highlands, explorers like Adela Breton (1903) and Carl Lumholtz (1902) began to report on a widespread tradition of shaft and chamber tombs and ceramic effigy figures found within them. Since looted tombs were the only source of the ceramic effigies in the absence of professional excavation, the states of Jalisco, Nayarit, and Colima became unified under the umbrella of an ancestral "tomb culture" (Corona Núñez 1960). This particular characterization has still not been entirely shaken off, despite excavations of hollow figures in residential and temple contexts (Beekman 2016, in press) and the intellectual poverty of a "cult of the dead."

The watershed for the interpretation of the cultures of far West Mexico came in the 1940s. Paul Kirchhoff (1943) defined the cultural area of Mesoamerica based on a list of traits largely from the time of the Spanish Conquest, and his initially inclusive definition encompassed all of the regions discussed here. Significant excavations began in Tzintzuntzan (Rubín de la Borbolla 1939, 1941) accompanied by surveys across the state that sought to delineate the nature of the empire (Goggin 1943; Lister 1947). Another seemingly positive development was the public exhibition of the hollow ceramic effigies from the far western highlands, particularly focusing on the enormous private collection of Diego Rivera (for example, Toscano 1946). While this event brought these materials to the attention of a much wider public and professional audience, it also highlighted differences between the visual culture of part of West Mexico and the recently recognized Olmec culture of the Gulf Coast (Drucker 1943a, 1943b; Stirling 1943). Initial comparisons quickly noted that descendants of Olmec designs and features could be identified in the visual culture of most Mesoamerican groups, with the exception of West Mexico (Covarrubias 1946). Rather than positing a multinodal origin for Mesoamerica based on this evidence, the source of Mesoamerican culture was soon equated with the Gulf Coast Olmec (Bernal 1969: 143). Any deviation from idealized Mesoamerican practices came to be attributed to South America (for example, shaft tombs [Long 1967], shamanism [Furst 1967], Purhépecha language [Swadesh 1956], clothing [Anawalt 1992]). Instead of recognizing that there could be multiple contributors to the hybrid culture area of Mesoamerica, Olmec origins (or affiliations) became the primary criterion for membership (Grove 2009). The result was a consistently pejorative approach to West Mexico. Schöndube (1975, 1980) wrote in terms of a "delayed Formative," extending until AD 600, followed directly by a Postclassic period ushered in by more recognizable Mesoamerican forces. J. Charles Kelley (1989) described northwest Mexico in similar terms with a "retarded Formative," slow to arrive out of a central Mexican hearth. And Townsend (1998) described the region as a living fossil from pre-Mesoamerican times, a time "before gods, before kings."

It was only in the 1970s and 1980s that the perspective began to change. A complex Postclassic Aztatlán trade network along the Pacific Coast became increasingly clearly delineated by surveys and intensive excavations (Beltrán Medina 2004; Jiménez Betts 2017; Meighan 1976; Mountjoy 2000; Olay Barrientos 2004). Excavations and absolute dates forced a realignment of the Chalchihuites culture sites of Zacatecas and Durango that divorced their sequences from those of central Mexican powers (Kelley 1985; Nelson 1997; Punzo Díaz 2016; Trombold 1990). Various investigators in Michoacán began to sort out the chronological and cultural sequence that gave a Formative and Classic period backstory to the Tarascan Empire, so that it no longer appeared to emerge from a vacuum (for example, Arnauld and Faugère-Kalfon 1998; Carot 2001; Pollard 2008). And in the far western highlands, excavations and settlement surveys removed the shaft tomb culture of Jalisco, Nayarit, and Colima from the realm of myth and into a social, political, and economic context (Beekman and Pickering 2016; Galván 1991; Weigand 1993, 2000).

Weigand was the most aggressive in promoting this changed perspective, and laid out what he called the "Simplicity Complex," in which investigators had become habituated to downplay signs of social or political complexity (Weigand 1985). His valid points, made in reaction to the excesses of prior decades, nonetheless formed part of a broader vision of West Mexico that in the end created a new trope of "unrecognized complexity" and a new myth in need of correction. Weigand's research had led him and colleagues to see a highly complex political system based in the Tequila valleys of central Jalisco, with large-scale intensified agriculture, semicentralized resource control and craft production, boundary centers guarding a central core, and peripheral centers elsewhere in the western highlands (Beekman 1996a, 1996b, 2000; Soto de Arechavaleta 1990; Weigand 1993, 2000; Weigand and Beekman 1998). This perspective was jarringly at odds with past characterizations of West Mexico, and although ultimately flawed due particularly to a poor control over chronology, at least one of the editors of this volume can testify that it drew him into working in the region.

The archaeology of West Mexico is no less exciting today. We have a growing sense of the interconnections within and beyond the region. Connections to Central/South America and the American Southwest are becoming more grounded through the use of both material and visual culture (for example, Hers 2013; Hosler 1994; Mathiowetz 2011; Mountjoy 2012; Mountjoy and Beltrán 2005). Research in most areas has diversified beyond the bare spatial and temporal distribution of cultures, and into topics of craft production, social organization, and ideology (for example, Beekman 2008; Beekman and Pickering 2016; Darras 1999; Heredia 2016; Heredia and Englehardt 2015; Hirshman 2008; López Mestas 2011; Olay Barrientos 2012). And a suite of new ethnographic and linguistic research is enriching our archaeological models by forcing us to address issues of identity, cosmovision, and myth alongside traditional concerns such as economics, politics, and mortuary ritual (for example, Coyle 2001; Guadalupe Becerra 1997; Jáuregui and Neurath 2003; Neurath 2002, 2008; Roskamp 1998; Stone 2004). The contributions to this volume, introduced and contextualized below, aim to demonstrate this dynamism through data-focused studies that both advance our understanding of West Mexico's pre-Hispanic cultural landscape, and complement archaeological research in other areas of Mesoamerica and the world.

Organization and Goals of the Present Volume

Contributors to this volume were selected primarily on the basis of their participation in recent archaeological projects producing new data. For this volume, we also sought to incorporate chapters that highlight regional and chronological variation, so as to convey the diversity of both current research in West Mexico and the region's ancient societies. As the region is nearly as large as the rest of non-Maya Mesoamerica, there was substantial cultural, political, and economic variation through time and space (Figure I.1 and Table I.1). In this sense, the individual contributions to this volume are best understood as part of a broader regional dialogue, rather than as a single, unified whole that seeks to advance a specific argument or perspective.

The fundamental goals of this volume are to contribute to contemporary debate by highlighting, engaging, and provoking questions pertinent to our understanding of the ancient West Mexican past and the dynamic cultural processes and practices of its constituent cultures, as well as historical trajectories of sociocultural change in the region as a whole. In doing so, we aim to create a fuller understanding of both West Mexico and the pre-Columbian Mesoamerican world by critically examining various specific case studies drawn from multiple spatial, temporal, and cultural contexts that highlight the cultural diversity of this understudied and poorly understood region and its multiple cultures. However, we do not wish to perpetuate the troublesome trope of the West as a singularly unique or exotic "other," discussed above. Rather, we seek to underline the ways in which the West was—and remains—an integral part of broader Mesoamerica, whose societies engaged in the same historical processes and cultural dynamics evident in other subregions. In that sense, the volume is not intended solely for an audience of regional specialists.

On the contrary, we hold that the data, methodologies, and interpretations presented here will assist in constructing archaeological explanations of similar phenomena in other contexts, both within Mesoamerica and on a more holistic level.

Structure of This Collection and Individual Contributions

To achieve these goals, and to underline regional diversity and variation, each of the chapters critically examines different aspects of sociocultural organization in ancient West Mexico. Areal coverage extends from southern Sinaloa to southern Michoacán. As the individual chapters demonstrate, the cultural manifestations present in each subregion or subarea are dynamic, at times showing close ties with the better-known cultures of Central Mexico, while at others the ties appear oriented to the north or within West Mexico, and not to "core" Mesoamerican regions or cultures. This broader temporal and geographic span allows for a richer

(after Beekman 2010: 46, table 1)

Date on calibrated timescale	Traditional Meso-american periods	Valley of Colima (Kelly 1980; Mountjoy 2012)	Tequila Valleys, Jalisco (updated from Beekman and Weigand 2008)	Suchil and Malpaso Valleys, Zacatecas (Kelley 1985; Nelson 1997)	Cuitzeo and Lerma Basin, Bajío (Darras and Faugère 2005; Hernández 2001)	Pátzcuaro Basin in Michoacán Highlands (Pollard 2008)
1500	Late Postclassic	Periquillo	Atemajac II		Acámbaro Tardío	Tariácuri
1400			Atemajac I			
1300	Middle Postclassic	Chanal			Acámbaro Temprano	
1200						Late Urichu
1100				Retoño	Perales Terminal	
1000	Early Postclassic			Calichal		
900	Epiclassic or Late		El Grillo			Early Urichu
800	Classic	Armeria		Alta Vista/La Quemada	Perales	
700		Colima				Lupe–La Joya
600	Classic			Canutillo		
500		Comala	Tequila IV		Choromuco	Jarácuaro
400						Loma Alta 3
300						
200	Late Formative	Ortices	Tequila III		Mixtlan	Loma Alta 1 and 2
100						
0					Chupícuaro Reciente 2	
100			Tequila II			
200					Chupícuaro Reciente 1	Chupícuaro
300	Middle Formative		Tequila I			
400					Chupícuaro Temprano	
500						
600						
700						
800						
900		Pantano				
1000	Early Formative	Capacha	Magdalena			
1100						
1200						
1300						
1400						
1500						
2000	Late Archaic					
2500						La Alberca

understanding of the varied cultural systems that once occupied the ancient West Mexican landscape. In this way, individual contributions seek to move beyond a monolithic or singular emphasis on the better-known areas within West Mexico (for example, the Pátzcuaro Basin). In addition, the chronological breadth of the volume also highlights contexts removed from—or at the spatial or temporal limits of—the earlier shaft tomb cultures or the Postclassic Tarascan Empire, which are perhaps more well known archaeologically.

The methodologies and analytic foci utilized in the individual chapters are also diverse, including theoretical analysis, chronological reassessment, intensive site survey and intrasite spatial configurations, large-scale regional survey, salvage work, excavations, and combinations of these methods. In this way, contributions attempt to approach traditional archaeological questions from new, more multidimensional perspectives. Chapters explore diverse topics in the archaeology of West Mexico, building on earlier research while also presenting new data that add significantly to scholarly understanding of this vast area in both synchronic and diachronic terms. In examining the developmental trajectories and sociocultural processes at play in its various constituent cultures, the volume and its chapters demonstrate that West Mexico is the focus of diverse research projects that undoubtedly have the potential to impact larger Mesoamerica archaeology as a whole.

Given the range of the individual contributions, as well as the spatial and temporal variability in the region, chapters are arranged in three broad sections, as suggested by the subtitle of this volume: Time, Space, and Diversity. Internally, each section is organized in rough chronological order. The first of these sections is comprised of the first four chapters, each of which treats general questions of chronology and temporal sequences in varied contexts, from the Formative period in Colima to the cultural sequence of southeastern Michoacán. These chapters expand our chronological understanding of cultural developments in the West, and present new data that serve to refine assessments of specific archaeological phenomena, such as Capacha ceramics or the Aztatlán colonization of the Pacific coast during the Postclassic period. They also raise theoretically important questions that may assist in understanding similar events in distinct contexts. For example, what are the ramifications of Capacha beyond Colima, and what do these imply about links between the West and other regions at this time? How can the study of Aztatlán colonization complement investigations of expansion by, for example, Teotihuacan or the Triple Alliance? In considering such critical queries, and others, these chapters demonstrate how archaeological research in West Mexico can impact the study of other regions and cultural processes.

The second section, Space, is more methodological in focus. Each of the three chapters in this section employs a relatively new methodology—spatial analysis (albeit in diverse forms)—to explore "old" questions, such as the sociopolitical structure of Los Guachimontones, Jalisco, or regional settlement patterns in northwestern Mexico. Although focused squarely on sites and regions within West Mexico, the methodologies deployed in these studies may be readily transferred to investigations in discrete contexts and across cultures. Further, and like the chapters of the first section, these contributions open new possibilities for the consideration of wider theoretical issues applicable in multiple circumstances, such as the reflection of sociocultural structure in archaeological materials, or the ways in which ancient societies actively constructed the built cultural environment.

Diversity, the third and final section, is comprised of three chapters, each of which seeks to augment and refine our conceptualization of ancient societies in West Mexico. These chapters treat specific cultural developments in more overtly theoretical frameworks. In doing so, they consider perspectives ranging from factional political strategies and segmentation to world systems and the reflection of ideology in material culture. Although foregrounded in current research in West Mexico, the contributions in all sections thus serve as templates for how researchers may productively approach fundamental issues in archaeology—such as interregional interaction or sociopolitical organization—from diverse theoretical and methodological perspectives. Moreover, the collection as a whole presents a range of new data that is of great potential value in cross-cultural contexts.

Although we intend the individual chapters to both stand alone and complement each other, serving as part of a wider, regional dialogue, we have organized the volume in this way so as to provide focus and balance to the collection. In addition, structuring the contributions in this way highlights cohesion among the varied sections and chapters, as well as the potential impact of this collection on cross-cultural archaeological research. Of course, no one volume can possibly do justice to the breadth of sociocultural developments through time in such a vast region, but these contributions attempt to mirror the cultural richness that flourished across the West, thereby presenting the diverse nature of what we may more profitably call "western Mesoamerica" in pre-Hispanic times. Below, we briefly introduce the constituent chapters of each of these subsections to more fully contextualize these individual contributions.

Time: Temporal Variability and Revised Regional Chronologies

For specialists and nonspecialists alike, the term Capacha evokes thoughts of the earliest cultural sedentary manifestations in West Mexico, shaft tombs, *bule*-type pottery, and very early dates. In addition, Capacha-phase pottery has recently been examined in terms of its association with the development of distillation in Mesoamerica (Zizumbo-Villarreal et al. 2009). Nonetheless, the Capacha phase, as Almendros's Chapter 1 explains, is far from being well understood, specifically as regards its dates. Although problems exist with the dates Kelly (1980) reported (for example, lack of secure contexts), researchers have too often taken them at face value and employed them as if they were well established (but compare Mountjoy 1994; Olay Barrientos 2012). As Almendros details, the placement of Capacha in the Early Formative period was questioned immediately after Kelly's publication, and even by Kelly herself. While the emerging consensus would place Capacha somewhat later in time, Almendros argues for a wider temporal span that encompasses older and newer data. Moreover, Almendros's research suggests that the temporal span of the phase is longer than previously thought, stretching from 1400 to 100 BC and overlapping with the subsequent Ortices phase. This may even be conservative, given the related materials emerging in Initial Formative coastal Oaxaca (Hepp 2015). In these ways, Almendros provides future directions for the study of this problematic yet important phase of pre-Hispanic cultural development in Colima. In addition, Almendros's contribution may serve as something of a cautionary tale for the archaeological essentialization of cultural phenomena, for as her chapter implicitly queries, if we do not really know what (or when) Capacha is, why is it so well known among Mesoamerican archaeologists?

Several of the contributions to this volume focus on the Teuchitlán Tradition defined by Weigand (1993) that forms the latest of the totalizing narratives for West Mexico. Much of this work examines the site of Los Guachimontones, one of several related sites excavated in the 1990s and 2000s (along with Huitzilapa, Llano Grande, and Navajas, and several in the Bolaños Canyon to the north), but the only one investigated with an eye toward boosting tourism. A decade of investigations at the site have served both to enhance our understanding of the Late Formative and

Early Classic periods in the central lake basins of highland Jalisco, and to dramatically increase the public profile of archaeology in the region. It is also a focal point for a new synthesis of political organization in the region.

Beekman's Chapter 2 is the first of the contributions to treat the site of Los Guachimontones, and the Teuchitlán Tradition more broadly. As Beekman explains it, a significant limitation for advancing our understanding of central Jalisco's prehistory has been the very loose approach to chronology. A series of highly speculative seriations of decontextualized ceramic effigies in public and private collections (Alsberg and Petschek 1968; Corona Núñez 1960; Kubler 1962; Townsend 1998 [his "Late Comala" style]) contradicted one another. A relative dating method using the extent of manganese deposits on object surfaces looked promising but did not receive sustained investigation (Long 1966). Absolute dates for the shaft tombs and related materials only began to trickle in during the 1960s and 1970s (Bell 1972; Delgado 1969; Furst 1966; Long 1966). While a general connection to the Late Formative and Early Classic periods seemed evident for most of the hollow ceramic figures associated with the shaft tombs, the poor selection of sample materials (many dates were on shell) and the premature reliance on a poorly developed dating method (obsidian hydration [Meighan et al. 1968]) did not allow for more precise dating. Weigand's later work on the Teuchitlán Tradition (Weigand 1979, 1985, 2000) relied upon a weakly supported and unilineal architectural sequence, while other investigators slowly developed a ceramic sequence covering the same period (Beekman 1996b, 2006; Beekman and Weigand 2000, 2008; Galván 1991). These competing chronologies probably confused the literature, particularly after old phase names continued in use in some publications even after the abandonment of the architectural sequence (Beekman and Weigand 2008). Beekman's chapter sorts out these issues and addresses the relative and absolute dating evidence for the pivotal site of Los Guachimontones. He outlines its growth out of a dimly visible Middle Formative community into a Terminal Formative ceremonial center, and its subsequent decline. Although he does not discuss its later occupation (the site's peak occupation actually came much later in the Late Postclassic period), Beekman's clarification of the chronological placement of some of its features significantly advances our understanding of the site. Moreover, his revised chronology serves to foreground and complement the later chapters by Sumano and Englehardt, and Heredia, which also explore Los Guachimontones and the Teuchitlán Tradition, respectively.

Moving southeast from the central valleys of Jalisco, long-term projects in the state of Michoacán have contributed greatly both to establishing the trajectory of social change in that region and to delineating the emergence of the Tarascan Empire in the Late Postclassic. Very few, however, have explored cultural developments outside of the spatial and temporal contexts of the Postclassic Tarascan "core." In Chapter 3, Punzo and colleagues present the results of a salvage project in southeastern Michoacán in the vicinity of Huetamo, where research has been sporadic, with long periods of hiatus. The construction of an irrigation dam opened an opportunity to carry out pedestrian and archaeomagnetic surveys and excavations at 10 of the 59 sites previously reported in the area. Results of this salvage project include a preliminary chronology based on ceramics and radiocarbon dates. The earliest occupation in the area dates to the Ancón phase (AD 0–200) and occupation continues, with a brief hiatus during the Tamarindos phase (AD 1250–1400) or Early Postclassic, up to the time of contact. This region of Michoacán was highly prized by the Tarascan Empire for its mineral wealth, specifically copper, and sites were situated in strategic locations along trade routes between the Pátzcuaro Basin and the lower Tierra Caliente. The survey identified important regional sites with *yácatas*, the quintessential architecture of the Tarascan Empire. In addition, the authors propose a ceramic sequence for the region as well as general characteristics of the lithic assemblage through time. They conclude with a discussion of how the location of these sites speaks to the power executed by local elites, expressed in the architecture and the prestige goods encountered through their archaeological survey. In doing so, Punzo and his colleagues offer valuable new data on a severely understudied area, as well as new perspectives on the construction of the Tarascan state.

The final chapter of this section, by Mountjoy and colleagues, concerns the Early- to Middle-Postclassic Aztatlán towns, found along the coasts of Sinaloa, Nayarit, and Jalisco, but also extending inland along trade routes through Durango northward and upriver toward Lake Chapala (Beltrán Medina 2004; Carpenter and Sánchez Miranda 2008; Jiménez Betts 2017; Ramírez Urrea 2006). Aztatlán sites and cultural remains were first defined by Carl Sauer and Donald Brand during their surveys of the archaeology and geography of northwestern Mexico (Sauer and Brand 1932). The elaborately decorated ceramics that initially defined these centers were quickly recognized as related to those of the "Mixteca-Puebla" style known from Postclassic central and southern Mexico (Ekholm 1942; Toro 1925), and resonated with the growing recognition that the region largely spoke Southern

Uto-Aztecan languages at the time of Spanish contact (Barragán Trejo and Yáñez Rosales 2001; Gerhard 1993; Valiñas 1994; Yáñez Rosales 1998, 2001). Strangely, the complex iconography of these ceramics received very little serious attention (Von Winning 1956 and 1976 being the sole exceptions). Only recently have researchers taken a closer look (Mathiowetz 2011; Mathiowetz et al. 2015; Pohl 2012), developing a much more detailed understanding of ideology and social ties among Aztatlán centers, and placing the Aztatlán centers within a wider context of interacting Mesoamerican centers. It has long been hypothesized that the Aztatlán centers formed a network of related but politically independent centers engaged in specialized production and exchange (for example, Publ 1985). Several authors (Bordaz 1964; Mathiowetz 2011; Meighan 1976; Mountjoy 2000; Scott and Foster 2000) have reviewed archaeological and ethnohistoric data that support the conclusion of valley-specific specialization in the production of metals, cacao, shellfish, tobacco, or cotton. In Chapter 4 of this volume, Mountjoy and his colleagues discuss the newly identified site of Arroyo Piedras Azules on the northwestern coast of Jalisco. Their data-centered chapter pinpoints an early Aztatlán occupation and concludes that outright colonization, rather than the expansion of a mercantile network, led to the foundation of the site in a rich *Spondylus* zone. This refinement of older, more generalized models is vital for advancing our understanding of Aztatlán and other developments. Further, this chapter both contextualizes and complements Mathiowetz's later contribution, which treats similar temporal and cultural contexts.

Space: Spatial Analysis, Old Questions, and New Methods

An improved spatial and chronological database is beginning to allow more specialized studies of Los Guachimontones and other sites in the Tequila valleys of central Jalisco during the Formative and Early Classic periods. There have been studies of ceramic function (Johns 2014) and obsidian exchange (Hoedl 2013) at the Terminal Formative site of Navajas, comparisons of obsidian production between Navajas and Early Classic Llano Grande (Wagner 2014), and compositional studies of ceramics at Navajas and Los Guachimontones (Jørgensen et al. 2016), all based on excavated materials. The extensive artifact collections at Los Guachimontones provide many opportunities for specialized studies in the long term, but the exposed architecture at the site has already received considerable attention. DuVall (2007) evaluated potential astronomical orientations at the site, and Hollon (2015) assessed issues of visibility and accessibility in the site core from a phenomenological perspective. In Chapter 5, Sumano and Englehardt take a different approach, contrasting spatial organization and architectural characteristics in what are believed to be functionally distinct sectors of the site—the site core vs. Loma Alta. Building on Beekman's chronological refinements, and employing the novel methodology of spatial syntax analysis, Sumano and Englehardt examine sociopolitical organization at Los Guachimontones, focusing on how sociocultural structure may be reflected in the built environment. They explore the nature of the site in terms of accessibility and what the constructed landscape or environment informs about the political structure system. While supporting a distinction between the two areas, the differences are subtle and suggest that a plurality of methods will be necessary to distinguish spatial variability within settlements like Los Guachimontones.

The Late Classic or Epiclassic period (beginning circa AD 500) has long been associated with major cultural and organizational changes in the western states. Schöndube (1975, 1980) saw it as the belated "Mesoamericanization" of West Mexico, accompanied by the appearance of rectangular platform mounds and recognizable deities, such as the Flayed God and the Storm God. Others have updated the proposal of Armillas (1969), and argued that climate in the north provoked an expansion along the northern frontier, followed by its retraction by the end of the period, bringing dislodged farming populations from Guanajuato and eastern Jalisco into neighboring areas (Beekman 2015; Beekman and Christensen 2003, 2011; Braniff 1989). Jiménez Betts rejects this temporal alignment altogether and argues that the same developments across northern and western Mexico represent an intrusion by Teotihuacan in the Early Classic (Jiménez Betts 1988, 1992, 2017). The chapter by Muñiz and Sumano focuses on the development of the cultural landscape at the northern extreme of the Chalchihuites culture in the Guadiana Valley of Durango (see Kelley 1971, 1976). The broadest definition of the Chalchihuites culture includes, besides the Guadiana Valley at its northern extreme, the better-known sites of Alta Vista in the Suchil Valley and La Quemada in the Malpaso Valley. These centers shared an emphasis upon rectangular architectural templates such as enclosed/sunken patios or halls of columns, likely originating in traditions located farther to the southeast. Muñiz and Sumano draw upon recent Instituto Nacional de Anthropología e Historia (INAH) survey data to analyze settlement distribution for the period of Chalchihuites occupation (AD 550–1350) through a consideration of architectural types and their distribution, site visibility and situation within the landscape, and ceramic analysis. Their placement of early Chalchihuites sites at the entry to the canyons, later Chalchihuites sites on the low mesas nearby, and historic sites within the canyons themselves, is a strikingly clear pattern in which they see changing efforts to appropriate the landscape. Their analysis of settlement patterns, and in particular intersite visibility from the nuanced perspective of landscape archaeology, is in the end a rarity in a region dominated by the single-site approach and complements investigations on the human construction of cultural landscape in other areas.

Diversity: Refining Our Theoretical Conceptualization of West Mexico

Verenice Heredia Espinoza's nuanced contribution is the first of the final section, which treats more theoretically oriented questions. Heredia took over the directorship of the archaeological investigations at the site of Los Guachimontones after the passing of Phil Weigand. She has overseen a new phase of work at the site simultaneously with her systematic regional survey of the surrounding valleys. Her Chapter 8 is the final treatment of the Teuchitlán Tradition offered in this collection, and combines data derived from her work at Los Guachimontones, as well as her regional survey projects, with a conceptual evaluation of the underpinnings of sociopolitical complexity in the central valleys of Jalisco. Building on and complementing the earlier, site-focused chapters by Beekman and Sumano and Englehardt, Heredia takes a more global view in her contribution. Working from a broader regional perspective, she reassesses the earlier interpretations of greater sociopolitical complexity by Weigand and Beekman (Beekman 1996a, 2000; Weigand 1985, 1993, 2000; Weigand and Beekman 1998) and critically dismantles proposed economic and political characterizations of the region (for example, the segmentary state) for which there is no longer, or never was, sufficient evidence. Heredia concludes that political organization in the central valleys of Jalisco steered more of a middle course between the "tomb culture" model of the 1960s and the "state-like society" proposed during the 1980s and 1990s. Her critical review stakes out a new baseline for the next generation of studies and emphasizes the need for systematic survey data at the regional scale and not just excavation projects. Heredia's chapter at once significantly advances our understanding of the Teuchitlán Tradition and serves as a paradigm for how—and how *not* to—conceptualize diachronic development and cultural phenomena at the regional scale.

Moving northeast from central Jalisco, the modern state of Aguascalientes is virtually unknown in its pre-Hispanic period. In Chapter 9, Caretta and Dueñas report on the survey and excavation at the Cerro de Santiago site in Aguascalientes, with a focus on its relations to other regions of Mesoamerica during the Epiclassic—a period of major cultural shifts in the West. Adopting a world-systems approach, the authors show that the site, and by extension the region north of the Río Verde Grande in the aforementioned state, was intimately in contact with the Bajío and the Jalisco highlands to the south. Ceramics and architecture attest to this increased economic interaction due to widespread processes of political reorganization throughout Mesoamerica during the Epiclassic. The authors suggest that, during the Epiclassic, new regional trade networks for prestige goods incorporated a larger region, in which Cerro Santiago was fully immersed. In embracing a world-systems perspective, Caretta and Dueñas offer a new model for understanding both the sociopolitical development and place of Cerro de Santiago within the region, and the various interaction spheres in which the site was integrated and that extended across central-north and western Mesoamerica during a crucial transitional period. However, rather than uncritically assuming an exploitative, "top-down" world-system with Teotihuacan or central Mexico at its core (an all-too-common interpretation), the authors instead argue for a type of decentralized world-system based on mutually favorable interaction and material exchange. This new conceptualization is a welcome perspective in a region whose archaeological interpretation is frequently dominated by totalizing narratives or essentializing characterizations of areal sociocultural phenomena (for example, the Teuchitlán Tradition). This chapter further demonstrates how cultural developments in Aguascalientes were both influenced by and impacted concurrent (and subsequent) processes in adjacent regions. Finally, like the chapter by Muñiz and Sumano, Caretta and Dueñas's contribution presents new data and interpretations that enhance archaeological understanding of a severely understudied area.

The final chapter, by Michael Mathiowetz, explores the Aztatlán phenomenon introduced in the earlier chapter by Mountjoy et al. Mathiowetz's contribution, however, takes a broader perspective, recognizing Aztatlán centers as the linchpin for Mesoamerican-Southwest interaction in the Postclassic. It has been obvious for many years that any interaction between Mesoamerica and its northern neighbors necessarily involved West Mexico, although this has not necessarily translated into increased research in the western states. Mathiowetz's chapter draws out the ideological linkage between the child solar deity Xochipilli/Piltzintli and a ritual economy of cotton in the Aztatlán area, and the ties between both and the Flower

World complex. He cites archaeological, ethnographic, and ethnohistoric evidence that coastal and inland sites may have specialized in the production of raw materials vs. finished products, respectively. His analysis also furnishes the basic components for the *chaîne opératoire* of textile production in the west—cotton and *maguey* production, weaving implements, sewing needles, and ceramic stamps for introducing dyed designs onto the finished product. In theoretical terms, Mathiowetz employs conceptual metaphor theory to connect the archaeological past with the ethnographic present. This approach is not only a novel method by which to access pre-Hispanic ritual-religious systems and recognize their material correlates in the archaeological record, it is also a welcome alternative to the direct historical approach that has too frequently been employed in interpretations of ideological systems in the ancient West. Although his chapter is slightly less reliant on explicitly archaeological data and field practice (as many others are), Mathiowetz's contribution nonetheless is a sophisticated take on the use and potential value of ethnographic analogy in archaeological research, as well as a critical contribution to our knowledge of Postclassic ideological systems in the West.

Mathiowetz's chapter forms an integral part of the collection, and an excellent point at which to end the regional dialogue running through the volume. Further, links to other contributions and cross-cultural contexts are evident. For example, his research highlights the importance of decentralized economic systems in the Early to Middle Postclassic periods, complementing the interpretations of Caretta and Dueñas on Epiclassic exchange in the preceding chapter. Medio period Casas Grandes, the Aztatlán centers, the Mixtec kingdoms, and central Mexican polities formed networks of small centers that increasingly moved goods along coastlines and river systems. It is, in fact, with the political consolidation of Late Postclassic Michoacán—touched on in the chapters by Punzo and colleagues—to the east that at least some of these links were blocked, likely explaining the decline of Aztatlán centers after AD 1350. Although dealing with distinct contexts, data, and cultural phenomena, Mathiowetz's chapter also complements the contribution by Sumano and Englehardt in demonstrating the reflection of social structure—in this case ideological—in material culture.

Finally, Stephen Kowalewski gives us his perspective on the contributions outlined here. As a seasoned Mesoamericanist with a long-standing—albeit secondary—interest in West Mexico, Kowalewski is in a uniquely knowledgeable position to assess and evaluate the progress being made, and his conclusions aid in binding the diverse contributions into a coherent whole. Kowalewski's concluding discussion also serves to further underline many of the major themes that run throughout the collection—to wit, the facts that the West enjoyed a long and complex relationship with Mesoamerica "proper," that it formed an essential part of this wider super-region, and that its constituent societies engaged in the same cultural practices and exhibited the same dynamic processes evident in other past societies in Mesoamerica and elsewhere.

Final Remarks

The case studies and treatments presented in this volume serve to elucidate the complexity and diverse nature of the cultures of pre-Columbian West Mexico, and provide both similarities and contrasts with better-known, "core" Mesoamerican regions. A notable lack of research in the region has perpetuated a conception of the west as "simple," "uncivilized" (see Bernal 1969; Weigand 1985, 1993: 69–79) or unrelated to other "typical" Mesoamerican regions. Subsequently, new tropes centering on regional singularity emerged. Paradoxically, by highlighting supposedly "distinctive" features (for example, shaft tombs, *guachimontones*), these revisionist treatments merely served to perpetuate the perception of regional isolation. Building on recent and ongoing archaeological research throughout western Mexico, this volume aims to counter both historical tendencies.

As research in the region progresses, it is becoming more and more apparent that West Mexico was not an isolated backwater or "mishmash" of cultures somehow disconnected from the rest of Mesoamerica (Beekman 2010). Research has revealed precocious sociopolitical complexity in the Formative period, urbanism prior to the Tarascan expansion, and the important role of the region in Postclassic trade routes, particularly along the Pacific coast and stretching toward the U.S. Southwest. Indeed, it is increasingly evident that ancient West Mexico was integrated into pan-Mesoamerican interaction spheres and cultural complexes from dates earlier than most current models consider. Although it is true that certainly distinctive material and cultural configurations are evident, both intraregionally and in relation to wider Mesoamerica, the same may be said of any macroregion. In this sense, West Mexico was decidedly not a unique region consistently defined by non-Mesoamerican features. Thus, it is not our intention to present this collection as "just another volume on (un[der]appreciated) West Mexico," or as a book aimed squarely at regional specialists, but rather as a showcase of the new data, methodologies, models, and interpretations that have been produced by recent investigations in the region, and the potential value of these perspectives for archaeological research in other contexts. In fact, we believe that it is no longer feasible to publish a volume simply on "West Mexico," à la Townsend's seminal 1998 tome—but we would consider this significant progress, and a testament to the quality of regional investigations in terms of refining archaeological narratives concerning the West.

Nor do we conceive of this collection as a "patchwork quilt" of shifting complexity around West Mexico. Although certainly wide-ranging in its scope, to our mind, several common threads emerge from the individual contributions that serve to bind the disparate contexts and foci presented in the chapters and sections of this volume into a more balanced and cohesive regional dialogue. First, all authors present recent research on pre-Hispanic West Mexico that moves beyond more extensively treated areas such as the Postclassic Tarascan core or the Bajío. Indeed, the fact that the scope of the volume is not explicitly limited to a single spatial or temporal context or specific case study—and that many chapters explore regions and/or temporal contexts infrequently treated in previous investigations—both highlights intraregional diversity through time and space and underlines the need to create new conceptual models capable of further increasing scholarly understanding of the complex, multifaceted nature of the societies that developed in ancient West Mexico. In this sense, the volume and its chapters seek to critically reevaluate the commonly held assumptions outlined above in order to refine extant theoretical models of regional sociocultural developments. The volume thus complements previous treatments by bringing to light new data, methodologies, and perspectives that will substantially contribute to contemporary views on ancient West Mexico, as well as the wider Mesoamerican world. That said, the interpretations presented in these papers also seek to move beyond the totalizing narratives that previously colored our understanding and perception of the region. It is hoped that these new data, approaches, and interpretations will contribute to the construction of archaeological models that relate to macroregional cultural processes in variable contexts, in order to better reflect culturally, spatially, or temporally specific configurations, modes, and dynamics, in both pre-Colombian Mesoamerica and beyond.

We are convinced that a consideration of West Mexico in the construction and application of such theoretical models will greatly enhance our knowledge of the alternative pathways to complexity that are evident in the chapters presented in this volume. In this sense, this volume also puts into relief the need for macroregional perspectives robust enough to provide models to explain these diverse pathways; such models and perspectives are readily applicable to archaeological research in multiple contexts. We are hopeful that in calling attention to the diachronic sociopolitical processes that gave rise to the cultural configurations—and by extension to diverse material manifestations—evident throughout the region, this volume will contribute to that goal. We would contend that the holistic approaches adopted by the authors and the application of methodologies to multidimensional datasets from disparate contexts not only advance comprehension of the ancient Mexican west, but also lay the groundwork for further research that will continue to add to our rapidly changing interpretation of this vast region and its place in the larger Mesoamerican world.

As Phil Weigand (1992: 23) presciently wrote, "We must have the flexibility to recognize that our conceptualization of West Mexico is rapidly changing, and will continue to change" as more and more research is conducted (translation ours). We agree with Weigand's sentiment, and further his characterization of the construction of archaeological knowledge as a process with no clear end in sight. Although we may be closer to a more complete picture of ancient West Mexico than we were when Weigand wrote those words 25 years ago, it is clear that much work remains to be done if we are to fill the lacunae in our understanding. We offer this volume as another step in this process.

References

Albiez-Wieck, Sarah

2011 *Contactos exteriores del Estado Tarasco: Influencias desde dentro y fuera de Mesoamérica.* Unpublished PhD dissertation, Department of Anthropology of the Americas, University of Bonn.

Alsberg, John L., and Rodolfo Petschek

1968 *Ancient Sculpture from Western Mexico: The Evolution of Artistic Form.* Nicole Gallery, Berkeley, CA.

Altman, Ida

2010 *The War for Mexico's West. Indians and Spaniards in New Galicia, 1524–1550.* University of New Mexico Press, Albuquerque.

Anawalt, Patricia Rieff

1992 Ancient Cultural Contacts between Ecuador, West Mexico, and the American Southwest: Clothing Similarities. *Latin American Antiquity* 3(2): 114–129.

Anguiano, Marina
1992 *Nayarit. Costa y Altiplanicie en el Momento del Contacto.* Universidad Nacional Autónoma de México, Mexico City.
Armillas, Pedro
1969 The Arid Frontier of Mexican Civilization. *New York Academy of Science Bulletin* 2(316): 697–704.
Arnauld, Marie Charlotte, and Brigitte Faugère-Kalfon
1998 Evolución de la ocupación humana en el centro-norte de Michoacán (Proyecto Michoacán, CEMCA) y la emergencia del Estado tarasco. In *Génesis, Culturas y espacios en Michoacán*, edited by Veronique Darras, pp. 15–34. Cuadernos de Estudios Michoacanos Series, Centre d'Etudes Mexicaines et Centraméricaines, México.
Arnauld, Marie Charlotte, Linda R. Manzanilla, and Michael E. Smith (editors)
2012 *The Neighborhood as a Social and Spatial Unit in Mesoamerican Cities.* University of Arizona Press, Tucson.
Aveni, Anthony F., Horst Hartung, and J. Charles Kelley
1982 Alta Vista (Chalchihuites), Astronomical Implications of a Mesoamerican Ceremonial Outpost at the Tropic of Cancer. *American Antiquity* 47(2): 316–335.
Barragán Trejo, Daniel, and Rosa Yáñez Rosales
2001 Investigaciones sobre las lenguas indígenas en Jalisco durante el siglo XX. *Estudios del Hombre* 13/14: 61–93.
Batres, Leopoldo
1903 *Visita a los Monumentos Arqueológicos de "La Quemada," Zacatecas.* Imprenta de la Vda. De Francisco Diaz de Leon, Mexico City.
Beekman, Christopher S.
1996a Political Boundaries and Political Structure: The Limits of the Teuchitlán Tradition. *Ancient Mesoamerica* 7(1): 135–147.
1996b *The Long-Term Evolution of a Political Boundary: Archaeological Research in Jalisco, Mexico.* PhD dissertation, Department of Anthropology, Vanderbilt University, Nashville, TN; University Microfilms, Ann Arbor, MI.
2000 The Correspondence of Regional Patterns and Local Strategies in Formative to Classic Period West Mexico. *Journal of Anthropological Archaeology* 19(4): 385–412.
2006 The Chronological Context of the Central Jalisco Shaft Tombs. In Mortuary Practices and Shaft Tombs in Western Mexico: An *Homenaje* to Phil Weigand, a Special Section of *Ancient Mesoamerica* 17(2): 239–249.
2008 Corporate Power Strategies in the Late Formative to Early Classic Tequila valleys of central Jalisco. *Latin American Antiquity* 19(4): 414–434.
2010 Recent Research in Western Mexican Archaeology. *Journal of Archaeological Research* 18(1): 41–109.
2015 Causes and Consequences of Migration in Epiclassic Northern Mesoamerica. In *Migration and Disruptions: Toward a Unified Theory of Ancient and Contemporary Migrations*, edited by Brenda Baker and Takeyuki Tsuda, pp. 73–96. University Press of Florida, Gainesville.

2016 Settlement Patterns and Excavations: Contexts of Tombs and Figures in Central Jalisco. In *Shaft Tombs and Figures in West Mexican Society: A Reassessment*, edited by Christopher S. Beekman and Robert B. Pickering, pp. 85–96. Gilcrease Ancient Americas Series, Gilcrease Museum, Tulsa, OK.

In press Unseating the Shaman: Reinterpreting the Hollow Figures of Western Mexico. In *Gods, Ancestors, and Human Beings: Anthropomorphic Representations in Highland Mexico*, edited by Brigitte Faugère and Christopher S. Beekman. University Press of Colorado, Boulder.

Beekman, Christopher S., and Alexander F. Christensen

2003 Controlling for Doubt and Uncertainty through Multiple Lines of Evidence: A New Look at the Mesoamerican Nahua Migrations. *Journal of Archaeological Method and Theory* 10(2):111–164.

2011 Power, Agency, and Identity: Migration and Aftermath in the Mezquital Area of North-Central Mexico. In *Rethinking Anthropological Perspectives on Migration*, edited by Graciela S. Cabana and Jeffery J. Clark, pp. 147–171. University Press of Florida, Gainesville.

Beekman, Christopher S., and Robert B. Pickering (editors)

2016 *Shaft Tombs and Figures in West Mexican Society: A Reassessment*. Gilcrease Ancient Americas Series, Gilcrease Museum, Tulsa, OK.

Beekman, Christopher S., and Phil C. Weigand

2000 *La Cerámica Arqueológica de la Tradición Teuchitlán, Jalisco*. El Colegio de Michoacán, Zamora; and Secretaría de Cultura del Estado de Jalisco, Guadalajara.

2008 Conclusiones, Cronología, y un Intento a Síntesis. In *Tradición Teuchitlán*, edited by Phil C. Weigand, Christopher Beekman, and Rodrigo Esparza, pp. 303–337. El Colegio de Michoacán, Zamora.

Bell, Betty

1972 Archeological Excavations in Jalisco, Mexico. *Science* 175(4027): 1238–1239.

Beltrán Medina, José C.

2004 Los concheros y el desarrollo cultural en Nayarit, la tradición Aztatlán. In *Introducción a la Arqueología del Occidente de México*, edited by Beatriz Braniff C., pp. 397–410. Universidad de Colima, Instituto Nacional de Antropología e Historia, Colima and Mexico City.

Berghes, Carl de

1996 *Descripción de las Ruinas de Asentamientos Aztecas durante su Migración al Valle de México, a través del Actual Estado Libre de Zacatecas*. Traducción, Notas y Estudio Introductoria de A. Lelgemann. Joyas Bibliográficas Zacatecanas II, Facultad de Humanidades, Universidad Autónoma de Zacatecas, Gobierno del Estado de Zacatecas, Centro Bancario del Estado de Zacatecas, Zacatecas.

Bernal, Ignacio

1969 *The Olmec World*. University of California Press, Berkeley.

Boas, Franz

1917 El Dialecto Mexicano de Pochutla, Oaxaca. *International Journal of American Linguistics* 1(1): 9–44.

Bordaz, Jacques

1964 *Pre-Columbian Ceramic Kilns at Peñitas, a Post-Classic Site in Coastal Nayarit,*

Mexico. Unpublished PhD dissertation, Department of Political Science, Columbia University, New York.

Braniff, Beatriz

1989 Oscilación de la Frontera Norte Mesoamericana: Un Nuevo Ensayo. *Arqueología* 1: 99–114.

Breton, Adela C.

1903 Some Mexican Portrait Clay Figures. *Man* 3: 130–133.

Campbell, Lyle R., and Ronald W. Langacker

1978 Proto-Aztecan Vowels: Part I. *International Journal of American Linguistics* 44(2): 85–102.

Campbell, Lyle R., Terrance Kaufman, and Thomas C. Smith-Stark

1986 Meso-America as a Linguistic Area. *Language* 62(3): 530–570.

Cardona Machado, Héctor

2016 *Las Narrativas de la Autoridad: Activación del Pasado en el Sitio Arqueológico Los Guachimontones, Teuchitlán, Jalisco*. Unpublished MA thesis, Centro de Estudios Arqueológicos, El Colegio de Michoacán, La Piedad.

Carot, Patricia

2001 *Le Site de Loma Alta, Lac de Zacapu, Michoacán, Mexique*. Paris Monographs in American Archaeology 9, BAR International Series 920. Archaeopress, Oxford.

Carpenter, John P., and Guadalupe Sánchez Miranda

2008 Entre la Sierra Madre y el Mar: La Arqueología de Sinaloa. *Arqueología* 39: 21–45.

Chavero, Alfredo

1887 *México a través de los siglos, Volumen 1*. Espasa y Compania, Editores, Barcelona.

Corona Núñez, José

1960 *Arqueología. Occidente de México*. Dirección General del Patrimonio Turístico, Guadalajara.

Corona Núñez, José, and Emilio Estrada

1994 *Origen Histórico de los Tarascos*. Universidad Michoacana de San Nicolás de Hidalgo, Morelia.

Covarrubias, Miguel

1946 El Arte "Olmeca" o de La Venta. *Cuadernos Americanos* 27: 153–179.

Coyle, Philip E.

2001 *Nayarí History, Politics, and Violence: From Flowers to Ash*. University of Arizona Press, Tucson.

Darras, Veronique

1999 *Tecnologías Prehispánicas de la Obsidiana: Los Centros de Producción de la Región de Zináparo-Prieto, Michoacán, México*. Collection Etudes Mesoamericaines II–17, Cuadernos de Estudios Michoacanos 9. Centre d'Etudes Mexicaines et Centraméricaines, Mexico City.

Darras, Veronique, and Brigitte Faugère

2005 Cronología de la Cultura Chupícuaro: Estudio del Sitio La Tronera, Puruaguita, Guanajuato. In *El Antiguo Occidente de México: Nuevas Perspectivas sobre el Pasado Prehispánico*, edited by Ediuardo Williams, Phil C. Weigand, Lorenza López Mestas, and David C. Grove, pp. 255–282. El Colegio de Michoacán, Zamora.

Darras, Veronique, Brigitte Faugère, Christophe Durlet, Catherine Liot, Javier Reveles, Rosalba Bérumen, Omar Cervantés, Cédric Caillaud, and Cybèle David

1999 Nouvelles Recherches sur la Culture Chupícuaro (Guanajuato, Mexique). *Journal de la Société des Américanistes* 85: 343–351.

Delgado, Diego

1969 *Arquitectura Funeraria Precolombina en el Estado de Jalisco*. Unpublished MA thesis, Department of Latin American Studies, University of California Los Angeles.

Drucker, Philip

1943a *Ceramic Sequences at Tres Zapotes, Veracruz, Mexico*. Bureau of American Ethnology, Bulletin 140. Smithsonian Institution, Washington, DC.

1943b *Ceramic Stratigraphy at Cerro de las Mesas, Veracruz, Mexico*. Bureau of American Ethnology, Bulletin 141. Smithsonian Institution, Washington, DC.

DuVall, Shina

2007 *Shared Symbolism in Powerful Places: Explorations into Cosmological Principles Displayed through the Ceremonial Public Architecture of the Teuchitlán Tradition in the Tequila Valleys of Jalisco, West Mexico*. Unpublished MA thesis, Department of Anthropology, University of Colorado–Denver.

Ekholm, Gordon F.

1942 *Excavations at Guasave, Sinaloa, Mexico*. Anthropology Papers of the American Museum of Natural History No. 38(4). AMNH, New York.

Foster, Mary L.

1969 *The Tarascan Language*. UCPL No. 56. University of California Press, Berkeley.

Freddolino, Marie Kimball

1973 *An Investigation into the Pre-Tarascan Cultures of Zacapu, Michoacán, Mexico*. Unpublished PhD dissertation, Department of Anthropology, Yale University, New Haven, CT.

Furst, Peter T.

1966 *Shaft Tombs, Shell Trumpets, and Shamanism: A Culture-Historical Approach to Problems in West Mexican Archaeology*. Unpublished PhD dissertation, Department of Anthropology, University of California, Los Angeles.

1967 Tumbas de Tiro y Cámara: Un Posible Eslabón entre México Occidental y los Andes. *Eco* 26: 1–5. Instituto Jalisciense de Antropología e Historia, Guadalajara, Jalisco.

Galván Villegas, Luis Javier

1991 *Las Tumbas de Tiro del Valle de Atemajac*. Instituto Nacional de Antropología e Historia, Mexico City.

Gerhard, Peter

1993 *The North Frontier of New Spain*. 2nd edition. University of Oklahoma Press, Norman.

Goggin, John M.

1943 An Archaeology Survey of the Rio Tepalcatepec Basin, Michoacán, Mexico. *American Antiquity* 9(1): 44–58.

Gorenstein, Shirley, and Michael S. Foster

2000 West and Northwest Mexico: The Ins and Outs of Mesoamerica. In *Greater Me-*

soamerica. The Archaeology of West and Northwest Mexico, edited by Michael S. Foster and Shirley Gorenstein, pp. 3–20. University of Utah Press, Salt Lake City.

Gorenstein, Shirley, and Helen Perlstein Pollard

1983 *The Tarascan Civilization: A Late Prehispanic Cultural System*. Vanderbilt University Publications in Anthropology No. 28, Nashville, TN.

Gorenstein, Shirley, David Chodoff, John Hyslop, Helen Perlstein Pollard, Michael Snarskis, and Lee Anne Wilson

1985 *Acámbaro: Frontier Settlement on the Tarascan-Aztec Border*. Vanderbilt University Publications in Anthropology No. 32, Nashville, TN.

Greenberg, Joseph Harold

1987 *Language in the Americas*. Stanford University Press, Palo Alto, CA.

Grove, David C.

2009 Morelos, el Occidente, y Mesoamerica en el Preclásico temprano. In *Las Sociedades Complejas del Occidente de México en el Mundo Mesoamericano*, edited by Eduardo Williams, Lorenza López Mestas, and Rodrigo Esparza, pp. 315–326. El Colegio de Michoacán, Zamora.

Guadalupe Becerra, Celina (editor)

1997 *Los Occidentes de México (Siglos XVI–XIX) El Archivo: instrumento y vida de la investigación histórica*. Universidad de Guadalajara, CMCA, El Colegio de Michoacán, Instituto Nacional de Antropología e Historia Zacatecas, Colegio de Jalisco, CIESAS, Archivo Histórico del Municipio de Colima.

Guerrero, Raúl G., and Ignacio M. del Castillo

1948 Etnografía y Lingüística de la Costa de Michoacán. In *El Occidente de México*, Cuarta Reunión de Mesa Redonda, pp. 127–128. Sociedad Mexicana de Antropología, Mexico City.

Haskell, David L.

2018 *The Two Taríacuris and the Early Colonial and Prehispanic Past of Michoacán*. University Press of Colorado, Boulder.

Hepp, Guy David

2015 *La Consentida: Initial Early Formative Period Settlement, Subsistence, and Social Organization on the Pacific Coast of Oaxaca, Mexico*. Unpublished PhD dissertation, University of Colorado, Boulder.

Heredia Espinoza, Verenice Y.

2016 Complexity without Centralization: Corporate Power in Postclassic Jalisco. In *Alternative Pathways to Complexity: A Collection of Essays on Architecture, Economics, Power, and Cross-Cultural Analysis in Honor of Richard E. Blanton*, edited by Lane F. Fargher and Verenice Y. Heredia Espinoza, pp. 79–103. University Press of Colorado, Boulder.

Heredia Espinoza, Verenice Y., and Joshua D. Englehardt

2015 Simbolismo Pan-Mesoamericano en la Iconografía Cerámica de la Tradición Teuchitlán, Jalisco. *Travaux et Recherches dans les Amériques du Centre* 68: 9–34.

Hernández, Christine L.

2001 Una Comparación entre la Fase Perales en el Noreste de Michoacán y la Fase Lerma en Acámbaro, Guanajuato. *Arqueología* 25: 23–45.

Hers, Marie-Areti (editor)
2013 Miradas renovadas al Occidente indígena de México. Conaculta, Universidad Nacional Autónoma de México, Instituto Nacional de Antropologia e Historia, Centro de Estudios Mexicanos y Centroamericanos, Mexico City.

Hirshman, Amy J.
2008 Tarascan Ceramic Production and Implications for Ceramic Distribution. *Ancient Mesoamerica* 19(2): 299–310.

Hoedl, Lucas M.
2013 *Shades of Black: Obsidian Distribution and Social Organization at the Teuchitlán Tradition Site of Navajas during the Late Formative Period.* Unpublished MA thesis, Department of Anthropology, University of Colorado–Denver.

Hollon, Kristie
2015 *Analysis of Social Space: Analyzing Lived Experiences in the Los Guachimontones Central Ceremonial Area.* Unpublished MA thesis, Department of Anthropology, University of Colorado–Denver.

Hosler, Dorothy
1994 *The Sounds and Colors of Power. The Sacred Metallurgical Technology of Ancient West Mexico.* Massachusetts Institute of Technology Press, Cambridge.

Hrdlička, Aleš
1903 The "Chichimecs" and Their Ancient Culture, with Notes on the Tepecanos and the Ruin of La Quemada, Mexico. *American Anthropologist* 5: 385–440.

Jáuregui, José, and Johannes Neurath (editors)
2003 *Flechadores de Estrellas. Nuevas Aportaciones a la Etnología de Coras y Huicholes.* Instituto Nacional de Antropología e Historia, Universidad de Guadalajara, Mexico City and Guadalajara.

Jiménez Betts, Peter
1988 Ciertas Inferencias de la Arqueología del Sur de Zacatecas. In *Primera Reunión Sobre las Sociedades Prehispánicas en el Centro Occidente de México, Memoria,* pp. 39–50. Cuaderno de Trabajo 1, Instituto Nacional de Antropologia e Historia, Mexico City.
1992 Una Red de Interacción en el Noroeste de Mesoamérica: Una Interpretación. In *Origen y Desarrollo en el Occidente de México,* edited by Bridget Boehm de Lameiras and Phil C. Weigand, pp. 177–204. El Colegio de Michoacán, Zamora.
2017 *Orienting West Mexico. The Mesoamerican World System 200–1200.* Unpublished PhD dissertation, University of Gothenburg.

Johns, Catherine
2014 *Ceramic Activity Analysis of Navajas Circle 5 and the Need for Practice Theory in Unusual Monumental Architecture.* Unpublished MA thesis, Department of Anthropology, University of Colorado–Denver.

Jørgensen, Mads S., Kong F. Cheong, Christopher S. Beekman, and Benjamin Turkel
2016 Next, the Museum? An Analysis of West Mexican Ceramics from Los Guachimontones and Navajas using a Handheld Portable X-ray Fluorescence (pXRF) Spectrometer. In *Shaft Tombs and Figures in West Mexican Society: A Reassessment,* edited by Christopher S. Beekman and Robert B. Pickering, pp. 109–128. Gilcrease Ancient Americas Series, Gilcrease Museum, Tulsa, OK.

Kelley, J. Charles

1971 Archaeology of the Northern Frontier: Zacatecas and Durango. In *Handbook of Middle American Indians, Vol. 11, Archaeology of Northern Mesoamerica*, edited by Robert Wauchope, Gordon F. Ekholm, and Ignacio Bernal, pp. 668–801. University of Texas, Austin.

1976 Alta Vista: Outpost of Mesoamerican Empire on the Tropic of Cancer. In *Las Fronteras de Mesoamerica, XIV Mesa Redonda*, pp. 21–40. Sociedad Mexicana de Antropología, Saltillo.

1985 The Chronology of the Chalchihuites Culture. In *The Archaeology of West and Northwest Mesoamerica*, edited by Michael S. Foster and Phil C. Weigand, pp. 269–287. Westview Press, Boulder, CO.

1989 The Retarded Formative of the Northwest Frontier of Mesoamerica. In *El Preclásico o Formativo: Avances y Perspectivas,* Martha Carmona Macias, coordinadora, pp. 405–423. Instituto Nacional de Antropología e Historia, Mexico City.

Kelly, Isabel

1980 *Ceramic Sequence in Colima: Capacha, an Early Phase.* Anthropological Papers of the University of Arizona 27. University of Arizona Press, Tuscon.

Kirchhoff, Paul

1943 Mesoamerica. *Acta Americana* 1: 92–107.

Kubler, George

1962 *The Art and Architecture of Ancient America.* Penguin Books, Harmondsworth, UK.

León, Nicolás

1888 *Anales del Museo Michoacano, Año Primero.* Imp. y Lit. del Gobierno en la Escuela de Artes, Morelia.

Lister, Robert H.

1947 Archaeology of the Middle Rio Balsas Basin, Mexico. *American Antiquity* 13(1): 67–78.

Long, Stanley V.

1966 *Archaeology of the Municipio de Etzatlán, Jalisco.* Unpublished PhD dissertation, Department of Anthropology, University of California, Los Angeles.

1967 Formas y Distribución de Tumbas de Pozo con Cámara Lateral. *Revista de la Universidad de los Andes* 1: 73–87.

López Mestas C., Lorenza

2011 *Ritualidad, Prestigio y Poder en el Centro de Jalisco durante el Preclásico Tardío y Clásico Temprano. Un Acercamiento a la Cosmovisión e Ideología en el Occidente del México Prehispánico.* Unpublished doctoral dissertation, Department of Social Sciences, Centro de Investigaciones y Estudios Superiores en Antropología Social, Guadalajara.

Lumholtz, Carl

1902 *Unknown Mexico.* Scribner's, New York.

Malvido Miranda, Elsa

2000 La Guerra contra las Momias en Nueva España. El Siglo XVIII, Jesuitas, Franciscanos, Autoridades Seculares e Inquisición. *Chungará* 32(2): 199–205.

Mathiowetz, Michael
2011 *The Diurnal Path of the Sun: Ideology and Interregional Interaction in Ancient Northwest Mesoamerica and the American Southwest.* PhD dissertation, Department of Anthropology, University of California Riverside. University Microfilms, Ann Arbor, MI.
Mathiowetz, Michael, Polly Schaafsma, Jeremy Coltman, and Karl Taube
2015 The Darts of Dawn: The Tlahuizcalpantecuhtli Venus Complex in the Iconography of Mesoamerica and the American Southwest. *Journal of the Southwest* 57(1): 1–102.
McCarty, Kieran, and Dan S. Matson
1975 Franciscan Report on the Indians of Nayarit, 1673. *Ethnohistory* 22(3): 193–221.
Medina González, José Humberto, and Baudelina L. García Uranga
2010 *A 100 Años de su descubrimiento. Alta Vista.* Instituto Nacional de Antropología e Historia, Mexico City.
Meighan, Clement
1976 *The Archaeology of Amapa, Nayarit.* Monumenta Archaeologica 2. Institute of Archaeology, University of California, Los Angeles.
Meighan, Clement W., Leonard J. Foote, and Paul V. Aiello
1968 Obsidian Hydration Dating in West Mexican Archaeology. *Science* 160(3842): 1069–1075.
Michelet, Dominique
1995 La Zona Occidental en el Posclásico. In *Historia antigua de México. Vol. III, el horizonte Posclásico y algunos aspectos intelectuales de las culturas mesoamericanas*, edited by Linda Manzanilla and Leonardo López Luján, pp. 153–188. Instituto Nacional de Antropología e Historia, Universidad Nacional Autónoma de México, and Grupo Editorial Miguel Angel Porrúa, Mexico City.
1998 Topografía y Prospección Sistemática de los Grandes Asentamientos del Malpaís de Zacapu: Claves para un Acercamiento a las Realidades Sociopolíticas. In *Génesis, Culturas y Espacios en Michoacán*, edited by Veronique Darras, pp. 47–59. Centre d'Études Mexicaines et Centraméricaines, Mexico City.
Michelet, Dominique, Grégory Pereira, and Gérald Migeon
2005 La Llegada de los Uacúsechas a la Región de Zacapu, Michoacán: Datos Arqueológicos y Discusión. In *Reacomodos Demográficos del Clásico al Posclásico en el Centro de México*, edited by Linda Manzanilla, pp. 137–154. Universidad Nacional Autónoma de México, Mexico City.
Mountjoy, Joseph B.
1994 Capacha: Una Cultura Enigmática del Occidente de México. *Arqueología Mexicana* II(9): 39–42.
2000 Prehispanic Cultural Development along the Southern Coast of West Mexico. In *Greater Mesoamerica. The Archaeology of West and Northwest Mexico*, edited by Michael S. Foster and Shirley Gorenstein, pp. 81–106. University of Utah Press, Salt Lake City.
2012 *El Pantano y otros sitios del Formativo Medio en el valle de Mascota, Jalisco.* Secretaría de Cultura, Gobierno de Jalisco, Universidad de Guadalajara Centro Uni-

versitario de la Costa, y el H. Ayuntamiento de Mascota, Jalisco, Acento Editores, Guadalajara.

Mountjoy, Joseph B., and José C. Beltrán

2005 Anthropomorphic Peg-Based Sculptures from the Banderas Valley of Coastal West Mexico. *Ancient Mesoamerica* 16: 155–168.

Nelson, Ben A.

1997 Chronology and Stratigraphy at La Quemada, Zacatecas, Mexico. *Journal of Field Archaeology* 24(1): 85–109.

Neurath, Johannes

2002 *Las Fiestas de la Casa Grande: Procesos Rituales, Cosmovisión y Estructura Social en una Comunidad Huichola.* Consejo Nacional para la Cultura y las Artes, Fondo Editorial Tierra Adentro, Instituto Nacional de Antropología e Historia, Universidad de Guadalajara, Mexico City and Guadalajara.

Neurath, Johannes (editor)

2008 *Por los Caminos del Maíz. Mito y Ritual en la Periferia Septentrional de Mesoamérica.* Fondo de Cultura Económica, Consejo Nacional para la Cultura y las Artes, Mexico City.

Olay Barrientos, Ma. de los Ángeles

2004 *El Chanal, Colima: Lugar que habitan los custodios del agua.* Instituto Nacional de Antropología e Historia, Mexico City.

2012 *Volcán de Fuego, Cuna del agua, Morada del viento. Desarrollo social y procesos de cambios en el valle de Colima. Una propuesta de interpretación.* Instituto Nacional de Antropología e Historia, El Colegio de Michoacán, Mexico City and Zamora.

Orozco y Berra, Manuel

1864 *Geografía de las Lenguas y Carta Etnográfica de México.* Imprenta de J. M. Andrade y F. Escalante, Mexico City.

Pereira, Grégory, Gérald Migeon, and Dominique Michelet

2005 Transformaciones Demográficas y Culturales en el Centro-Norte de México en Vísperas del Posclásico: Los Sitios de Cerro Barajas (Suroeste de Guanajuato). In *Reacomodos Demográficos del Clásico al Posclásico en el Centro de México*, edited by Linda Manzanilla, pp. 123–136. Universidad Nacional Autónoma de México, Mexico City.

Pickering, Robert B., and Christopher S. Beekman

2016 A Historical Overview of Shaft Tomb Archaeology in Western Mexico. In *Shaft Tombs and Figures in West Mexican Society: A Reassessment*, edited by Christopher S. Beekman and Robert B. Pickering, pp. 1–22. Gilcrease Ancient Americas Series, Gilcrease Museum, Tulsa, OK.

Pohl, John M. D.

2012 La Tradición Aztatlán de Nayarit-Jalisco y el Estilo Nahua-Mixteca de Cholula. *Arqueología Mexicana* 19(115): 60–65.

Pollard, Helen P.

1977 An Analysis of Urban Zoning and Planning at Prehispanic Tzintzuntzan. *Proceedings of the American Philosophical Society* 121: 46–69.

1980 Central Places and Cities: A Consideration of the Protohistoric Tarascan State. *American Antiquity* 45(4): 677–696.

1993 *Tariacuri's Legacy: The Prehispanic Tarascan State*. University of Oklahoma Press, Norman.

2008 A Model of the Emergence of the Tarascan State. *Ancient Mesoamerica* 19(2): 217–230.

Pollard, Helen P., and Laura Cahue

1999 Mortuary Patterns of Regional Elites in the Lake Patzcuaro Basin of Western Mexico. *Latin American Antiquity* 10(3): 259–280.

Publ, Helmut

1985 *Prehispanic Exchange Networks and the Development of Social Complexity in Western Mexico: The Aztatlán Interaction Sphere*. Unpublished PhD dissertation, Department of Anthropology, University of Southern Illinois, Carbondale.

Punzo Diaz, José Luis

2016 Revisando la Cronología en la Frontera Norte de Mesoamérica, Estado de Durango, México. *Arqueología Iberoamericana* 29: 38–43.

Ramírez Urrea, Susana

2006 Relaciones Interregionales en la Tradición Aztatlán (Postclásico Temprano y Medio). In *Transformaciones Socioculturales y Tecnológicas en el Sitio de La Peña, Cuenca de Sayula, Jalisco*, edited by Catherine Liot, Susana Ramírez Urrea, Javier Reveles, and Otto Schöndube, pp. 435–444. Instituto Nacional de Antropología e Historia, Centro Universitario de Ciencias Sociales y Humanidades, Universidad de Guadalajara, Guadalajara.

Retes, Miguel

1845 Apuntes de un Viaje. Santiago Ixcuintla (Departamento de Jalisco). *Museo Mexicano* 5: 1–6.

Roskamp, Hans

1998 *La historiografía indígena de Michoacán: el Lienzo de Jucutácato y los títulos de Carapan*. Research School CNWS, Leiden.

Roth-Seneff, Andrew, Robert V. Kemper, and Julie Adkins (editors)

2015 *From Tribute to Communal Sovereignty. The Tarascan and Caxcan Territories in Transition*. University of Arizona Press, Tucson.

Rubín de la Borbolla, Daniel F.

1939 Antropológica Tzintzuntzan-Ihuatzio: Temporadas I y II. *Revista Mexicana de Estudios Antropológicas* 3: 99–121.

1941 Exploraciones Arqueológicas en Michoacán: Tzintzuntzan, Temporada III. *Revista Mexicana de Estudios Antropológicas* 5: 5–20.

Santoscoy, Alberto

1902 Diversos errores acerca de las lenguas indígenas del antiguo Obispado y del actual Arzobispado de Guadalajara. *Anales del Museo Nacional de Antropología* Tomo VII: 311–328.

Sauer, Carl

1948 *Colima of New Spain in the Sixteenth Century*. University of California Press, Berkeley.

Sauer, Carl, and Donald Brand

1932 *Aztatlán: Prehistoric Mexican Frontier on the Pacific Coast*. Ibero-Americana 1. University of California Press, Berkeley.

Schöndube Baumbach, Otto

1975 La Evolución Cultural en el Occidente de Mexico: Jalisco, Colima y Nayarit. In *XLI Congreso Internacional de Americanistas, Volumen 1*, pp. 332–337. Mexico City.

1980 Capitulos V–XI. In *Historia de Jalisco, tomo I*, edited by José María Muría, pp. 111–258. Gobierno de Estado de Jalisco, Guadalajara.

Scott, Stuart D., and Michael S. Foster

2000 The Prehistory of Mexico's Northwest Coast. A View from the Marismas Nacionales of Sinaloa and Nayarit. In *Greater Mesoamerica. The Archaeology of West and Northwest Mexico*, edited by Michael S. Foster and Shirley Gorenstein, pp. 107–135. University of Utah Press, Salt Lake City.

Smith, Michael E.

2010 The Archaeological Study of Neighborhoods and Districts in Ancient Cities. *Journal of Anthropological Archaeology* 29(2): 137–154.

2011 Empirical Urban Theory for Archaeologists. *Journal of Archaeological Method and Theory* 18(3): 167–192.

Soto de Arechavaleta, María dc los Dolores

1990 Áreas de Actividad en un Taller de Manufactura de Implementos de Piedra Tallada. In *Nuevos Enfoques en el Estudio de la Litica*, edited by María de los Dolores Soto de Arechavaleta, pp. 215–242. Universidad Nacional Autónoma de México, Mexico City.

Stirling, Matthew W.

1943 *Stone Monuments of Southern Mexico*. Bureau of American Ethnology, Bulletin 138. Smithsonian Institution, Washington, DC.

Stone, Cynthia L.

2004 *In Place of Gods and Kings: Authorship and Identity in the Relación De Michoacán*. University of Oklahoma Press, Norman.

Swadesh, Morris

1956 Problems of Long-Range Comparison in Penutian. *Language* 32(1): 17–41.

1966 Porhe y Maya. *Anales de Antropología* 3: 173–204.

Toro, Alfonso

1925 Una Nueva Zona Arqueológica en Sinaloa. *Anales del Museo Nacional de Arqueología, Historia, e Etnografía* Epoca 5, Tomo III (1): 56–58.

Toscano, Salvador (editor)

1946 *Arte Precolombino del Occidente de México*. Secretaría de Educación Pública, Mexico City.

Townsend, Richard F.

1998 Before Gods, Before Kings. In *Ancient West Mexico: Art and Archaeology of the Unknown Past*, edited by Richard F. Townsend, pp. 107–135. Art Institute of Chicago, Chicago, IL.

Trombold, Charles D.

1990 A Reconsideration of Chronology for the La Quemada Portion of the Northern Mesoamerican Frontier. *American Antiquity* 55(2): 308–324.

Valiñas C., Leopoldo

1994 Transiciones Lingüísticas Mayores en Occidente. In *Transformaciones Mayores en el Occidente de México*, edited by Ricardo A. Palafox, pp. 127–166. Universidad de Guadalajara, Guadalajara.

Von Winning, Hasso

1956 Offerings from a Burial Mound in Coastal Nayarit, Mexico. *The Masterkey* 30(5): 157–170.

1976 Escenas Rituales en la Cerámica Polícroma de Nayarit. In *Actas del XLI Congreso Internacional de Americanistas, Volumen II*, pp. 387–400. Instituto Nacional de Antropología e Historia, Mexico City.

Wagner, John P.

2014 *Circles of Glass and Grain: Economic Differences between Core and Semi-peripheral Zones, a Study of Public Center Lithics from the Tequila Valleys of West Mexico*. Unpublished MA thesis, Department of Anthropology, University of Colorado–Denver.

Warren, J. Benedict

1985 *The Conquest of Michoacán: The Spanish Domination of the Tarascan Kingdom in Western Mexico, 1521–1530*. University of Oklahoma Press, Norman.

Weigand, Phil C.

1968 The Mines and Mining Techniques of the Chalchihuites Culture. *American Antiquity* 33(1): 45–61.

1978 The Prehistory of the State of Zacatecas: An Interpretation: Part I. *Anthropology* 2(1): 67–87.

1979 The Formative-Classic and Classic-Postclassic Transitions in the Teuchitlán-Etzatlán Zone of Jalisco. In *Los Procesos de Cambio: XV Mesa Redonda, vol. 1*, pp. 413–423. Sociedad Mexicana de Antropología, Mexico City.

1985 Evidence for Complex Societies during the Western Mesoamerican Classic Period. In *The Archaeology of West and Northwest Mexico*, edited by Michael Foster and Phil C. Weigand, pp. 47–91. Westview Press, Boulder, CO.

1992 Introducción. In *Origen y Desarrollo de la Civilización en el Occidente de México: Homenaje a Pedro Armillas y Ángel Palerm*, edited by Brigitte Boehm de Lameiras and Phil C. Weigand, pp. 13–26. El Colegio de Michoacán, Zamora.

1993 *Evolución de una Civilización Prehispánica*. El Colegio de Michoacán, Zamora.

2000 The Evolution and Decline of a Core of Civilization: The Teuchitlán Tradition and the Archaeology of Jalisco. In *Greater Mesoamerica: The Archaeology of West and Northwest Mexico*, edited by Michael S. Foster and Shirley Gorenstein, pp. 43–58. University of Utah Press, Salt Lake City.

Weigand, Phil C., and Christopher S. Beekman

1998 The Teuchitlán Tradition: Rise of a Statelike Society. In *Ancient West Mexico: Art and Archaeology of the Unknown Past*, edited by Richard Townsend, pp. 35–51. Art Institute of Chicago, Chicago, IL.

Weigand, Phil C., and Acelia García de Weigand

1996 *Tenamaxtli y Guaxicar: Las Raíces Profundas de la Rebelión de Nueva Galicia*. El Colegio de Michoacán, Zamora.

Weitlaner, Robert J.

1948 Situación Lingüística del Estado de Guerrero. In *El Occidente de México*, Cuarta

Reunión de Mesa Redonda, pp. 129–133. Sociedad Mexicana de Antropología, Mexico City.

Wilcox, David R.

1986 The Tepiman Connection: A Model of Mesoamerican-Southwestern Interaction. In *Ripples in the Chichimec Sea: New Considerations of Southwestern-Mesoamerican Interactions*, edited by Frances J. Mathien and Randall H. McGuire, pp. 135–153. Southern Illinois University Press, Carbondale.

Williams, Eduardo

2004 Prehispanic West Mexico: A Mesoamerican Culture Area. Electronic document, http://www.famsi.org/research/williams/, accessed October 3, 2016.

Yáñez Rosales, Rosa H.

1998 La Construcción del Nahuatl como Lengua General y la Supresión de las Lenguas Indígenas Locales: El Caso del Occidente Novohispano. In *El Occidente de México: Arqueología, Historia y Medio Ambiente. Perspectivas Regionales. Actas del IV Coloquio Internacional de Occidentalistas*, edited by Ricardo Ávila, Jean Pierre Emphoux, Luis Gómez Gastélum, Susana Ramírez, Otto Schöndube, and Francisco Valdez, pp. 55–64. Universidad de Guadalajara, Instituto Francés de Investigación Científica para el Desarrollo en Cooperación (ORSTOM), Guadalajara.

2001 *Rostro, Palabra y Memoria Indígenas: El Occidente de México: 1524–1816.* Centro de Investigaciones y Estudios Superiores en Antropología Social, Mexico City.

Zizumbo-Villarreal, Daniel, Fernando González-Zozaya, Ma. de los Ángeles Olay-Barrientos, Laura Almendros-López, Patricia Flores-Pérez, and Patricia Colunga-García Marín

2009 Distillation in Western Mesoamerica before European Contact. *Economic Botany* 63(4): 413–426.

PART 1

TIME

Revised Regional Chronologies

1

The Cultural Sequence during the Formative Period in the Valley of Colima

LAURA ALMENDROS LÓPEZ

This chapter presents a review of current knowledge regarding the Formative period in the Colima Valley where two phases—Capacha and Ortices—are recognized on the basis of studies conducted in the region by Isabel Kelly. Focusing primarily on the Capacha phase and the chronological problematics associated with it, the study examines existing interpretations of this cultural development and corroboration provided by more recent research in the area. Our findings at the El Diezmo–Adonaí site emphasize the need to apply new methodological approaches in order to obtain a better understanding of the chronology of the Capacha phase, which to this point has been defined solely on the basis of funerary contexts and their unique ceramic contents. Because in future studies we plan to pursue our interest in conducting chronometric analyses involving the application of the Bayesian statistical model, a substantial part of this text centers, on the one hand, on the properties of radiocarbon dating, and, on the other, the possibility of utilizing the Bayesian model to interpret the dates obtained using this method.

When we think of the Formative period in Colima (in West Mexico), Capacha comes immediately to mind with its well-known incised pottery: the famous *bules* or double pots. We are reminded of Isabel Kelly's studies in the 1960s and 1970s that discovered and identified this ceramic material, especially her 1980 publication, *Ceramic Sequence in Colima: Capacha, an Early Phase*, which presented the data on this early phase identified in the Colima Valley. This is the pioneer study in which she analyzed the pre-Hispanic cultural sequence in this region and specifically posited possible relations between Capacha and other nearby areas of western Mesoamerica in the present-day states of Michoacán and Jalisco (Mexico), and even with such far-off regions as northwestern South America.[1] But Kelly's study concludes with a somewhat baffling statement: "To recapitulate, Capacha is Capacha. It is not Mesoamerican, yet not quite South American, although perceptible ingredients link it with the northwest part of South America" (Kelly 1980: 37).

The question now is what do we know about Capacha today, four decades after Kelly's initial research? In answer to this question, the following section presents a review of the key aspects of the Formative period in Colima, especially in relation to the Capacha phase with some comments on the Ortices phase, on the problems that persist, and on some new issues that have emerged with the results of more recent studies. At the same time, I suggest future directions for research on the chronological period of pre-Hispanic development known as the Formative period.

Capacha: What, Where, and When?

The title of Kelly's 1980 monograph indicates that she saw Capacha as one phase within a ceramic sequence found in Colima. This suggests that in her view, what characterizes a particular phase is the presence of different criteria from the phases that preceded and followed it. Furthermore, this entails adding the element of time. In this regard, it is Capacha pottery that distinguishes it from the preceding phase and marks it as the first expression of this ceramic style in Colima. In addition, this material shows sufficiently significant contrasts with the succeeding phase, called Ortices. Further below, I will show that Kelly also took chronological differences into account when distinguishing the two Formative phases, Capacha and Ortices.

Turning to the literature on this topic, we find that various researchers—including myself—use terms like culture, horizon, complex, or tradition when referring to Capacha (Almendros et al. 2014; López Mestas 2011; Morales et al. 2013 Mountjoy 1994, 2010, 2012, 2015; Olay 2015; Olay et al. 2010). However, when we analyze the concrete characteristics recognized as being related to Capacha, it becomes clear that we lack the elements necessary to apply any other term beyond what Kelly defined as a "phase." This affirmation is based on the understanding that defining cultures or traditions requires a much broader array of elements, as Tylor (1995) sustained when he wrote that a complex of this type includes the beliefs, art, customs, habits and abilities that are common to a given group of human beings.

In the case of Capacha in Colima, the data available today are only marginally more ample than those that Kelly presented in 1980, so there is no

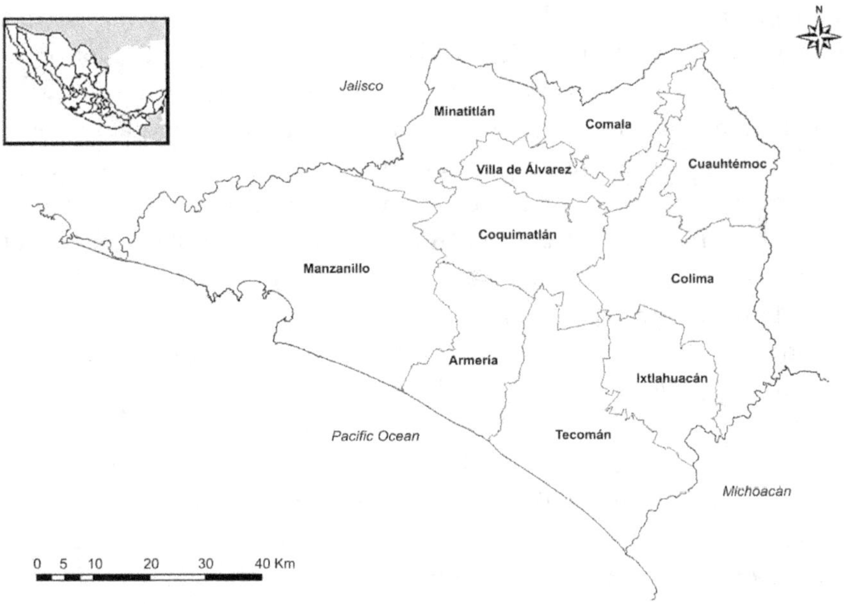

Figure 1.1. Map of Colima detailing current municipal divisions (map by Georgina Ortiz).

justification for using terms like "culture" or "tradition." In contrast, a phase refers—much more modestly—to observations of discontinuities with an earlier expression that are sufficiently significant to justify identifying a phase. In the definition offered by López Austin and López Luján (2002: 15), history consists of a series of interrelated events that form epochs which then contribute to defining a historical process. In terms of their temporal dimension, these are called periods. Hence, "[t]he passage from one period to another may manifest itself in an abrupt rupture, a gradual dissolution, or an overlapping. The stages within periods tend to be called phases."

In this sense, phase has a temporal connotation, but also requires identifying a criterion that makes it possible to define a discontinuity; for Capacha, this is ceramic material. Therefore, in order to correctly place the Capacha phase we must consider two factors: one spatial, the other temporal. Unfortunately, in the case of Capacha it is clear that we can only respond tentatively to these two aspects because of the large gaps in current knowledge of this cultural development.

Virtually all the archaeological contexts found up to now that have been associated with Capacha are located in the Colima Valley, with only a few in the coastal area of the state of Colima (Figure 1.1). Thus, this cultural development is manifested, above all, in a region that begins on the hillsides of the famous Colima Volcano (the Volcán de Fuego) and forms an ample valley that runs from north to south, surrounded by a sierra that includes the mountains called La Cuesta de La Salada, La Mina, La Mesa, and La Media Luna, among many others. On the other side, this mountain range descends to the coastal plain of Colima that borders the Pacific Ocean around Tecomán (INEGI 1981). Outside the valley, evidence of Capacha-type pottery is scanty at best, both along the coast and in areas closer to the Colima volcanoes.

Although it is clear that this valley was settled during the Capacha phase, we cannot yet define a specific settlement pattern there because the distribution of deposits in the valley is extremely complicated and the available evidence of occupied areas is still severely limited. With respect to chronological placement, the primary topic of our research on Capacha, we face similar restrictions. In general, it seems clear that this phase developed during a broad period in Mesoamerican prehistory known as the Formative or Preclassic, which is subdivided into Early, Middle, and Late expressions. However, when we attempt to define more precisely the epoch in which the Capacha phase developed, we confront the severe problems of chronological uncertainty that have made it so difficult to define the cultural characteristics of Capacha and compare this phase to contemporary developments in other areas of Mesoamerica. Here we attempt to clarify this situation.

A Controversial Date

It was Isabel Kelly who coined the term *Capacha* to refer to what she considered the first identifiable phase in the ceramic and cultural sequence of Colima in pre-Hispanic times; but the question is, on what did she base this chronological ascription? A few years after Kelly's 1980 publication appeared, the archaeologist Paul Schmidt Schoenberg wrote the following comments in the Mexican journal *Anales de Antropología*:

> In terms of absolute dating, various dates obtained by obsidian hydration are presented, spanning the period from 806 to 234 BC. The only carbon-14 (C14) date was determined using the organic contents of Capacha-style potsherds gathered from the surface at site No. 4, which produced results of 1450 BC (uncorrected) and 1870–1720 BC (corrected). This is not particularly convincing but taking

> into account similarities in the ceramics and a carbon-14 date from El Opeño (1500 BC, uncorrected), at this juncture this second date seems more logical than the later ones obtained via obsidian hydration. (Schmidt 1983: 210)

In her chapter on chronology, Kelly attempted to address the thorny issue of stratigraphy that persists today because of the scarce evidence of habitation zones (the entire excavated area is associated with funerary contexts), and the problematic evidence for different phases in the stratigraphy. In this area, remains of clearly distinct temporal phases can be found not only scattered on the surface but even in a single layer of an excavation. Indeed, examples of this confusion abound in the Colima Valley (Almendros and González 2009).

Because of this stratigraphic problem, Kelly commented that it was extremely difficult to distinguish subphases within the Capacha phase, thus identifying an issue that continues to generate complications and controversy in the Colima area today. And this problem is made even more acute by the fact that much research is of the salvage archaeology type or, worse yet, rescue operations with time restrictions that preclude undertaking the extensive excavations that would offer the only real means of finally resolving this complex stratigraphic problem.

Kelly's section on chronology revolves primarily around the question of absolute dating, which she discusses in terms of two procedures: obsidian hydration and C14. It is interesting to note that hers was likely one of the first studies to utilize data of this type in western Mexican archaeology. And we can by no means set these data aside because they may provide the key for constructing the chronology of the Capacha phase. The seven dates obtained by obsidian hydration dating (OHD) in her work came from 14 specimens of this volcanic glass. Kelly states that the first two dates were from obsidian that was associated definitively with Capacha. Those samples produced the most recent of the seven dates found; namely, 8.5 microns (260 BC) and 8.4 microns (234 BC). In contrast, the other five dates were determined using 11 specimens of obsidian that, as Kelly observed (1980: 27), "[were] allegedly found with Capacha pottery [and] produced some of the oldest readings of the entire Colima series: two of 10.6 microns (806 BC), and one each of 10.4 (754 BC), 9.8 (598 BC), 9.6 (546 BC), and 9.5 (520 BC); the remaining specimens gave more recent dates."

These results, obtained with the OHD method in analyses conducted at UCLA in the early 1970s, suggested a chronology running from 806 to 234

BC, dates that would correspond to the Middle-to-Late Formative. Finally, a C14 date was obtained from a sample sent to the *Geochron Laboratories* in Massachusetts instead of UCLA. That date was 3400 ± 200 BP, or 1450 BC, but was corrected to 1870–1720 BC. Although the author does not specify the correction tool applied, the time period leads to the inference that the Suess curve was used to calibrate the C14 dates (Mestres 2003).

Kelly acknowledges that this date is very early, but mentions the similarity of nearby pottery remains to ceramics from northwestern South America, on the one hand, and, on the other, at the El Opeño site, and cites the C14 date obtained from a tomb at the latter site in Michoacán: 1500 BC, corrected to 1280 ± 80 BP. She closes the section on chronology with these words: "Because of ceramic similarities between Capacha and Opeño (Chapter 4: Opeño Phase), the two independently derived dates provide mutual support for the early placement of both phases" (Kelly 1980: 28).

Thus, Kelly's scheme was established, once and for all, even though she was not convinced of the reliability of the date obtained, one that has generated innumerable doubts, but has been used throughout the literature, as well as in traditional chronological charts. In fact, this date has been used to posit cultural interpretations in other areas, thus endowing this early chronology with validation and legitimation, despite the fact that it has not been corroborated by new empirical data.

In this regard, an observation from Schmidt's (1983) review related to the need for future research is of interest:

> Before we become too enchanted with relations with Capacha, a good procedure would be to localize it stratigraphically, obtain additional absolute dates, and improve the sample, of both sites and contents, in the West. [But], we should not become so empirical as to repress speculation [for] this is quite legitimate, provided, as in this case, it is recognized as such; without this . . . without intuition, empiricism leads us nowhere. (Schmidt 1983: 210; translation P. Kersey)

At this point, I would simply repeat Schmidt's caution against allowing this old and controversial date to become the basis for an interpretation of Capacha. Archaeologists continue to repeat this date in the hopes of eventual improvements to our knowledge, but this is difficult in a region where archaeological investigations are neither long term nor well supported.

Evaluating Capacha

Conducting an evaluation of all the works that have used Capacha data—especially those related to chronology and analogies with ceramic materials from this phase—would be a huge undertaking, but it is important to analyze, albeit briefly, some of the most significant proposals that have emerged.

The investigator who has worked the most on the Capacha phase is Mountjoy, albeit in an area considered outside of the nuclear region in Colima. Based on his work in the Mascota Valley of Jalisco, he has addressed various aspects of this phase, and chronology is without question one of them. For years he has expressed doubts as to the absolute dating proposed by Kelly (1980), based on the various issues already discussed, and he has shown this through his discoveries of contexts in the Mascota Valley associated with ceramics similar or even identical to those of Capacha.

Mountjoy, who has always supported a placement in the Middle Formative, identifies this period in his own study area in four main sites. He has separated the material into two subphases, the first (at Los Coamajales and Los Añiles) with dates from 1000 to 900 BC, and the second (at El Pantano and El Embocadero II) around 800 BC. It is important to underline that these are absolute C14 dates, all from representative cultural contexts and calibrated. Mountjoy (2012) proposes these two subphases for Capacha based on the ceramic material, but also on the characteristics of the excavated contexts, which are quite different from one another. He proposes then that in Colima the earliest evidence for Capacha must range between 1200 and 900 BC (Mountjoy 2012: 176), while the most recent pertains to 900–500 BC. Based on these chronological data and the associated materials, Mountjoy proposes that far western Mexico was colonized during this period (Mountjoy 2015), a hypothesis that carries major implications for Archaic period sites in the area.

Turning now to the Colima Valley, the area deemed the "nuclear zone" of the Capacha phase, several researchers have explored contexts in this time frame. Unfortunately, published chronological data are scarce, so the following section focuses on the information produced by our own research, conducted within the framework of the project entitled The Formative Period in Colima: Occupational Continuity. These data were gathered principally during explorations of the El Diezmo–Adonaí site in the municipality of Colima. The main goals of our work were to respond to the questions that have emerged in relation to Capacha and to define this phase on its own merits, not as derived from other areas or developments. These are not tasks to be taken lightly because they may entail rejecting assumptions that time and repetition have transformed into certainties. Taking temporality as our central axis, and based on the succession of occupations, or phases, in the Colima Valley, the principal hypothesis we sustain is, simply, that those occupations were marked by continuity. All the literature available on this issue, all the traditional chronological tables, and even most academic discussions, include the argument that a temporary hiatus, or abandonment, occurred in the Colima Valley between the Capacha phase and the later Ortices phase. Obviously, this assumption was based on the broad acceptance of the Capacha chronology based on the C14 date (Kelly 1980) discussed above. Often, this abandonment was attributed to a volcanic eruption and its environmental consequences—scorched forests, contaminated water, and the virtual annihilation of fauna—that forced the human groups living there to abandon the valley. But this assumption has not been subjected to a specific study of Capacha sites, and no published materials exist to support it.

This proposed temporal hiatus has been attributed, as Olay (2012: 265–266) explains, to two of the volcanic eruptions of the Volcán de Fuego between 3500 and 2500 BP or 1500 and 500 BC. This proposal is based on studies carried out by a group of geologists on these eruptive episodes. They have obtained different dates for, among others, the eruptions associated with the Capacha-Ortices era (Komorowsky et al. 1994). This could explain the temporal distance between Kelly's dates for Capacha and Ortices, which could oscillate between 11 and 13 centuries (Olay 2012: 268). In this respect, and before treating the evidence for Capacha that we have obtained from the excavations at El Diezmo–Adonaí, we carried out a study of the ceramics of both phases, assuming that we should be able to identify so many centuries of abandonment given the lack of contact between the populations of both phases. That is to say, we tried to look for ceramic traits that would clearly separate the two phases.

If the more-than-500-year hiatus were true, then the differences in the ceramics would be substantial and obvious. But our observations reveal precisely the opposite: while differences exist in the ceramic remains among the three phases we compared—Capacha, Ortices, and Comala (see Figure 1.2)—there are also clear, evident characteristics that suggest continuity. These are based on a shared substrate that we identified in terms of both form and decoration, as well as in relation to certain manufacturing techniques (Almendros et al. 2014).

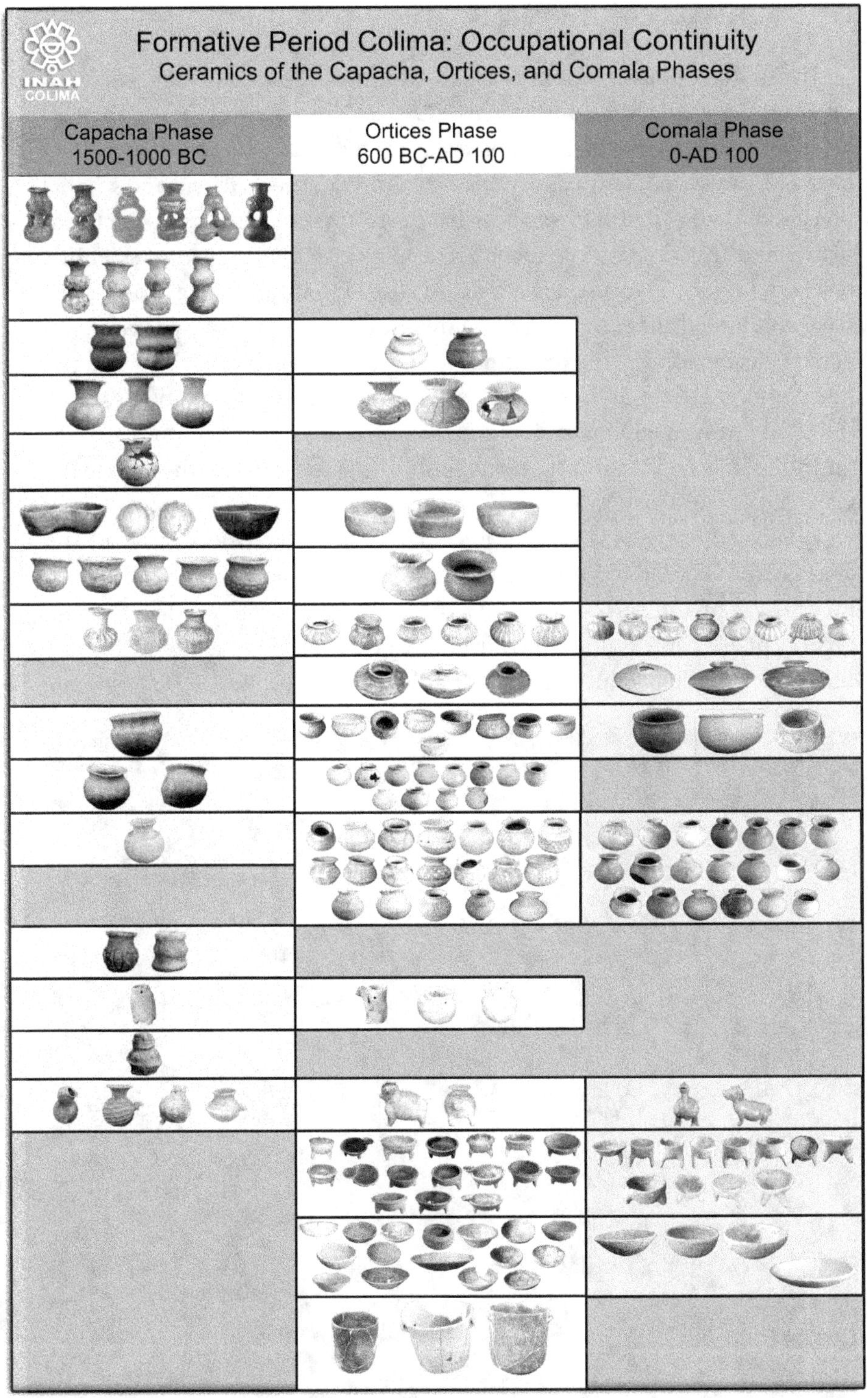

Figure 1.2. Comparative table of ceramic forms of the Capacha, Ortices, and Comala phases (cf. Almendros et al. 2014: 127). Evident similarities in vessel form, decoration, and manufacturing techniques among Capacha-, Ortices-, and Comala-phase ceramics suggest occupational continuity throughout the Formative period.

Turning to our excavations at the El Diezmo–Adonaí site that have been ongoing since 2011, the appearance of Capacha funerary contexts there allowed us to interpret a distribution area of remains as a cemetery of the Capacha phase, with later occupations pertaining to the Armería and Chanal phases (AD 600–1500). It was possible to shift work at this site, which was discovered in a context of rescue archaeology, into a long-term scientific research project by investing significant effort into fundraising. To date, we have conducted three seasons of fieldwork that include the application of geophysical studies, the excavation of 10 interments, and the obtaining of one C14 date. That C14 sample was sent to the Center for Isotope Research (CIO) at Groningen University in Holland for dating. It was a molar from an individual in Burial 9 that was found in association with two clearly Capacha vessels that had been deposited beside the skull (Figure 1.3). It was analyzed by the AMS technique (Accelerator Mass Spectrometry) and generated a date of 2050 ± 35 BP (Figure 1.4: sample no. GrA-52579; Almendros et al. 2013). Together with this sample, we sent a molar from an Ortices phase context unearthed from Burial 22 at the El Zalate site (Figure 1.5). That sample produced a date of 2140 ± 30 BP (Figure 1.6: sample no.

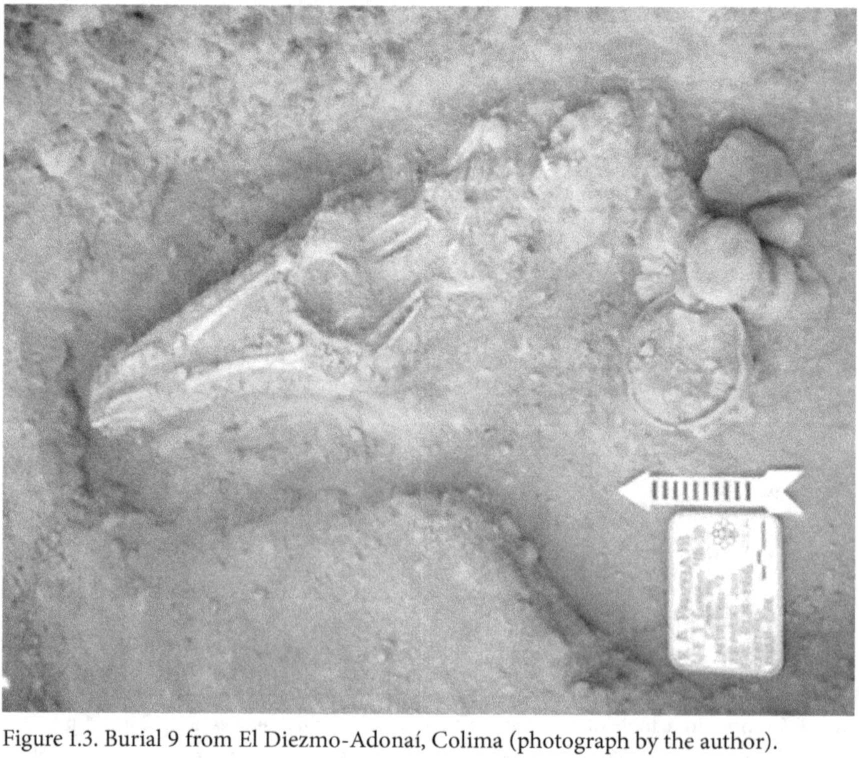

Figure 1.3. Burial 9 from El Diezmo-Adonaí, Colima (photograph by the author).

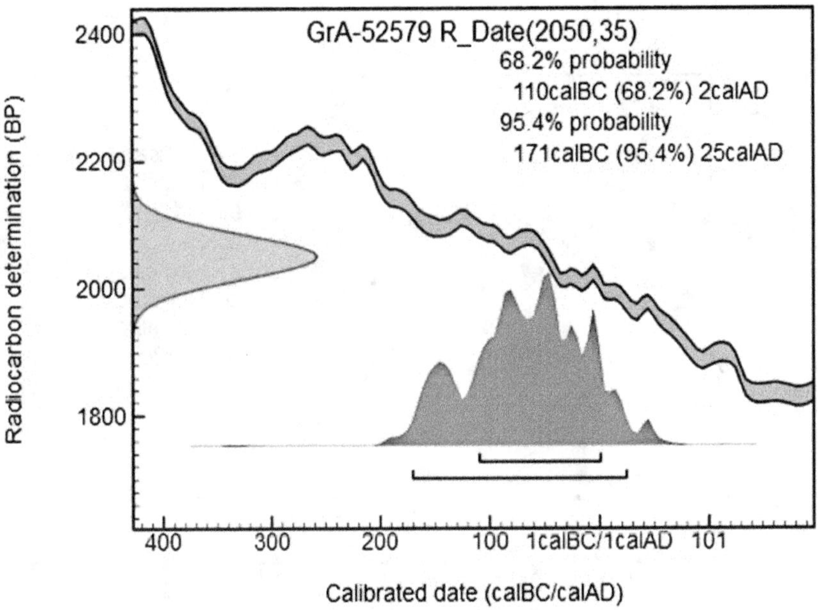

Figure 1.4. Calibrated dates obtained from molar in Burial 9 at El Diezmo-Adonaí (Ramsey 2015; Reimer et al. 2013).

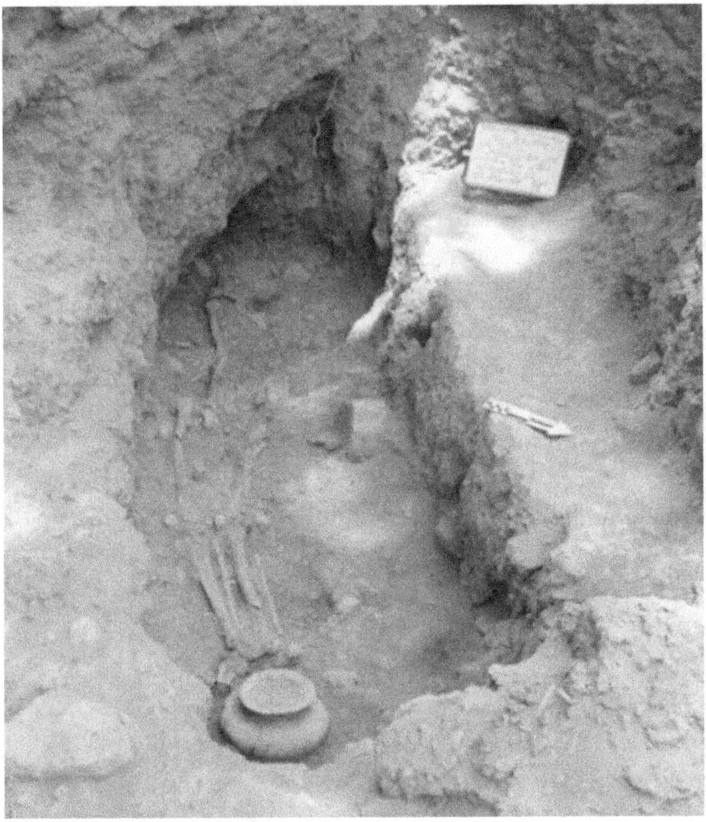

Figure 1.5. Burial 22 from El Zalate, Colima (photograph by the author).

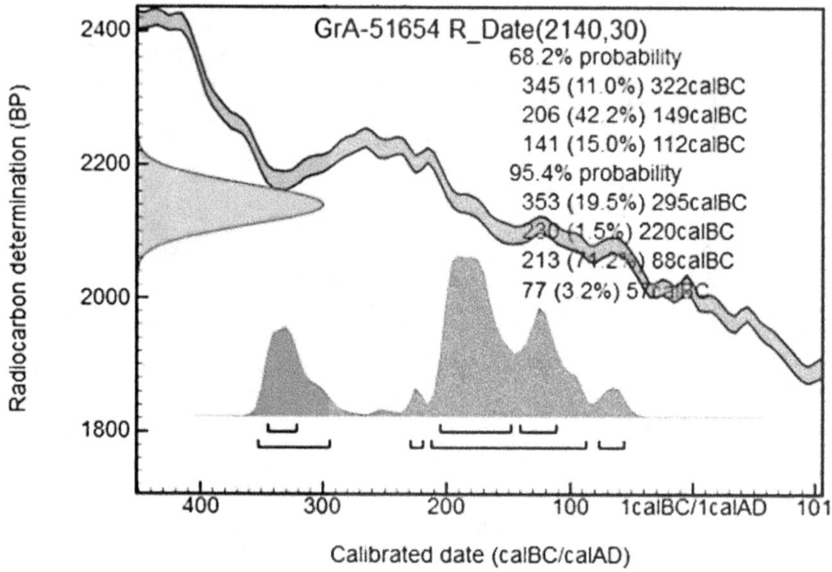

Figure 1.6. Calibrated dates obtained from molar in Burial 22 at El Zalate (Ramsey 2015; Reimer et al. 2013).

GrA-51654). Clearly, these two dates serve to muddle the controversy that surrounds Capacha's chronology even more, because after calibration there is a 95.4 percent probability that these samples correspond to the interval between 171 BC and AD 25, dates that are more recent than those attributed to the Ortices phase. Far from resolving the doubts surrounding the dates for Capacha, this sample intensifies them, especially when we consider that the date for the sample from El Diezmo–Adonaí is much more recent than we expected and so may be deemed as unreliable or simply false.

In this regard, we stress that we have no intention of repeating the error of basing an entire interpretation on one date. In contrast to Kelly's date, the reliability of ours is very high because the sample was extracted from a rigorously controlled archaeological excavation and was processed and analyzed in a laboratory using modern C14 methods—not to mention the nature of the material analyzed, and its half-life. What this new date does establish is the need to work more diligently on research into the chronology of the Capacha phase and the occupational sequence in the Colima Valley, conscious of the fact that uncertainty is currently the principal characteristic of the schemes that pretend to delineate the temporality of early cultural developments in West Mexico.

Here we think the proposal by various investigators—that there must be subphases within Capacha—is warranted, but the problem remains of how to identify these subphases in the scarce number of contexts that have been excavated. Without a doubt, there exist some preliminary data such as the different levels of burials that Alcántara (2005) identified in the cemetery of Las Fuentes, to the south of the Colima Valley, and where surely the analysis of the rich offerings combined with absolute dates will throw much light on our problem. In the same manner, although with fewer examples, the site of El Diezmo–Adonaí shows different levels of burial deposition that will be able to provide new data to test Mountjoy's interesting proposal of Capacha subphases.

Before addressing other aspects of the Formative in Colima, and to continue with the theme of chronology, we claim that until we have a series of absolute dates from Capacha contexts that allow us to address its length and subphases, we will not be in a position to confidently improve our understanding. There are areas in western Mexico that fortunately have well-supported projects with absolute dating, but Colima is not one of them.

There are other methodological approaches now that would allow us to get closer to a chronology of the phase and its interpretation. One of them is proposed clearly by Beekman and Weigand (2008), who emphasize the identification of ceramic complexes, beyond a few diagnostic types, that are then placed in relative order, and at some later date supported with absolute dates from secure contexts. In our case, we propose the combination of different data, both absolute and relative, to obtain a better interpretation of what happened in the past, incorporating contextualized radiocarbon dates, stratigraphic information, and typological data with the objective of using Bayesian analysis to arrive at a more confident interpretation. To that end, we will now speak to this last point.

A New Methodological Focus: Radiocarbon Dating and Bayesian Statistics

In relation to the Capacha phase, we have very few assumed "certainties"—we know little about its material and funerary culture and regional distribution, and virtually nothing regarding habitation zones, settlement patterns, economic practices, or origins in the Colima area, much less of its possible contacts with other regions. This is because Capacha is always analyzed as a function of exogenous developments. This broad uncertainty opens up a veritable panoply of lines of research, but we consider that priority should be given to anchoring the chronology, which is one of the fundamental objectives of our discipline—determining the time frame in which the archaeological events we strive to understand occurred. It is only on the basis of a solid chronological ascription that we can advance our knowledge of the spaces occupied by the groups we study and of their economic and cultural characteristics, not to mention the obvious need to resolve chronological issues if we genuinely wish to ascertain the role that those groups played in micro- and/or macroregional cultural dynamics.

It is within this framework that we insist upon the need to obtain an ample series of radiocarbon dates, since C14 is currently the method that provides the greatest possibility of refining the time intervals in which the archaeological events we wish to date took place. In the future, this project will focus on obtaining absolute dates—based primarily on radiocarbon—in order to be able to define not only the period to which materials considered to be of the Capacha type, and the human groups that produced them, correspond, but also the relation between the Capacha and Ortices phases or, perhaps, the lack of any such relation.

But first it is necessary to understand what really stands behind C14, or radiocarbon, dating. Great advances have been made in the half century since Libby and his team (1952) discovered this isotopic dating method, although few archaeologists delight in entering the world of dating and calibrations, much less that of the application of statistical models.

Using an organic sample from an archaeological context, we can obtain a conventional C14 age expressed in "years before the present" (BP), together with an error measure (standard deviation), shown as ± "x" years (López Sáez 2008; Mestres 2003, 2008; Santos et al. 2015). The question is whether what we obtain at this point is an absolute date. Our answer is yes . . . and no. While such dates are absolute, they need to be defined and refined. When C14 first began to be used—and, unfortunately, still today in some research—archaeologists considered that the average date determined corresponded to the one on which the event occurred. In reality, however, the probability that the date calculated in this way is the "real" one is very low. This is because a factor of relativity exists in absolute dating, a process that is enveloped within a whole series of uncertainties that must be eliminated one-by-one if we are to approach more closely the concrete date we seek; that is, the one that corresponds to the archaeological event in question.

The first key aspect here is calibration, which is required because C14 years and calendar years are not of exactly the same length (Barceló 2008: 24) due to the fact that the C14 content in the atmosphere has not been constant during the time period for which this method is effective. This means, for example, that the date for the Capacha context at El Diezmo–Adonaí (Figure 1.4) indicated above has a probability of 68.2 percent (1 sigma) of falling between 110 BC and AD 2. While logically this is earlier than the date with a probability of 95.4 percent (2 sigmas), we cannot rule out the possibility that it falls within the first range. When we observe the date from the Ortices context at the El Zalata site (Figure 1.6), the pronounced variation among them means that none can be ruled out.

Upon examining the characteristics that Mestres (2003, 2008) mentions as requirements for valid radiocarbon dates, we see that the first is chemical in nature; that is, it entails both eliminating contamination from the sample and calibrating the dating process. But in addition to these criteria, it is important to point out that the amount of carbon in samples will vary because the different organisms that we may attempt to date could have undergone differential exchanges of C14. Here, it is important to understand, for samples of marine or aquatic origin as well as those taken from organisms that may have had some type of marine influence (for example, the bones of an individual whose diet consisted largely of marine organisms), that the latter will produce an earlier date than a sample taken from a contemporary organism that did not consume marine products.

These aspects and complications associated with radiocarbon dating are of a chemical and analytical order, according to Mestres's (2003, 2008) proposal regarding the validity of C14 dating. But there is a third, no less important, order to be considered: the archaeological one. To achieve validity in this category, a date must be representative of the context for which we want to assign a date. This involves two characteristics: association and synchrony (Mestres 2003: 24). First, the sample to be dated must be associated with a human activity; second, synchrony means obviously that the death of the organism from which the sample is taken must correspond in time to the archaeological event that interests us. Therefore, to the uncertainties of a chemical and procedural nature, we must add those of an archaeological character that may be introduced at any point between the phases of the contexts up until the selection of the most suitable samples. We must consider as well their representativeness in relation to the historical moment to be dated. Thus, the excavation process and good stratigraphic recording play a fundamental role in narrating the history of any sample. By this I mean that the taphonomic and postdepositional processes that the sample chosen may have suffered will allow us to make corrections, and determine the significance—or lack thereof—of performing the dating procedure.

Clearly, the selection of archaeological events goes hand-in-hand with the key topic that we set out to investigate several years ago; namely, understanding the development of Capacha and its possible relation to Ortices. This requires not only defining precise dates and obtaining associated chronometric data, but combining this evidence with other data from the context, including stratigraphy, ceramic associations, settlement patterns, and funerary architecture, among others.

In order to achieve a good conjunction and interpretation based on such data, Barceló (2007: 23) proposes applying a statistical approach. In his words: "Statistics as such is today understood as the logic through which we can climb up one rung on the ladder that leads from data to information." But different statistical approaches exist, primarily the "classic" and "Bayesian" types. In recent years, the latter has gained wide acceptance in archaeology because, while both approaches set out from an assumption of verisimilitude, unknown parameters are handled as fixed values in the first, but in the second they are considered as random (Barceló 2008; Jover et al. 2014). Largely for this reason, researchers who adopt the Bayesian approach coincide in identifying the feature that makes it especially attractive: its ability to incorporate data from the context that one wishes to analyze (Barceló 2008; Beramendi et al. 2009; Capuzzo 2014; Jover et al. 2014; Ramsey 2015).

Thus, the qualitative homogeneity between relative information, understood as stratigraphic and associative data, and absolute data, such as radiocarbon dates, is the required condition that all authors recognize in relation to the potential application of an effective Bayesian statistical approach to resolving archaeological problematics (Barceló 2008; Bayliss et al. 2007; Capuzzo 2014; Jover et al. 2014; Steier et al. 2001; Williams 2012). The conclusions of many studies that have adopted Bayesian inference contain this warning of the need to ensure that data are both reliable and comparable.

We may well have come late to the application of such studies compared to areas where they are no longer deemed "novel" but, rather, "routine." Whatever the case, what matters is to begin this process and ensure that our research on the Formative period in Colima is based on data that are increasingly robust, reliable, and scientific. We must rid ourselves of the stigma that the research we carry out is judged as "marginal" by finding responses through our own processes and comparisons with others, based on the homogeneity and quality of information; that is, on the archaeological data that will allow us to progress in our general understanding of the Formative period in West Mexico.

Other Aspects of the Formative Period in Colima

In addition to the spatial and chronological aspects of the Capacha and Ortices phases in the traditional cultural sequence of Colima, other interesting issues have emerged that are the subject of ongoing research. Perhaps the most important of these involves funerary contexts since, as we have seen, the vast majority of the sites found to date are associated with burials.

In recent years, monumental architecture and its association with the Ortices phase has gone hand-in-hand with research that probes a possible relation with the Teuchitlán Tradition best known from central Jalisco (see Beekman, this volume). In this regard, there have been many changes in the discourses on West Mexico, where monumental construction was thought to only appear with the arrival of Central Mexican influences around AD 600. But this long-standing belief is increasingly being challenged by new studies in different zones in the northern Colima Valley and on the slopes of the volcano. Here, we could mention the sites of Comala, La Herradura, and Potrerillos (Olay and Sánchez 2015), all of which present remains of ceremonial architecture that dates to the Late Formative and Classic periods.

Returning to funerary elements, there is the eternal question of whether or not the Capacha phase included shaft tomb–style burials. Our research has yet to identify any association between these tombs, which constitute the diagnostic feature of the Formative and Classic periods in West Mexico, with Capacha, at least as defined on the basis of its best-known diagnostic remains, the double *olla*. In this sense, we can call attention to the evidence presented by Mountjoy in Mascota with the association of materials assigned to Capacha with shaft and chamber tombs. Without a doubt, this is one of the great questions for researchers in Colima (Olay 2012). Nevertheless, with the exception of looters' reports of Capacha material in shaft tombs, which cannot be confirmed, we have no evidence of this in Colima. In El Diezmo–Adonaí, we have used ground-penetrating radar (GPR) without apparent success (Barba et al. 2013), since we have to date identified no subsurface features corresponding to a hollow space. Another type of funerary architecture has been proposed; namely, the perimeter walls that delimit burial areas corresponding to the Capacha phase (Ramos de la Vega et al. 2005). While we have no definitive evidence of such walls at El Diezmo–Adonaí, a fragment of a wall of rustic construction was unearthed in the southern sector of the excavated area. It consists of an irregular accumulation of medium-sized stones that we have followed for about 3 meters, but it has not appeared in other areas of the site. Thus, the nature of this feature needs to be determined. We can say that it is found at a distance from the burials that coincides with common cemetery sizes (as reported by Kelly 1980, 30 × 30 m), there is currently a gap between the feature and known burials.

Stone "burial markers" are another architectural feature associated with funerary contexts, namely, the Capacha and Ortices phases. In general, and in the absence of detailed analyses of this kind of architectural element, we can say that most such markers consist of mounds of medium-sized stones combined, in some cases, with larger rocks. Those associated with Ortices materials may include broken *metates* that were reutilized in these constructions.

We can address this at El Diezmo–Adonaí. Our extended excavations of a level that includes this architectural feature revealed what appears to be a Capacha cemetery, although we do not yet have all the data necessary to categorically affirm the existence of a pattern in the orientation of the stone piles or the precise characteristics of their construction. Nor can we (1) discern if they were built prior to, or at the time of, the burials; (2) determine their functionality; or, (3) discern if they constitute a typology associated with a chronological aspect or questions of social differentiation among the individuals interred there. But findings do show, again in general, that the burials in both of these Formative phases contain skeletons arranged in an extended position accompanied by offerings usually placed at their feet or head. Also visible in both phases is the presence of cranial deformation and individual as well as multiple interments (Alcántara 2005; Olay et al. 2010).

Both phases are characterized by burial pits dug into the earth—sometimes down to the bedrock—but, in contrast to the Ortices phase, Capacha has no shaft tombs. Another little-studied topic mentioned earlier involves functional analyses of ceramic materials, but this procedure faces a severe limitation: the fact that pottery remains come almost exclusively from funerary contexts. While these materials may often be interpreted as everyday household articles that are later included in mortuary offerings, it is also common to find that the materials interred with the dead were designed especially for that purpose.

With respect to the ceramics, we can confirm the presence of the most common Capacha form, which is the double *olla* or *bule*, which is present as a complete vessel in the offerings and in fragments found throughout the excavations. In addition, the vessels include the typical "sunburst" decorative element, although in different varieties, basically made with a central circular depression surrounded by radiating lines. Decoration can also include triangles filled with punctations and incised lines.

The great majority of the ceramic fragments correspond to the monochrome type presented by Kelly (1980). Nevertheless, we have examples with red-pink and zoned decoration. As for forms, besides the most common form already mentioned, we find simple wide-mouthed *ollas*, bowls of various sizes, neckless jars or *tecomates*, and miniature *ollas*. They are found both in the offerings and in fill contexts.

We can also take into consideration the materials that Mountjoy found associated with the Capacha sites in Mascota, where pinkish paint appears in the most recent phase. In the case of El Diezmo–Adonaí, we have a broken example of the double *olla* that was found isolated and outside of any funerary offering. It has the "sunburst" design with the application of a red-pink paint and was found in the uppermost levels of the excavation. Without doubt, this is a something we need to investigate to see whether it can be associated with the different occupations at the site, and it may support the interpretation of subphases proposed by Mountjoy (2012).

We also found a vessel whose form is reminiscent of the "water cap" form of the Capacha contexts identified by Mountjoy for the later subphase in Mascota. This vessel is associated with a secondary burial and located in the highest levels of the excavation. Although important, this requires further analysis.

Continuing with the issue of ceramics, it is important to bring to the table the interesting discussion of Guy Hepp (2015) working at the site of La Consentida on the coast of Oaxaca. Hepp has been able to identify a number of characteristics similar to Capacha among the ceramic and lithic material at that site, suggesting the movement of people from the Early Formative period. The rich deposits at that site have allowed him to address various issues relating to their economic base and social characteristics, which will need to be carefully assessed in relation to Capacha in Colima.

Among the elements that we must continue to investigate in Capacha are whether these people were dependent on agriculture, with a fully sedentary lifestyle, and with a hierarchical social organization, characteristics found widely among Middle and Late Formative period groups elsewhere in Mesoamerica. Nonetheless, as we can see from the case of La Consentida, detailed studies at the site level are necessary before we can begin to generalize effectively.

Note

1. Particularly the Valdivia and Machalilla phases in Ecuador, with chronologies between 3800 and 1000 BC.

References

Alcántara, Andrés Saúl

2005 *Las Fuentes. Un Panteón Preclásico en Colima.* Tesis de Licenciatura. Escuela Nacional de Antropología e Historia, Mexico City.

Almendros, Laura, and Fernando González

2009 El Occidente de México: La Reocupación del Valle de Colima. *Boletín Americanista* 59: 137–154.

Almendros, Laura, Rafael Platas, and Mariza Cuevas

2013 Entierros en el Sitio El Diezmo–Adonaí. *Arqueología Mexicana* 120: 10–12.

2014 Continuidad y Discontinuidad en la Cultura Material del Periodo Formativo en Colima. *Boletín Americanista* 69: 111–134.

Barba, Luis, Laura Almendros, Agustín Ortiz, Maritza Cuevas, Jorge Blancas, and Rafael Platas

2013 Estudios geofísicos aplicados al estudio de contextos arqueológicos. El Diezmo–Adonaí, un caso en el valle de Colima. *Memoria del VIII Foro Juan Carlos Reyes Garza. Colima y su región. Arqueología, antropología e historia.* CONACULTA Y Secretaría de Cultura de Gobierno del Estado de Colima, 123–145.

Barceló, Joan Anton

2007 *Arqueología y Estadística: Introducción al Estudio de la Variabilidad de las Evidencias Arqueológicas.* Universidad Autónoma de Barcelona, Barcelona.

2008 La Incertesa de les Cronologies Absolutes en Arqueologia: Probabilitat i Estadística. *Cypsela* 17: 23–33.

Bayliss, Alex, Christopher Bronk Ramsey, Johannes van der Plicht, J., and Alasdair Whittle

2007 Bradshaw and Bayes: Towards a Timetable for the Neolithic. *Cambridge Archaeological Journal* 17: 1–28.

Beekman, Christopher S., and Phil C. Weigand

2008 Conclusiones, Cronología, y un Intento a Síntesis. In *Tradición Teuchitlán*, edited by Phil C. Weigand, Christopher Beekman, and Rodrigo Esparza, pp. 303–337. El Colegio de Michoacán, Zamora, México.

Beramendi-Orosco, Laura E., Galia Gonzalez-Hernandez, Jaime Urrutia-Fucugauchi, Linda R. Manzanilla, Ana M. Soler-Arechalde, Avto Gogichaishvili, and Nick Jarboe

2009 High-Resolution Chronology for the Mesoamerican Urban Center of Teotihuacan derived from Bayesian Statistics of Radiocarbon and Archaeological Data. *Quaternary Research* 71(2): 99–107.

Capuzzo, Giacomo

2014 *Space-Temporal Analysis of Radiocarbon Evidence and Associated Archaeological Record: From Danube to Ebro Rivers and from Bronze to Iron Ages.* Unpublished doctoral dissertation, Departament d'Antropologia Social i Cultural, Universitat Autónoma de Barcelona, Barcelona.

Hepp, Guy David

2015 *La Consentida: Initial Early Formative Period Settlement, Subsistence, and Social Organization on the Pacific Coast of Oaxaca, Mexico*. Unpublished PhD dissertation, Department of Anthropology, University of Colorado–Boulder.

Instituto Nacional de Estadística y Geografía (INEGI)

1981 *Síntesis Geográfica*. INEGI, Mexico City.

Jover Maestre, Francisco Javier, Juan Antonio López Padilla, and Gonzalo García-Donato Layrón

2014 Radiocarbono y Estadística Bayesiana: Aportaciones a la Cronología de la Edad del Bronce en el Extremo Oriental del Sudeste de la Península Ibérica. *SAGVNTVM: Papeles del Laboratorio de Arqueología de Valencia* 46: 41–69.

Kelly, Isabel T.

1980 *Ceramic Sequence in Colima: Capacha, an Early Phase*. Anthropological Papers of the University of Arizona 27. University of Arizona Press, Tucson.

Komorowsky, Jean-Christophe, Carlos Navarro, Abel Cortés, and Claus Siebe

1994 The Unique Tendency for Recurrent Collapse at Colima Volcanoes (Mexico): The Challenge of Reconciling Geologic Evidence with 14C Chronology, Implications for Hazard Assessment. In *Proceedings of the International Conference on Volcano Instability on the Earth and Other Planets*. Geological Society of London, London.

Libby, Willard F.

1952 *Radiocarbon Dating*. University of Chicago Press, Chicago, IL.

López Austin, Alfredo, and Leonardo López Luján

2002 La Periodización de la Historia Mesoamericana. *Arqueología Mexicana* 11: 6–15.

López Mestas C., Lorenza

2011 *Ritualidad, Prestigio y Poder en el Centro de Jalisco durante el Preclásico Tardío y Clásico Temprano. Un Acercamiento a la Cosmovisión e Ideología en el Occidente del México Prehispánico*. Unpublished doctoral dissertation, Department of Social Sciences, Centro de Investigaciones y Estudios Superiores en Antropología Social, Guadalajara.

López Sáez, José Antonio

2008 Reconstrucciones Paleoambientales a partir de Estudios Paleopalinológicos y Precisiones sobre el uso del Método de Datación por 14C. In *Técnicas y Aplicaciones Multidisciplinares de los Isótopos Ambientales*, edited by Paloma Alcorlo, Ramón Redondo Ortega, and José Toledo, pp. 107–124. Cuadernos de Apoyo 22. Universidad Autónoma de Madrid, Madrid.

Mestres, Joan Salvador

2003 La Química i la Cronologia: La Datació per Radiocarboni. *Revista de la Societat Catalana de Química* 4: 10–25.

2008 El Temps a la Prehistòria i el seu Establiment a través de la Datació per Radiocarboni. *Cypsela* 17: 11–21.

Morales, Juan, Avto Gogichaichvili, Ma. de los Ángeles Olay Barrientos, Claire Carvallo, and Bertha Aguilar Reyes

2013 Archeointensity Investigation on Pottery Vestiges from Puertas de Rolón, Capacha Culture: In Search for Affinity with Other Mesoamerican Pre-Hispanic Cultures. *Studia Geophysica et Geodaetica* 57(4): 605–626.

Mountjoy, Joseph

1994 Capacha: Una Cultura Enigmática del Occidente de México. *Arqueología Mexicana* 2(9): 39–42.

2010 Caracterización de la Cerámica del Formativo Medio y Tardío Encontrada en la Costa de Jalisco y Nayarit y el Altiplano Adyacente. In *El Sistema Fluvial Lerma-Santiago durante el Formativo y el Clásico Temprano: Precisiones Cronológicas y Dinámicas Culturales,* edited by Laura Solar, pp. 43–56. Instituto Nacional de Antropología e Historia, Mexico City.

2012 *El Pantano y Otros Sitios del Formativo Medio en el valle de Mascota, Jalisco*. Secretaría de Cultura del Estado de Jalisco, Guadalajara.

2015 La Colonización del Lejano Occidente de México por Agricultores Sedentarios durante el Formativo Medio, 1200 a 400 a.C. *Revista Occidente* 1–18. Electronic document, https://www.academia.edu/37290710/LA_COLONIZACI%C3%93N_DEL_LEJANO_OCCIDENTE_de_M%C3%A9xico_por_agricultores_sedentarios_durante_el_Formativo_medio_1200_a_400_a.C, accessed August 9, 2019.

Olay Barrientos, Ma. de los Ángeles

2012 *Volcán de Fuego, cuna del agua, morada del viento: desarrollo social y procesos de cambio en el valle de Colima: una propuesta de interpretación*. INAH and El Colegio de Michoacán, Colima and Zamora.

2015 El Occidente Mesoamericano: Una Lectura sobre su Pasado. *Revista Occidente* 1–20. Electronic document, http://www.mna.inah.gob.mx/documentos/Revista_occidente/Olay_montado_final.pdf, accessed October 30, 2016.

Olay Barrientos, Ma. de los Ángeles, Saúl Alcántara, and Laura Almendros

2010 La Tradición Capacha: Evaluación y Perspectivas desde el Valle de Colima. In *El Sistema Fluvial Lerma-Santiago durante el Formativo y el Clásico Temprano: Precisiones Cronológicas y Dinámicas Culturales,* edited by Laura Solar, pp. 19–42. Instituto Nacional de Antropología e Historia, Mexico City.

Olay Barrientos, Ma. de los Ángeles, and Ligia Sofía Sánchez Morton

2015 La Tradición Teuchitlán en Colima: Notas para una Discusión. *Revista Occidente* 1–18. Electronic document, http://www.mna.inah.gob.mx/documentos/Revista_occidente/Olay_morton_grados_2.pdf, accessed October 30, 2016.

Ramos de la Vega, Jorge, Lorenza López Mestas, Jean-Pierre Emphoux, and Javier Revelez

2005 Investigaciones Arqueológicas en la Cuenca del Río Salado, Colima. In *El Occidente y Centro de México. IV Coloquio Pedro Bosch Gimpera* (II), edited by Ernesto Vargas, pp. 57–74. UNAM Instituto de Investigaciones Antropológicas, Mexico City.

Ramsey, Christopher Bronk

2013 OxCal 4.2. Web Interface Build. Electronic document, https://c14.arch.ox.ac.uk/oxcalhelp/hlp_contents.html, accessed October 13, 2016.

2015 Bayesian Approaches to the Building of Archaeological Chronologies. In *Math-*

ematics and Archaeology, edited by Joan Antón Barceló and Igor Bogdanovic, pp. 272–292. CRC Press, Boca Raton, FL.

Reimer, P. J., E. Bard, A. Bayliss, J. W. Beck, P. G. Blackwell, C. B. Ramsey, and P. M. Grootes

2013 IntCal13 and Marine13 Radiocarbon Age Calibration Curves 0–50,000 Years Cal BP. *Radiocarbon* 55(4): 1869–1887.

Santos Arévalo, Francisco J., Lidia Agulló García, Aurora Diéguez Ferrari, and Isabel Gómez Martínez

2015 Datación Radiocarbónica y Arqueología: La Experiencia del Centro Nacional de Aceleradores (Sevilla). *Menga: Revista de Prehistoria de Andalucía* 6: 95–111.

Schmidt Schoenberg, Paul

1983 Kelly, Isabel: Ceramic Sequence in Colima: Capacha, an Early Phase. *Anales de Antropología* 20(2): 208–212.

Steier, Peter, Werner Rom, and Stephan Puchegger

2001 New Methods and Critical Aspects in Bayesian Mathematics for 14C Calibration. *Radiocarbon* 43(2A): 373–380.

Tylor, Edward B.

1995 [1871] La Ciencia de la Cultura. In *El Concepto de Cultura: Textos Fundamentales,* compiled by J. S. Kahn, pp. 29–48. Editorial Anagrama, Barcelona.

Williams, Alan N.

2012 The Use of Summed Radiocarbon Probability Distributions in Archaeology: A Review of Methods. *Journal of Archaeological Science* 39(3): 578–589.

2

The Early Segment of the Chronological Sequence at Los Guachimontones

CHRISTOPHER S. BEEKMAN

The site of Los Guachimontones is located in the Tequila valleys of central Jalisco. With its nuclear sector and the secondary sectors of Talleres, Loma Alta, and Texcalame, it was the largest known archaeological site in western Mexico prior to the Epiclassic period. The built space at the site is composed of two types of formal public architecture: ballcourts and circular groups of platforms around a circular altar and known as either circles or *guachimontones*. Most residential architecture takes the form of rectangular patio groups familiar to Mesoamericanists, although some small circles are found among the residential areas as well. Nearly a century after the site's discovery (Breton 1903), it became the focus of the Proyecto Arqueológico Teuchitlán under the direction of Phil Weigand. In this first phase of operations, the project carried out excavations from 1999 to 2010 oriented toward restoration for purposes of tourism. The excavations exposed the weaknesses of Weigand's approach to archaeological chronology, in which he had relied first on the assumption that public architecture became more elaborate over time (Weigand 1979, 1985, 1993), and that ceramic analysis could be reduced to identifying major decorated types or "flags" (Weigand personal communications, 1992 onward). When the first radiocarbon dates for the Proyecto Arqueológico Teuchitlán (PAT) established that the largest and most elaborate circular group was also one of the earliest, it became clear that the architectural sequence was conceptually flawed and empirically incorrect (Beekman and Weigand 2008). As part of the ongoing efforts to reassess the collections from Los Guachimontones (Figure 2.1) for purposes of research, we must reevaluate the historical sequence at Los Guachimontones using additional categories of evidence. Based on a general review of the ceramics, it is clear that the site was occupied from the

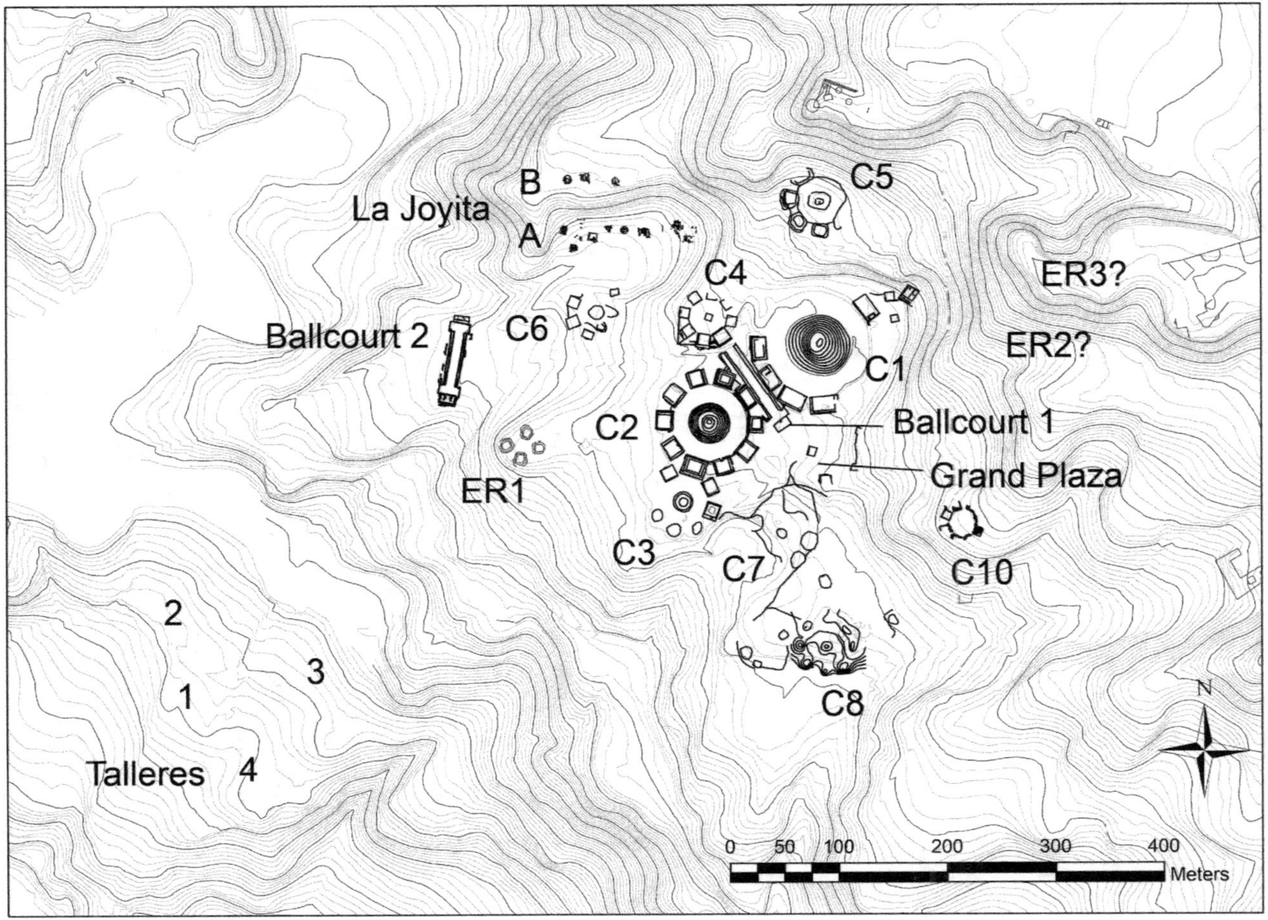

Figure 2.1. Map of the Nuclear and Talleres sectors of Los Guachimontones (courtesy of the Proyecto Arqueológico Teuchitlán). Operations ER2 and ER3 are not located in early PAT reports and are approximate based on text descriptions.

Table 2.1. Chronological sequence used in this chapter, with defining sites and publications

Phase	Proposed Dates	Published Ceramic Descriptions	Example Sites
Atemajac II	AD 1400–1600	Beekman and Weigand 2000; Davenport 2011; Galván 1979, 1983; Nance et al. 2013	Bugambilias Arriba, Anona, Tiana, Las Cuevas, El Guaje
Atemajac I	AD 900–1400	Davenport 2011; Glassow 1967; Nance et al. 2013	Huistla, Anona, Tiana, Las Cuevas, El Guaje
El Grillo	AD 450/500–900	Aronson 1993; Montejano 2007; Smith 2008	Tabachines box tombs, La Higuerita, Oconahua
Tequila IV	AD 200–450/500	Beekman 2001; Beekman and Weigand 2000; Galván 1991; López 2011; Ramos and López 1996	Tabachines shaft tombs, Llano Grande, Huitzilapa circle
Tequila III	100 BC–AD 200	Beekman and Weigand 2000; Blanco et al. 2010; Galván 1991; Johns 2014; López 2011; Ramos and López 1996	Tabachines shaft tombs, Navajas, Huitzilapa tomb
Tequila II	300–100 BC	Beekman and Weigand 2000, Blanco et al. 2010; Galván 1991	Tabachines shaft tombs
Tequila I	ca. 1000–300 BC	Heredia and Beekman 2018	Sites identified by the Proyecto Arqueológico Ex-Laguna de Magdalena (PAX)
Magdalena	ca. 1500–1000 BC	Heredia and Beekman 2018	Sites identified by the PAX

Middle Formative through the Late Postclassic period, encompassing three major ceramic traditions (with subdivisions) named Tequila, El Grillo, and Atemajac (Table 2.1). The first period received the most attention during Weigand's tenure as the director of the PAT, and it is the period that has been most affected by the use of the architectural chronology.

Earlier PAT personnel report having examined 65,000 sherds from Los Guachimontones architectural groups Circles 1, 10, Ballcourts 1 and 2, residential group La Joyita A, and Loma Alta Ballcourt 1 (Blanco et al. 2010). Few of these data exist in project files, but summary charts appear in some *informes*, or reports. The contexts analyzed are not recorded, so it is not possible to say whether these ceramics came from floors, fill, or other contexts. From our own efforts to identify useful contexts however, it seems clear that fill and trench contexts must have been incorporated to have reached the quantity reported. That analysis concluded that there were no discernible changes in the ceramics across the period attested to by radiocarbon dates at the site—that is, from 300 BC to AD 400. That analysis used a typology defined by proposed vessel functions, and sherds were recorded by type without additional information about rims, modes, diameters, decoration, forms, and so on. The apparent continuity in functional categories may be due to continuity in activities over time rather than the absence of chronological differences in the materials themselves. Their conclusions were in any case directly contradicted by Weigand, who in a separate discussion of artifacts from the site (Beekman and Weigand 2008, 2010) described a sequence of artifact changes over time. Since Weigand was a coauthor on both publications, it is hard to know which conclusions to follow. Therefore, a new and more detailed analysis was necessary to assess the site's chronology.

Problems with Working with the Collections from Los Guachimontones

Prior excavations at the Los Guachimontones site were oriented toward restoration for tourism purposes, emphasizing primarily the occupation from the Late Formative and Early Classic periods. Excavations thus treated later materials within the built space from these periods as overburden to be removed with limited recording. Excavations often proceeded beginning with one or more trenches that crossed or outlined architecture, and upper stratigraphic layers were treated summarily or grouped together in excavation. Excavators commonly collected all artifacts from the ground surface down to structures or floors as a single lot without distinguishing by stratigraphic context. The stratigraphy discerned in these trenches was recorded in section drawings, and this information was used to guide later excavations. Our current understanding of excavation recording is that *Capas* designated natural stratigraphic layers (with either numerical or text labels such as *húmica*, *negra*, and so on), while *Niveles* designated distances below surface (typically 20 centimeter units, as in 20–40 centimeters, although sequential numbers appear to have been used on occasion). Capas in at least some cases were assigned after excavation upon observation of the exposed stratigraphy.

Stratigraphic observations in the sections by Weigand during the first few seasons of excavation include much descriptive detail because of their

Table 2.2. Ceramic analyses to date using the combined type–variety/modal approach

Project	Site	# of Sherds Analyzed	Weight of Sherds Analyzed in g
Proyecto Arqueológico Teuchitlán (PAT) 2012, 2014, 2015	Los Guachimontones (incl. Loma Alta)	22,312	191,306
Tequila Valleys Regional Archaeological Project (TVRAP) 2003, 2004, 2011	Navajas	20,967	115,693
Proyecto La Primavera (PLP) 1993, 1994	La Venta Corridor	11,116	65,648
Proyecto Arqueológico ex–Laguna de Magdalena (PAX) 2013	Magdalena Basin	6,174	92,316
TVRAP 2000, 2003	Llano Grande	1,546	23,514
Total		59,408	455,290

Source: Compiled by author.

importance for architectural interpretation. As the project grew after 2001, stratigraphic descriptions became increasingly schematic, and very few features (postholes, trash or storage pits, and so on) were recorded anywhere in the excavated areas of the site. In the course of reviewing project reports, there were few locations where artifacts were collected in a natural stratigraphic sequence in a manner that would be of value for chronological purposes. Excessive attention during excavation was given to the recovery of artifacts and carbon samples from fill contexts, despite the weaknesses of such locations for dating, probably because of project interest in construction techniques.

Beginning in 2012, investigators from the University of Colorado–Denver, California State University–Los Angeles, El Colegio de Michoacán, University College of London, Trent University, and the University of Calgary have analyzed ceramics from use contexts at Los Guachimontones. Since 1993, I have developed a system of ceramic analysis based on materials from other locations in the Tequila valleys—surface collections and test excavations from the La Venta Corridor and the Magdalena Lake Basin (this last with Verenice Heredia), and from extensive excavations at Llano Grande and Navajas (Table 2.2). This has resulted in a type-variety system to collect data on surface treatment, forms, and pastes supplemented by a modal analysis to capture information on more sensitive changes in form. The type-variety component owes a great deal to the prior work of Javier Galván (especially 1991), and his initial analysis of tomb lots from the cemetery at Tabachines. Johns (2014) has recently completed a study of the manufacture characteristics of the different wares in order to better understand their possible functions. Because of our experience with collections from the extremes of the Tequila valleys, we now recognize regional variations in the occurrence of different wares and types. These antecedents are available in various publications (Beekman 1996a, 1996b, 2006; Beekman and Weigand 2000, 2008; Johns 2014). This chapter presents our ongoing application of this system to the early part of the sequence at Los Guachimontones.

Dating any sizable multicomponent settlement must address several problems. All parts of the site are likely to have seen some activity if only due to their close proximity to one another. Construction stirs up previously buried materials and incorporates them into the fill of new construction or scatters them among those contexts that would have been exposed in that period. Therefore, trace quantities of materials from a given phase cannot be considered to date an architectural group, and we should use multiple methods to date the construction of each group. Several methods are incorporated into the chronological analysis presented here:

Architectural stratigraphy. Weigand has already reviewed the evidence for the integrated construction sequence for Circles 1, 2, 3, 4, and Ballcourt 1, to develop a relative sequence (Weigand 2008).

Radiocarbon dates (Preliminary analysis of half the available dates in Beekman and Weigand 2008).

Analysis of excavated ceramic debris.

Analysis of figure and figurine fragments. This category and the preceding represent aggregated activities over time.

Analysis of the vessels and complete figures/figurines from offerings and burials. These reflect short-term events. One problem with using the existing collections was that the same vessels were attributed to different contexts in different project documents, and Lefae (2015, 2017) has reviewed the reports and original photos to correct these provenience problems.

Analysis of stratigraphic pits. We identified 20 pits with two or more strata that we could use to evaluate the relative order of ceramics and to test the sequence.

Architectural Stratigraphy

Weigand (1999, 2008; Weigand and García de Weigand 2000) observed during excavations at Los Guachimontones that the interface between architectural units enabled him to develop a construction sequence for the major architectural groups. This evidence unfortunately only extended to part of the nuclear sector, but it provides a relative framework for the analysis. We noted additional possible relationships between groups in the Loma Alta sector during our own analysis.

Weigand proposed that Los Guachimontones Circle 1 and Ballcourt 1 were built at approximately the same time since they shared multiple platforms on the side where they interdigitated. On the other side of the ballcourt, the lateral platform interdigitated with the platforms of Circle 2 when the ballcourt was expanded, and hence Circle 2 followed the construction of the ballcourt. Similarly, the platform forming the far northwestern end of the ballcourt was initially small, but it was later expanded and reoriented away from the ballcourt when it was incorporated into a new Circle 4. Circle 3 similarly appears to have been added after Circle 2 since their one shared platform is of a comparable size to those in Circle 2 but oversized compared to those in Circle 3. Although our consolidation and review of the plans, sections, and descriptions is continuing (DeLuca 2014, 2017), these relationships remain plausible.

Not discussed by Weigand was the arrangement of architecture in Loma Alta (Figure 2.2). Ballcourt 1 is notable for the manner in which it connects awkwardly with Circle A in that sector. The platform shared by Ballcourt 1 and Circle A shares the orientation of the ballcourt, even though it creates a disjunction in how that structure fits within Circle A. I suggest that both Circle A and the unnumbered circle north of Circle A were already in place, forcing the ballcourt to be constructed in an awkwardly defined space. Circle B appears to have been constructed after Circle A because they share a large platform that matches the scale of the other platforms on Circle A—the logic here is similar to the observation that the construction of Circle 3 in the nuclear sector postdates Circle 2.

The two architectural sequences may be linked by noting that Ballcourt 1 of Loma Alta shares the same orientation of 325° with the larger Ballcourt 1 of the nuclear sector. They may have been built at the same time, although the Loma Alta Ballcourt 1 is smaller and may be a derivative of the one in the nuclear sector. All these proposed relationships are represented in Figure 2.3 below. The relationships between these architectural groups only

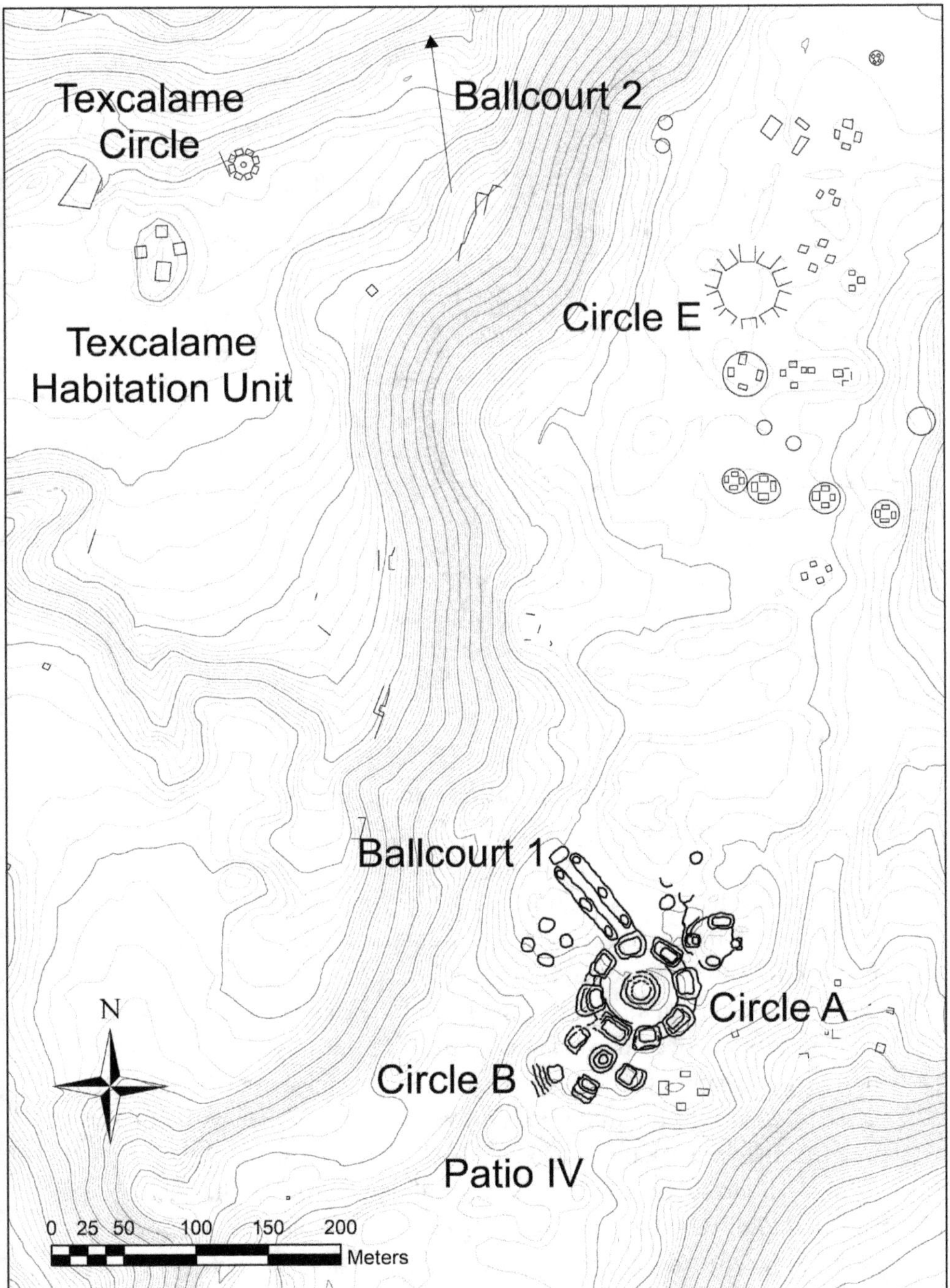

Figure 2.2. Map of the Loma Alta and Texcalame sectors (courtesy of the Proyecto Arqueológico Teuchitlán).

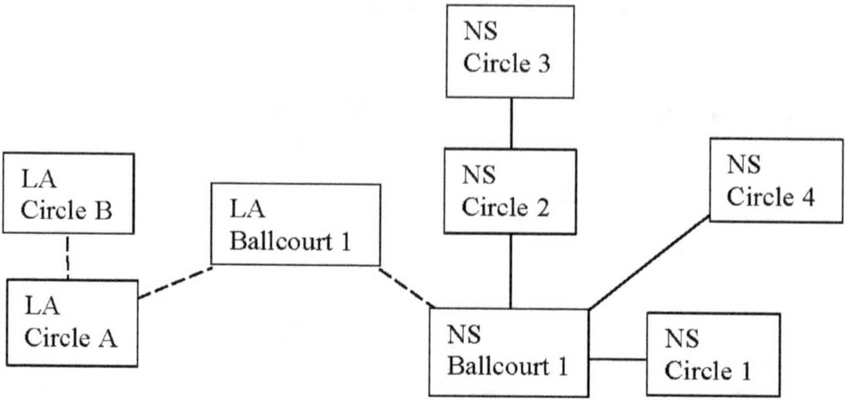

Figure 2.3. Schematic representation of relative stratigraphic relationships between architectural groups from the Nuclear Sector (NS), and the Loma Alta (LA) Sector (photograph by the author).

suggest the relative order in which they were constructed, and do not provide any information on how long activity continued in each space, which can best be evaluated through more chronologically sensitive datasets.

Radiocarbon Dates

Radiocarbon samples from the nuclear and Loma Alta sectors were collected and submitted for analysis by the PAT. Some have been reported elsewhere and were integrated with the Los Guachimontones architectural sequence at that time (Beekman and Weigand 2008, 2010). The central problem with those radiocarbon determinations was that many were taken from architectural fill contexts or from burned organics embedded within the clay used for *aplanado*, the fired face of the clay body of the platforms. Fill contexts can of course date to any period prior to sealing the construction, while organics within clay could potentially have been in place within the clay source itself and do not necessarily represent any particular moment of the construction history. The choice of these contexts stemmed from a reliance upon architecture instead of ceramics as the basis for the chronological sequence. The result is that the samples do not isolate and date significant transitions in the architectural history, such as construction or abandonment, as would have been the case had caches or interface contexts been dated more aggressively. While the dates have value, they may have a large range of error for dating construction. The prior analysis

(Beekman and Weigand 2008, 2010) attempted to take this into account and evaluate the dates within their stratigraphic position. Here I develop an updated version of the original analysis, which was not published in full, and include the second round of radiocarbon dates provided by Weigand but never analyzed in print. I use the original calibrations of the radiocarbon age determinations, and present all ranges at 1 sigma. I assume that initial construction dates for any platform of a *guachimontón* date the construction of that *guachimontón* as a whole. Continuing activity at an architectural group is represented by dates on platform expansions, but this information was provided by Weigand and has not yet been confirmed through the reanalysis of site stratigraphy.

Beta 192096 (390–350, 300–220 BC) provides the earliest dated evidence of activity at Los Guachimontones, in the form of material from the lower part of the oven in the La Joyita A residential area. A review of the *informes* suggests that this is probably from the "Gran Horno," but there is another oven reported from the group. With a range broadly defined as 400–200 BC, the date seems quite possible, although it is difficult to evaluate without any other dates from contexts linking it to the rest of the ceremonial center. No dating evidence speaks to how late the group was occupied.

Circle 6 has multiple dates in association with the construction sequence of the altar and, by extension, the circle as a whole. Tombs 4 (Beta 192086—160–60 BC) and 6 (Beta 192085—AD 60–160) are said to predate the initial construction of the altar, although it is not clear what this means. Presumably the altar was immediately built over them. The two dates do not agree between themselves, but the later date is unlikely because of the dates from the expansion of the altar. More information on the specific materials being dated might suggest the reason for the wide spread in dates. I propose that the first date is associated with the initial construction, and the later date suggests that the excavator may have been mistaken about the stratigraphic placement of Tomb 6 (Cach 2008 is not clear). The dates from Offerings 2 (Beta 192087—125 BC–AD 25) and 5 (Beta 192084—50 BC–AD 150) are reportedly within the next construction stage of the altar. Since we know that they date offerings, these are more likely to be good representatives of the time period of the expansion. The two dates overlap in the range 50 BC–AD 25, when the circle was likely expanded, and provide an upper limit on the initial construction of the circle. We can propose on this basis that Circle 6 was likely built before the mid-first century BC and that the expansion occurred quickly thereafter. This circle cannot be linked to the others in the overall site construction sequence, and so its relative position in the sequence can only be inferred from the dates. The contents of the offerings from the two stages suggest that the original construction falls within the Tequila II phase, while the expansion dates to early Tequila III (see below). This provides a date for the transition between these two subphases in line with previous work at Navajas (Beekman and Weigand 2008).

A good date for the construction of the largest and most complex, Circle 1, does not exist. The age determination Beta 215271 from the fill of Platform 1 should be quite close at 160–40 BC. Other dates pertain to later expansions in the construction sequence. The date from the expansion of Platform 7 of Circle 1 (Beta 192097—40 BC–AD 45) indicates that Circle 1 should have been built before this time. The dated *aplanado* on Platform 2 of Circle 1 (Beta 192093—20 BC–AD 85[1]) reportedly decorates an expansion of the building, and dates the organic material in the finished surface to the same general period of expansion as Platform 7. This is earlier than the date (Beta 192092—AD 95–195) from the *otate* superstructure, which is expected: the perishable superstructure would have been renovated periodically. The age determination Beta 215272 comes from a hearth in the patio in front of Platform 2 and dates to AD 60–240, extending occupation still later. Whether a hearth in the patio of the public architecture signifies ongoing occupation or decline requires further assessment, however. The date (Beta 192098—AD 30–120) from the Lunate Plaza behind Platform 7 of Circle 1 places its construction in the late first century AD, although we have not as yet located any reports describing its excavation.

Circle 2 should be the best dated of all the architectural groups, but the PAT relied upon poor contexts for dating. Built sometime after Circle 1, Circle 2 has several dates relevant to the original construction, and they are all reportedly fill dates. Beta 215280 (360–290, 250–230 BC), Beta 215273 (370–100 BC), Beta 192095 (AD 1–80), Beta 215278 (AD 60–240), and Beta 215279 (AD 240–420) reportedly come from the fill of several different platforms, and show a disappointingly wide range. Beta 215274 (340–320, 210–40 BC) may also be a fill date, as it is said to come from burned material at the base of a wall from (inside?) Platform 6. There is a date reportedly associated with a burned offering within Platform 10—Beta 192099 (AD 15–115), which should be the most trustworthy as it is on a cultural deposit presumably made at the time of construction. But the wide spread of fill dates makes dating the construction of Circle 2 very problematic, as it would seem to push construction until the third century; fortunately, the dates from Circle 3 (which follows Circle 2 in relative order of construction) force an earlier construction date. Further research should focus on clarifying these contexts—the latest "fill" dates may actually have been from unrecognized activity contexts.

Two dates come from within an extension to Platform 8, Circle 2, Beta 192091 (185–35 BC) and Beta 192089 (AD 150–290). These dates provide a *terminus post quem* for the expansions, and the second date is from a maize offering that should be closest to the date when the construction was sealed. Beta 215269 (10 BC–AD 170) comes from a hearth and associated trash on the lateral walls between Platforms 8 and 9. Depending on the exact physical location of these last two dates, they may pin down the expansion of Platform 8 more specifically to AD 150–170. A date from the *aplanado* of Platform 6 (Beta 192102—90 BC–AD 50) should be the latest in terms of construction sequence, but probably dates older carbon in the clay. Radiocarbon dating of carbon in clay should be avoided, although optically stimulated luminescence would not suffer from this problem. Beta 215270 (AD 1010–1150) from within the double wall of Platform 7 must be intrusive, or I am interpreting the stated recovery location incorrectly.

Circle 3 is the next in the sequence according to the construction data. The sample (Beta 192094—AD 80–220) from the layer deep within the fill may be close to the date of the circle's initial construction. We might split the difference between Circle 2 and Circle 3's construction to have the former built by AD 100 and the latter after AD 100. Even if this date pertains to an expansion of Platform 3 (as the description of the sample states as a possibility), such an addition seems to have taken place relatively soon after the original construction, which is narrowly bracketed between this date and the construction of Circle 2. Beta 192100 (AD 110–210) dates fill from a more secure expansion of this same Platform 3 to the second century AD or later, closely following and even overlapping the evidence for the initial construction. Again, we might split the difference to place the dating of the original construction of Circle 3 to AD 100–175 and its expansion to AD 175–220.

Circle 4, according to the construction sequence, was built when Ballcourt 1 was expanded and Circle 2 was built, or prior to AD 100. Two dates from Circle 4 are disappointing, but another two look promising. Beta 215276 (360–280, 260–240 BC) is burned material from the fill of Platform 3, while Beta 215277 (1270–940 BC) is from the *aplanado* of Platform 2. Both are far too early. Beta 215283 (40 BC–AD 130) and Beta 215284 (AD 30–220) are from within the patio floor in front of Platform 1. Circle 4 may have been built at any time after approximately 40 BC, but, as noted, prior to AD 100.

Circles 7 and 8 each have one date to help pin down their period of use. Beta 215282 (380–160 BC) from Circle 7, Platform A, is reportedly below a vessel within the platform that may be an offering. Alternatively, this could simply be another fill date. This context should receive further attention to clarify the situation. Beta 215281 (AD 220–400) comes from Circle 8, the "exterior platform . . . above the floor," and would seem to place this circle among the latest in use at the site. The only date obtained for Loma Alta is Beta 215285 (7950–7600 BC), from the fill beneath the Ballcourt 1 floor, and is clearly of little help.

In sum, the most parsimonious explanation of the sequence of dates is as follows. Dated evidence for occupation was apparent by the third or fourth centuries BC at La Joyita A, establishing a residential component prior to the construction of the public architecture. Circle 7 may have been built any time after 360 BC. Circle 6 was constructed circa 160–50 BC, and the altar was expanded with new offerings circa 50 BC–AD 30. Circle 1 was constructed between 160 and 40 BC since expansions were already under way by that time, and occupation continued into the second or third century. The Lunate Plaza was built by the late first century AD. The expansion of Ballcourt 1 and the constructions of Circles 2 and 4 should have occurred at the same time, to judge from the reported integration of the architecture, but there is still considerable leeway in the dating. Circle 4 could have been built as early as 50 BC, but the two useful dates from Circle 4 suggest a date closer to AD 100. Circle 2's construction is likely to have occurred AD 15–115, with expansions from AD 150 to 170. Circle 3 was likely built circa AD 100–175, and added to circa AD 175–220. Circle 8 was apparently in use in the third and fourth centuries AD This sequence traces the overall site occupation differently from what a strict enumeration of the radiocarbon dates suggests, because the dates are evaluated using what contextual information is available. We must keep in mind that the samples were not chosen to coincide with architectural or artifact transitions, and the resulting sequence remains loose.

Ceramic Debris

The analysis approached the collections from a small-site methodology perspective, in that each architectural group was assumed to have a relatively discrete period of occupation or use that would allow the recognition of more short-lived ceramic types. The assemblages for each group could be compared and potentially ordered into overlapping periods of time to reconstruct an overall sequence. Groups rather than structures were considered the ideal unit of analysis because: (1) evidence for activity in one structure was considered evidence for the group as a whole, (2) the architectural groups were treated as operations organizing the excavations, and (3) artifact bags are sometimes labeled with a code whose meaning has not yet been deciphered. It has therefore not always been possible to assign collections to specific structures even though their tags might describe them as coming from "floors" or other useful contexts.

First, architectural groups were selected for analysis. The architectural repertoire consists of formal circles, ballcourts, residential groups, and a mix of smaller architectural units that tend to be domestic or at least nonpublic in their associated activities. Our initial efforts focused on the major architectural groups, such as the circles and the intensively excavated residential groups of La Joyita A and B. The first season of analysis deemphasized Loma Alta until we had a better understanding of the nuclear sector of the site. In our third round of analysis in 2015, Loma Alta received increased attention, samples were increased from the nuclear sector, and a series of "minor" contexts such as *Estructura Residencial* (ER)-1, ER-2, and the Gran Plaza were targeted for analysis.

Second, more specific contexts were identified. A database assembled by Juan José Cortés Guzmán was used to identify and select individual bags from each architectural group. Some contexts were rejected for any analysis: fill, upper strata that had been lumped during excavation, and ambiguous contexts whose identity could not be established. This removed the majority of bags from consideration. Another group of bags retained sufficient information that they could be useful for analysis in the future, but may require a much more detailed study of PAT *informes*, as no master list had been created that explained the contexts and their codes. Of those bags that remained, analysis targeted contexts that were described as floor surfaces, caches, burials, hearths, and so on, that pertained to the occupation/use of the architectural group. The intention was to analyze the equivalent of at least one box of ceramics, or on average 20 bags, from each architectural group. This was often exceeded. Beekman selected all bags for analysis, except for those from La Joyita A and B, which were selected by Catherine Johns in the pursuit of combined chronological and activity analysis.

Ceramic analysis proceeded along the lines pursued in previous studies in the La Venta Corridor (1993, 1994), Llano Grande (2000, 2003, 2011), Navajas (2003, 2011), and the Magdalena Basin (2013). This involves the use of a mixed type-variety classification supplemented by a modal analysis, using categories that have continued to be expanded and refined over the course of the above investigations. Johns coordinated the analysis in 2012 and 2014, while Beekman did so in 2015. To keep data presentation from becoming too cluttered, Table 2.3 is the only place where I report both counts and weights, and the primary measurement used in the other charts will be weight in grams.

To briefly summarize the groups or wares, Tabachines was used for the finest semihemispherical bowls in simple cream, reduced black, or red on cream, the same color combinations as found among the small and solid figurines (Beekman 2006; Beekman and Weigand 2000; but see Johns 2014 for the most refined descriptions). The paste is very fine, with white silt-sized silicates and a prominent black core, with occasional hematite probably added during clay processing. These are the best candidates for ritual vessels for their thin walls and rapid firing. The *Estolanos* group includes vessels with thicker walls, made with a similar paste but with a uniformly crushed white temper, and the same decorative colors as Tabachines. It was used for bowls frequently with convergent walls, zoomorphic vessels, and the famous hollow figures best known from the shaft tombs (Beekman and Pickering 2016). At Los Guachimontones, the *Estolanos* paste was used for San Juanito, San Sebastián, and Ameca-Etzatlán style figures. The Colorines group was used for utilitarian vessels of all sizes, and includes a diverse range of inclusions of multiple-size classes. The Colorines paste forms a continuum that we have separated into Fine and Coarse Colorines. The latter was generally used for larger, closed vessels with less care invested in their painted decoration, while Fine Colorines tended to be for smaller open vessels. Hence the continuum in the fineness of the paste corresponds largely to that expected anyway between thicker-walled vessels that needed more support and thinner-walled vessels that did not. The Colorines vessels can be plain, or they can be decorated with red paint, or rarely reduced black, but the buff color of the paste frequently required a white slip for the application of the red decoration. This is particularly apparent in the Fine Colorines bowls. Finally, the Arroyo Seco group includes larger vessels with thick walls. They appear to be of everyday use based on their size and the simplicity of their decoration, but the surface is well polished, and the paste is very fine and shows intense processing. The vessels are generally open bowls with vertical or slightly divergent walls. They are fine, but large and simple, and Johns proposes that they were used for corporate public feasting.

Table 2.3. Number of ceramics analyzed by architectural group, by count, and by weight

	Magdalena-Tequila I		Tequila II-III-IV		El Grillo		Atemajac I-II	
Context	Count	Weight	Count	Weight	Count	Weight	Count	Weight
Nuclear Sector								
Circle 1			412	4,027	14	152	11	236
Patio			70	729	9	60	2	66
Pyramid			1	3				
Platform 1							5	58
Platform 12			31	553	2	77		
Unlocated			310	2742	3	15	4	112
Circle 2	1	7	205	1,855	47	512	383	8,198
Banq. betw. Plt. 8 & 9	1	7						
Platform 5			6	31			27	305
Unlocated			199	1,824	47	512	356	7,893
Circle 3			312	1,908	10	67	83	603
Altar			86	531	4	28	19	203
Platform 2			22	189			1	6
Platform 3			158	878	3	10	63	394
Platform 8			34	180	1	8		
Plaza Exterior			2	68				
Unlocated			20	72	2	21		
Circle 4	7	82	1241	8,326	28	124		
Platform 3			491	3,311	5	28		
Platform 4	4	73	483	3,590	3	14		
Unlocated	3	9	267	1,425	20	82		

(*continued*)

Table 2.3—*Continued*

Context	Magdalena-Tequila I		Tequila II-III-IV		El Grillo		Atemajac I-II	
	Count	Weight	Count	Weight	Count	Weight	Count	Weight
Circle 5			326	3,525	21	222	26	540
Platform 4			21	201	2	65	2	11
Unlocated			305	3,324	19	157	24	529
Circle 6			420	5,281	67	836	623	8,204
Patio			115	1,621	42	602	164	2,114
Altar			266	3,054	8	54	257	3,850
Unlocated			39	606	17	180	202	2,240
Circle 7	2	9	331	4,146	1	11	2	31
Altar			75	1,059	1	11		
Unlocated	2	9	256	3,087			2	31
Circle 8			448	4,040	8	67	116	1,860
Altar			71	621			6	39
East Platform			13	121				
Unlocated			364	3,298	8	67	110	1,821
Circle 10			217	2,250				
Banquette			6	463				
Unlocated			211	1,787				
Ballcourt 1	8	77	561	5,899	21	300	131	2,146
Court	4	30	38	322	9	97	2	46
East End Platform	1	8	159	1,593	6	75	6	202
North Lateral Platform			19	213				
South Lateral Platform			4	58				
Unlocated	3	39	341	3,713	6	130	123	1,898

Ballcourt 2			159	2,480			25	399
Ballcourt, East Plaza			80	629	1	9	4	105
Early Structure 1			19	97			1	66
Early Structure 2			23	231	1	9	2	30
Unlocated			38	301			1	9
La Joyita A	1	3	6673	59,582	156	1350	198	2,431
Estructura 1			251	3,187	17	72	2	40
Estructura 2	1	3	1362	12,051	34	256	85	948
Estructura 3			493	4,537	28	241	15	205
Estructura 4			547	3,639	6	47	6	79
Estructura 5			208	2,555	5	51		
Estructura 6			70	739	1	4		
Estructura 7			1136	10,299	2	8	12	184
Estructura 8			317	2,358	7	96	5	22
Estructura 9			76	665	4	26	2	11
Patio			197	1,447	2	8	4	16
Unlocated			2016	17,248	50	541	67	926
La Joyita B			1742	11,819	28	224	80	712
Estructura 1			823	4,802	17	111	74	596
Estructura 2			322	2,673	2	31	1	13
Estructura 3			226	1,808	1	4	2	53
Unlocated			420	2,939	5	32	4	33
ER-1			9	337	3	136	9	714
ER-2	1	8	3	42				

(*continued*)

Table 2.3—*Continued*

Context	Magdalena-Tequila I		Tequila II-III-IV		El Grillo		Atemajac I-II	
	Count	Weight	Count	Weight	Count	Weight	Count	Weight
Circle 1, Plaza Exterior			535	5,128	11	259	1	21
Structure A			35	280	1	13		
Structure C			20	319	3	31	1	21
Unlocated			480	4,529	7	215		
Grand Plaza			549	4,503	32	148	6	118
Talleres Sector								
Talleres 1			39	666	7	110		
Loma Alta Sector								
Circle A			584	3,021	6	17	1	3
Patio			285	1,614	4	11	1	3
Structure 3			120	517	2	6		
Unlocated			179	890				
Circle B			943	6,189	1	1	3	83
EE			305	2,802			3	83
EN			378	1,853				
Patio			17	56				
Unlocated			243	1,478	1	1		
Circles A and B			626	3,280	1	6	1	5
Circle E			171	920	1	3	12	213
Patio			143	762	1	3	12	213
Unlocated			28	158				

Ballcourt 1	67	427				
Patio	1	10				
South End Platform	10	31				
Unlocated	56	386				
Ballcourt 2	2347	11,452	5	26	2	11
North End Platform	80	287	1	2	1	2
Unlocated	2267	11,165	4	24	1	9
Patio IV	624	6,867				
Structure 1	228	3,315				
Structure 2	1	7				
Patio	97	991				
Unlocated	298	2,554				

Source: Compiled by author.

Notes: Groups that were excavated by the PAT, but that we have not yet analyzed, are the following:

Nuclear Sector

Ballcourt 2, Postclassic Structures 1 and 2—presumed Atemajac phase

ER-3-No reported date

Talleres Sector

Talleres II—presumed Atemajac phase

Talleres III—presumed Atemajac phase

Talleres IV—presumed El Grillo and Atemajac phases

Texcalame Sector

El Texcalame—composed of a small circle considered to be habitational (presumed Tequila II–IV phases) and a large residential structure (reported to include all periods)

Modes may refine the chronology, but here we will mention only a few that have been useful for chronological purposes. Based on their distributions in the shaft tombs from Tabachines (Galván 1991), oval vessels, miniature vessels, and jar necks of composite silhouette are indicators of the Tequila II phase. *Tecomates* do not show a consistent pattern at Tabachines, but may be Tequila II at Los Guachimontones—their presence will be monitored in future analyses. We have long recognized that bowls with concave bases are early in the sequence, probably Tequila II and III. Finally, modes that pertain to Tequila III and IV include the *campana* jar neck and rim, shaped like an upside-down bell, and square vessels with rounded corners. Due to their highly variable occurrences, modes must typically be multiplied by some factor in order to be easily compared between contexts. This factor varies for each mode, as the goal is only to compare different occurrences of the same mode.

Results

Blanco and colleagues (Blanco et al. 2010) used their results to propose relatively sharp differences between the *guachimontón* circles and groups of more residential function, based particularly in the ratios between finewares and utilitarian wares. Our results from more secure contexts and across a wider number of groups are more varied. The *guachimontón* architectural form does not show a single ceramic profile in relation to residential groups as represented by La Joyita, as summarized in Tables 2.4 and 2.5, respectively. All architectural groups share a basic pattern in which the majority of the ceramics are of the utilitarian Coarse Colorines ware, with the equally utilitarian Fine Colorines in second place. Tabachines and Estolanos wares are chronologically sensitive and thus more complex to interpret, but they are much less frequent in all contexts. The Arroyo Seco ware, also chronologically sensitive and perhaps not even made locally, shows up in still smaller amounts.

The two ballcourts stand out as having the most distinct assemblage. They both have the highest percentages of Coarse Colorines ware (averaging 82 percent), the lowest percentages of Tabachines ware (1 percent), and a low frequency of the specifically bichrome decorated types from the Coarse Colorines (9 percent) and Estolanos wares (4 percent), that is, ceramics are more utilitarian and less decorated. Some indicators suggest that there is a lower frequency of open relative to closed vessels, although this needs to be further examined by ware. One consequence of this kind of

Table 2.4. Distribution of functionally distinct wares and selected modes in the nuclear sector, by architectural group

Nuclear sector	Circle 1	Circle 2	Circle 3	Circle 4	Circle 5	Circle 6	Circle 7	Circle 8	Circle 10	Avg. Circles	Ballcourt 1	Ballcourt 2	Avg. Ballcourts
Wares													
Coarse Colorines as % of total	53	79	51	58	59	44	70	67	69	58	79	90	82
Fine Colorines as % of total	30	15	40	29	34	48	19	25	10	31	15	10	14
Arroyo Seco as % of total	0	0	0	0	0	0	1	1	0	0	0	0	0
Estolanos as % of total	6	3	4	6	5	8	6	3	1	5	3	0	2
Tabachines as % of total	11	2	5	7	2	0	4	4	21	6	2	0	1
Wares 18, 19, or 21 presence		X		X			X				X		
Modes													
Rim 23 Vertical olla neck (T I-II?) as % of all rims x 10		70		5		19				10			
Rim 132 (T I-II) as % of all rims x 100													
Rims 9, 34, 35, 139 + Other 31 Composite silhouette (T II) as % of all rims	9	7	1	7	5	7	7	6	10	6	3		1
Other 16 Pinched vessels (T II) in g			8		11	126				48			

(continued)

Table 2.4—*Continued*

Nuclear sector	Circle 1	Circle 2	Circle 3	Circle 4	Circle 5	Circle 6	Circle 7	Circle 8	Circle 10	Avg. Circles	Ballcourt 1	Ballcourt 2	Avg. Ballcourts
Other 34 Miniature vessels (T II) in g													
Rims 5, 6, 68 neckless jar (T II?) as % of all rims			7		1		3		4	2	2		1
Base Form 2 Concave/dimpled base (T II-III) presence/absence	X						X			X			
Rim 69 (T III-IV) as % of all rims x 10	17	70	3	123	13	7	7	59		33	14		7
Rim 30 Campana (T III-IV) as % of all rims x 100						1794		234		225	173		87
Other 33 Square vessel with rounded corners (T III-IV) presence/absence													

Source: Compiled by author.

Notes: Percentages of wares are based on weight. Mode frequencies are presented in a variety of percentages or modified percentages to best draw out the distinctions in quantities at the different architectural groups. Very rare attributes are shown in terms of presence/absence. The frequency of these modes should not be compared to one another, only to separate occurrences of the same mode.

Table 2.5. Distribution of functionally distinct wares and selected modes in the nuclear and Talleres sectors, by architectural groups that may have had functions outside of the public sector

Nuclear and Talleres sectors	ER-1	Talleres 1	C1, Exterior Plaza	Ballcourt 2, East Plaza	La Joyita B	La Joyita A	Gran Plaza	ER-2
Wares								
Coarse Colorines as % of total	87	83	82	80	73	72	66	64
Fine Colorines as % of total[a]	13	0	14	13	21	20	28	0
Arroyo Seco as % of total	0	0	0	0	1	0	0	0
Estolanos as % of total	0	10	3	2	3	5	2	0
Tabachines as % of total	0	7	1	5	2	3	4	20
Presence of Wares 18, 19, or 21						X		X
Modes								
Rim 23 Vertical olla neck (T I-II?) as % of all rims x 10					15	11		
Rim 132 (T I-II) as % of all rims x 100			266			7		
Rims 9, 34, 35, 139 + Other 31 Composite silhouette (T II) as % of all rims	134		1		6	5	15	
Other 16 Pinched vessels (T II) in g						22		
Other 34 Miniature vessels (T II) in g						8		
Rims 5, 6, 68 neckless jar (T II?) as % of all rims			1		4	4	3	
Base Form 2 Concave/dimpled base (T II-III) presence/absence					X	X		
Rim 69 (T III-IV) as % of all rims x 10			13	179	19	17	47	
Rim 30 Campana (T III-IV) as % of all rims x 100			1297		379	15	138	
Other 33 Square vessel with rounded corners (T III-IV) presence/absence								

Source: Compiled by author.

Notes: Percentages of wares are based on weight. Mode frequencies are presented in a variety of percentages or modified percentages to best draw out the distinctions in quantities at the different architectural groups. Very rare attributes are shown in terms of presence/absence. The frequency of these modes should not be compared to one another, only to separate occurrences of the same mode.

[a]The Fine Colorines ware was not yet separated out from Coarse Colorines until after the Talleres 1 analyses were completed, and the La Joyita A and B analyses were one third of the way to completion.

assemblage is that the ballcourts lack many of the distinctive wares and modes that are most useful for dating, so the ballcourts tend to be weakly placed in the sequence. The assemblage does suggest an interpretation of the ballcourts as more "corporate" (Blanton et al. 1996) spaces, with activities focused less on aggrandizement and more on integrative public events.

The means for the circle assemblages show a reduced percentage of Coarse Colorines (58 percent) and an increased presence of Fine Colorines (31 percent) relative to the ballcourts. The fine Tabachines (6 percent) and Estolanos (5 percent) wares are more common. Bichrome decorated types make up a larger percentage of their respective wares (63 percent of Coarse Colorines, 67 percent of Fine Colorines, 27 percent of Estolanos, and 59 percent of Tabachines wares). Thus relative to the ballcourts, the circle assemblages include more finewares and more decorated wares. However, there is a great deal of variation among the circles, with Coarse Colorines comprising 44–79 percent of a given circle's assemblage, and Circle 2 with the same general assemblage as the adjoining Ballcourt 1, reflecting their proximity and similar construction date. Tabachines wares are so rare in one circle as to go unregistered (Circle 6). The highest percentage of Tabachines occurs not in the largest and most prominent circles, but in Circle 10, the smallest circle, built in a marginal location perhaps more conducive to private ritual. The utilitarian Coarse Colorines occurs in its lowest percentage in this same circle. Although Arroyo Seco was common at Tabachines (Galván 1991) and made up over half the assemblage at Navajas (Johns 2014), it only occurs in trace percentages at its contemporary Los Guachimontones, primarily in small- to mid-sized circles that otherwise show signs of late occupation (Circles 7 and 8). The distribution of Estolanos ware varies from 1 to 8 percent of the assemblages, but shows no clear pattern related to circle size or location.

The architectural groups in Table 2.5 may be distinct in function from the public circles and ballcourts. ER-1, ER-2, La Joyita A and B, Talleres 1, and the C1 Plaza Exterior are all proposed domestic structures based on the size and morphology of the platform group. The East Plaza resembles these groups, but its close proximity to Ballcourt 2 suggests a ritual- and/or game-related function. The Gran Plaza is an open space flanked by two structures and with a central altar, nestled behind Circles 1 and 2. In general, they all show an elevated occurrence of the utilitarian Coarse Colorines ware, but this is not straightforward for the Talleres and La Joyita groups. The Fine Colorines ware was only defined after the analysis of Talleres 1, and one third of the way through the analysis of La Joyita A and B.

This is a hindrance to analysis, but we can partly compensate by combining both Colorines wares for comparisons. The total percentages of the two Colorines wares in La Joyita and the other groups would thus look much like those of the circles in Table 2.4. More promising is the C1 Exterior Plaza, which primarily differs from the circles in its higher percentage of Coarse Colorines at the expense of Fine Colorines ware. The difference is more pronounced when compared to the immediately adjoining Circle 1—this group may be more of a storage space for vessels used in feasting than a residence. It is also very similar to the ballcourt assemblages. The Gran Plaza and Ballcourt 2 East Plaza assemblages are most similar to those of the circles, suggesting more ceremonial activities, and their elevated quantity of Coarse Colorines utilitarian wares may point to public feasting. Here and at C1 Exterior Plaza, we will need to review vessel diameters to assess this possibility. The assemblages from ER-1, ER-2, and Talleres 1 have interestingly divergent quantities of some wares, but they are also the groups with the smallest samples.

Chronological factors play a role in the makeup of these assemblages. Wares 18, 19, and 21 were defined in the Magdalena Basin for Early and Middle Formative types, but only occur at Los Guachimontones in trace amounts in a few groups. The earliest radiocarbon date that appears to pertain to a Tequila II context centers just before 300 BC, suggesting that Tequila II begins circa 350–300 BC. Tabachines and Estolanos wares are assigned to Tequila II and III, and, combined, form 5 to 22 percent of the assemblages from the circles. Arroyo Seco ware is assigned to Tequila III and IV and occurs only as trace percentages in Circles 7 and 8. This would suggest only a limited Tequila III–IV occupation, but Arroyo Seco is geographically focused to the south and east of Los Guachimontones and these are likely imports. The higher percentages of Coarse Colorines ware in those circles thought to be late suggest an overall chronological trend toward higher percentages of that utilitarian ware, a process also seen at the site of Tabachines (Galván 1991). Any functional analysis must therefore take chronology into account to develop ceramic profiles for the architectural groups.

Interestingly, location may also be important for interpreting the relative proportions of different functionally distinct wares. As noted, Coarse Colorines wares make up higher percentages of assemblages from the later Tabachines tombs as the finewares drop out (Galván 1991). Conversely, the circles with the lowest percentages of Coarse Colorines ware are also the most centrally located. So we might suggest an indirect relationship in which the earliest circles were generally the most centrally located circles, which in turn had a lower ratio of utilitarian activities.

Modes potentially provide a more refined understanding of chronology (Tables 2.4 and 2.5). Two rims that were identified solely in Tequila I–II contexts during the PAX project are Rims 23 (a tall jar neck) and 132 (a distinctive trapezoidal rim that led to a composite silhouette jar neck). Both are of low frequency and only occur together here in La Joyita A. Other occurrences are in Circles 2, 4, 6, La Joyita B, and Circle 1 Exterior Plaza. Based on their distribution within the Tabachines shaft tombs, oval vessels that have been pinched in at the sides, miniature vessels, and jar necks with a composite silhouette are all indicators of the Tequila II phase. Composite silhouette jar necks are widely distributed, but are most common in ER-1 and the Gran Plaza. The other two vessel types are probably specialized forms (occurring across Circles 3, 5, and 6), but all occur together in La Joyita A. Neckless jars (*tecomates*) showed no clear pattern at Tabachines, but may be Tequila II here at Los Guachimontones. Bowls with dimpled bases occur in Circles 1, 7, and La Joyita A and B, supporting the early date for those groups. Finally, late modes that are associated with Tequila III–IV at Tabachines are Rim 69 (from flaring bowls with a thickening on the interior lip), Rim 30 (Galván's *campana* jar neck), and square vessels with rounded corners. Square vessels did not occur anywhere in our sample, but the rims occur together or in greater quantities in Circles 6 and 8, Ballcourt 1, La Joyita A and B, the Gran Plaza, and in Circle 1 Exterior Plaza. The failure of these rims to co-occur outside of these groups may mean that they were associated with different activities, or that our sample sizes were too small. In the end, the chronological analysis of the bulk ceramic debris indicates that the architectural groups did not have neatly sequential and discrete occupations as between Circles 1 and 5 at Navajas (Beekman 2008). Rather, circles had longer and overlapping occupations, or perhaps multiple occupations separated by periods of disuse.

The ceramic analysis from the Loma Alta sector has advanced to the point where we can address issues of function and chronology in that part of the site (Table 2.6). First and most noticeably, there is no evidence for the wares associated with the Magdalena or Tequila I phases. This sector seems to have been occupied later. Typical wares from Tequila II and III are present in abundance. The best late marker at Tabachines and Navajas was Arroyo Seco, but it is nearly absent. The modes suffer from similar problems. Early modes are rare to absent, with only trace amounts of the *tecomates* and composite silhouette jar necks so widespread in the nuclear

Table 2.6. Distribution of functionally distinct wares and selected modes in the Loma Alta sector, by architectural group

Loma Alta	Circle A	Circle B	Circles A and B	Circle E	Ballcourt 1	Ballcourt 2
Wares						
Coarse Colorines as % of total	70	58	74	78	63	65
Fine Colorines as % of total	26	39	21	17	26	26
Arroyo Seco as % of total	1	0	1	0	0	0
Estolanos as % of total	3	2	3	2	4	7
Tabachines as % of total	1	1	1	3	8	1
Wares 18, 19, or 21 presence						
Modes						
Rim 23 Vertical olla neck (T I-II?) as % of all rims x 10						
Rim 132 (T I-II) as % of all rims x 100						
Rims 9, 34, 35, 139 + Other 31 Composite silhouette (T II) as % of all rims	12				12	3
Other 16 Pinched vessels (T II) in g						
Other 34 Miniature vessels (T II) in g						
Rims 5, 6, 68 neckless jar (T II?) as % of all rims	2	1				4
Base Form 2 Concave/dimpled base (T II-III) presence/absence		X				
Rim 69 (T III-IV) as % of all rims x 10	14	67	38		111	39
Rim 30 Campana (T III-IV) as % of all rims x 100	1,255	444				
Other 33 Square vessel with rounded corners (T III-IV) presence/absence						

Source: Compiled by author.

Notes: Percentages of wares are based on weight. Mode frequencies are presented in a variety of percentages or modified percentages to best draw out the distinctions in quantities at the different architectural groups. Very rare attributes are shown in terms of presence/absence.

sector. The late bowls (Rim 69) appear to be widespread in the public architectural groups, and the *campana* rim (Rim 30) jars are numerous but limited to the circles. It may be that the activities overall in Loma Alta were less ceremonial and more quotidian, to judge from the low occurrence of these special forms (compare Sumano and Englehardt, this volume).

In terms of functional differences, the Loma Alta circles share the variability present within the circles from the nuclear sector. On balance, however, the Loma Alta circles show an increase in Coarse Colorines at the expense of Estolanos and Tabachines wares, which might support a less ceremonial function or match the later chronological trend noted above. The small Circle B differs notably in the greater presence of the Fine Colorines ware, and the dearth of Tabachines and Estolanos wares still suggests a late date. The excavations into the platform shared by Circles A and B found its assemblage to be very close to that of Circle A in all respects, rather than presenting some intermediate mixture of the activities from both circles. The Circle E assemblage has more in common with domestic groups when modes are considered. The two ballcourts share very similar assemblages that differ notably from the ballcourts in the nuclear sector. Those in Loma Alta have much lower quantities of the utilitarian Coarse Colorines ware, and these numbers are also lower than found in the Loma Alta circles. Fine Colorines, Estolanos and Tabachines are more common in the Loma Alta ballcourts, and the latter two finewares are more frequent than seen in the Loma Alta circles. The Loma Alta ballcourt assemblages look more like that for the Gran Plaza in the nuclear sector. The relative relationship of ballcourts to circles is therefore quite different in Loma Alta, and the ballcourts may have been used for more socially competitive versions of the game, such as between lineages. Note that this distinction in activity does not correspond to the relative size of the courts within a larger hierarchy, but to their physical position within the overall site. The only proposed residential assemblage from Loma Alta is that from Patio IV, proposed as an elite residential group by different authors (Beekman 2016a, 2016b; Smith 2008) based on size and morphology. Its assemblage resembles a mixture of the other residential groups in its more numerous Colorines wares and less common finewares overall, but with a greater emphasis upon Fine over Coarse Colorines. An elite residence does seem very plausible on this basis, with fewer ritual wares and increased fancy serving vessels. What is notable is how this assemblage is very different from the other group defined as an elite residence by the early PAT excavators at the

Circle 1 Exterior Plaza. The latter is radically different from Patio IV, with a Coarse-to-Fine ratio of 7:1 compared to Patio IV's 2:1. The latter is far more likely to have been an elite residence.

Figure and Figurine Debris

The figures and figurines are incorporated into the analysis in the same manner as the ceramic debris. Most fragments could be assigned to paste, and those that are chronologically sensitive can be used to date the architectural groups. For reasons of space, the detailed enumeration of figures and figurines by architectural group can be consulted in Beekman (in press). There are some discrepancies to work out—the very high percentage of Fine Colorines paste used for figurines in Circle 6, for example. But in brief, the figure-and-figurine debris data support Tequila II or III occupations in Circles 1, 2, 4, 5, 6, 10; La Joyita A and B; and Loma Alta Circles A, B, the shared A and B platform, Ballcourt 2, and Patio IV. Unfortunately, it is likely that this distribution is because figure and figurine fragments were pulled from the collection bags for these groups, but were not removed from the collections from other operations. Useful statistics will not be possible until the other collections have been reviewed more systematically. We can point out, however, that our analyses of ER-2, C1 Exterior Plaza, Talleres 1, Ballcourt 2 East Plaza, and three out of the four ballcourts found no figurines, and our analyses of La Joyita A and B found truly minuscule quantities when compared to the total volume of bags that were processed in the laboratory. We can propose on this basis that figurines commonly associated with household ritual may not have held this function at Los Guachimontones. Circle E's lack of figurines, then, is in line with its other suggestions of domestic activity, while the presence of figurine and figure fragments in the Gran Plaza assemblage suggest that it had some distinct functions.

The combined ceramic and figurine debris add to the chronological assignment of the architectural groups, but two points must be kept in mind. First is that these materials came from activity contexts, and do not date the initial construction of the buildings. Second is that trace amounts of chronologically sensitive materials do not necessarily represent activity, but are rather materials that would have been disturbed or moved around in the course of normal community activities.

Offerings/Burials

Although we had previously assessed the assignment of whole vessels and figurines from the PAT inventory to particular burials and caches, it was apparent that there were contradictions. Some vessels were attributed to different burials in the inventory than were attributed on labels stored directly with the vessel. In 2015, LeFae reviewed all of these items and compared them directly to the original excavation reports, using photographs, descriptions, and so on, to locate all items (LeFae 2015, 2017). The combined offerings from each cache or burial context were then evaluated to identify the phase(s) when all vessels could have been deposited together. For reasons of space, the specific offerings and burial assemblages can be consulted in Beekman (in press). Note that there are a few cases in which clearly intrusive materials are associated with the offering or burial, but there is insufficient information in the older PAT *informes* to clarify the nature of the intrusion. The contribution of these data to the chronology of the architectural groups can be summarized as follows:

Circle 1, Plaza Exterior (1 offering)—Tequila II.
Circle 3 (2 burials, 1 offering)—Occupation in Tequila III, with possible occupation in Tequila II and/or IV.
Circle 4 (1 in situ vessel)—Occupation unspecified, Tequila II–IV.
Circle 6 (5 tombs, 7 offerings)—Occupation in Tequila II and III–IV, and again in Atemajac I–II.
La Joyita A (3 burials, 3 offerings)—Occupation in Tequila II, with possible Tequila III–IV occupation.
Talleres 2 (2 offerings)—Occupation in Atemajac I–II.
Talleres 3 (7 burials, 1 offering)—Tequila II, possibly Tequila III and/or IV, El Grillo.

These assignments coincide with the evidence already discussed. Of the burials, tombs, offerings, and so on, reviewed, only two appear to have ceramics indicating contradictory phases—Circle 6: Tomb 2 and Talleres 3: Entierro 8. Both are considered to be much later intrusions of single vessels.

Stratigraphic Pits

Few stratigraphic columns were evident in the bag database at hand, and we selected cases in which two or more layers in stratigraphic order could be analyzed to test the relative order of ceramics, and to date occupational activity in the different architectural groups. For reasons of space, the detailed results are not presented here but can be consulted in Beekman (in press).

Circle 1 (2 columns)—Occupation begins in Tequila II or III, occupied at least into Tequila III.
Circle 2 (2 columns)—Occupation in Tequila phase unspecified, but mixed layers include Tequila II–III materials.
Circle 3 (3 columns)—Occupation begins in Tequila II, ending unknown, but mixed layers include Tequila III–IV materials.
Circle 7 (4 columns)—Occupation begins in Tequila II or III, occupied at least into Tequila III.
Circle 8 (4 columns)—Occupation begins in Tequila II or III, occupied at least into Tequila III.
Circle 10 (1 column)—Occupation begins in Tequila II–III, no later.
Loma Alta, Circles A and B (1 column)—Occupation begins in Tequila III, ending unknown.
Loma Alta, Patio IV (3 columns)—Occupation in Tequila II and III, ending unknown.

In sum, we analyzed 20 stratigraphic sets with 65 layers. Nine sets show a total of 24 mixed stratigraphic layers (that is, including El Grillo or Atemajac in with earlier materials). Two sets (in Circles 7 and 8) show reversed layers with Atemajac phase materials below Tequila phase materials, although the excavation reports record no such disturbance. None shows contradictions with the proposed sequence of subphases *within* the Tequila phase. The additional contributions of the offerings and stratigraphic columns are incorporated into the stratigraphic chart and displayed in Figure 2.4.

Summary

The sequence for the nuclear and Loma Alta sectors is summarized in Figure 2.4. The chart generally shows sherd evidence for activity within many groups earlier than anticipated based on the radiocarbon dates for their construction. This may be due to the presence of settlement around the earliest groups long since demolished or buried underneath later constructions. The Tequila I phase is sparsely represented, and while no construction is known to have taken place this early, the distribution of Tequila I sherds suggests that a small settlement did exist. They interestingly form a

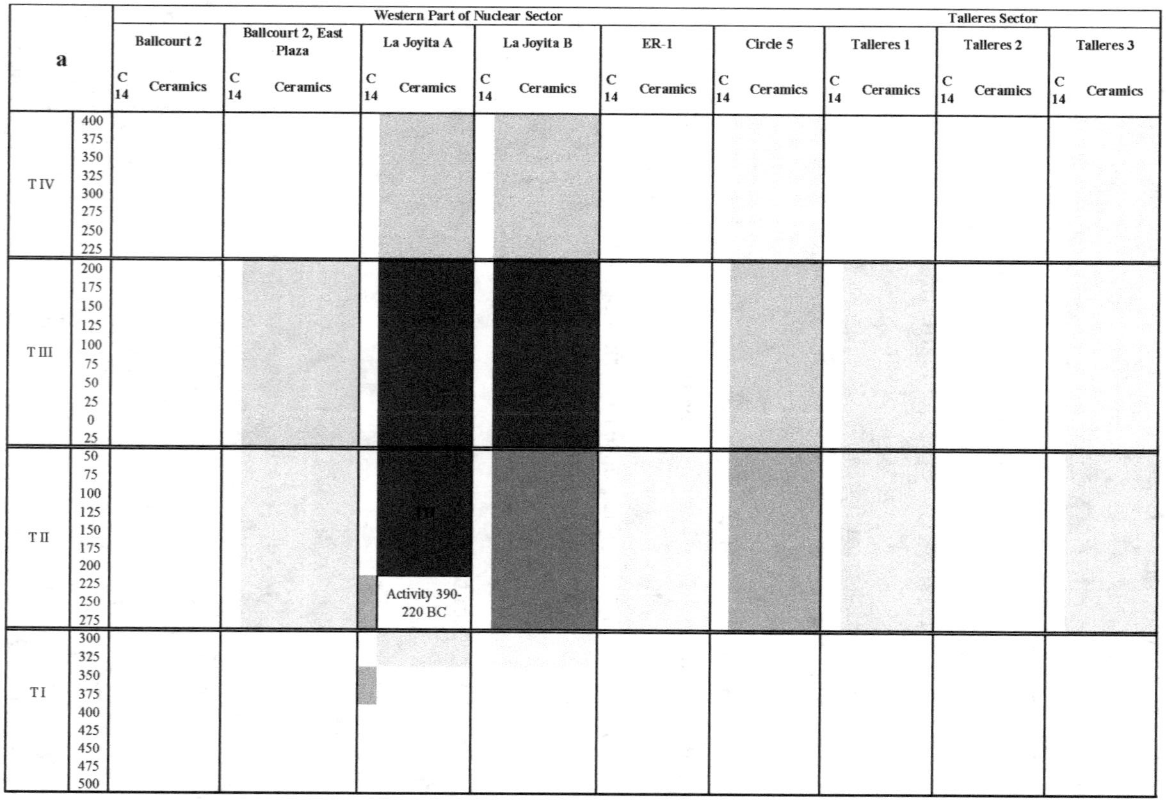

Figure 2.4a. Stratigraphic chart of the western part of the Nuclear Sector and the Talleres Sector, incorporating architectural stratigraphy, radiocarbon dates, ceramic and figurine debris, offerings, and stratigraphic columns. C14 date ranges are marked in the narrow columns between wider columns dedicated to the ceramic evidence. Each point of evidence for occupation during a particular phase (ware, mode, offerings, excavation evidence, etc.) adds an additional shade of gray for that phase block in the chart (compiled by the author).

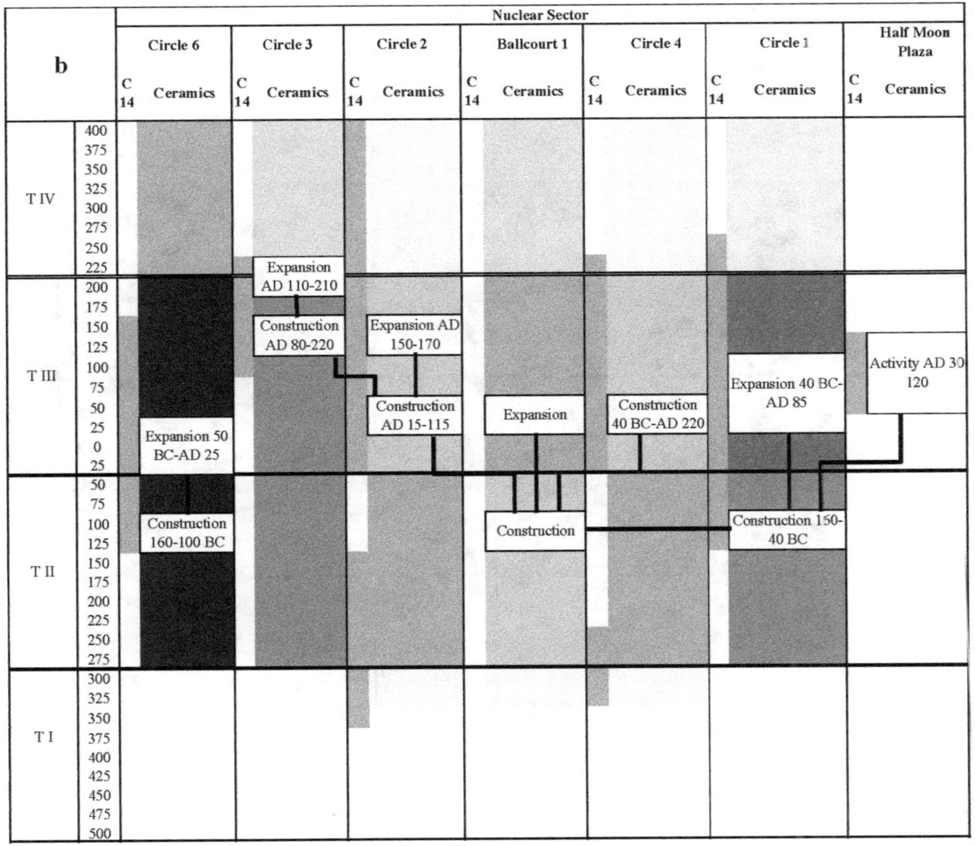

Figure 2.4b. Stratigraphic chart of the Nuclear Sector, incorporating architectural stratigraphy, radiocarbon dates, ceramic and figurine debris, offerings, and stratigraphic columns. C14 date ranges are marked in the narrow columns between wider columns dedicated to the ceramic evidence. Each point of evidence for occupation during a particular phase (ware, mode, offerings, excavation evidence, etc.) adds an additional shade of gray for that phase block in the chart (compiled by the author).

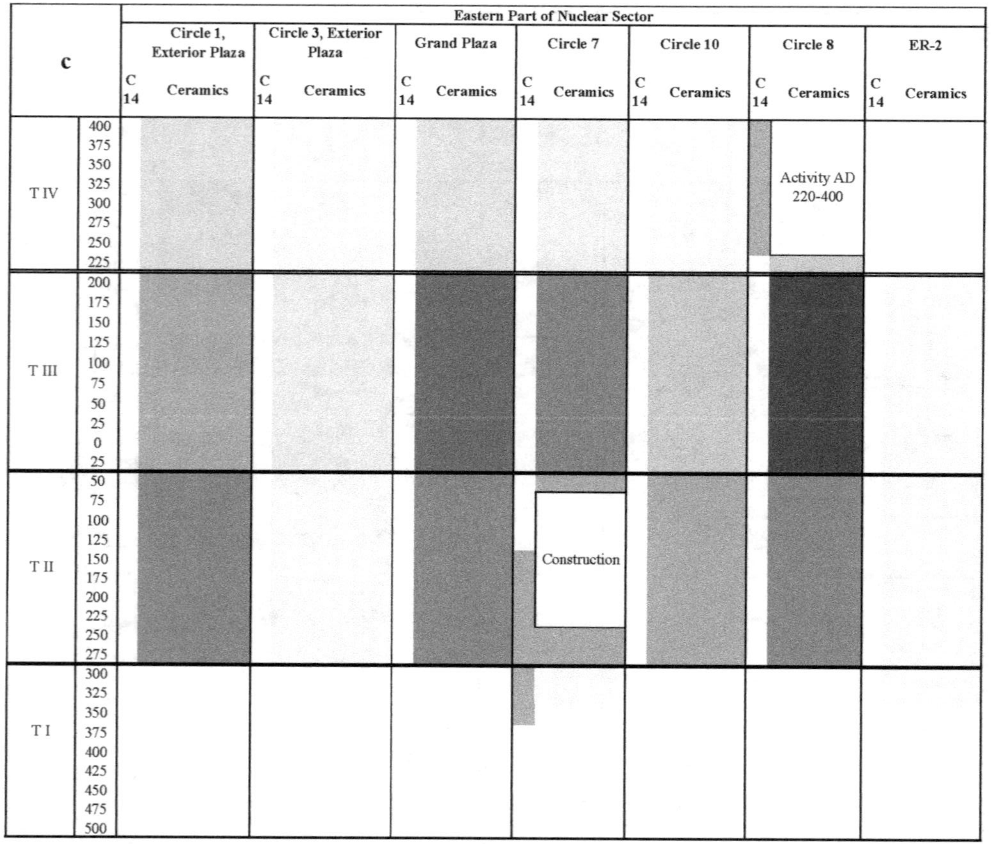

Figure 2.4c. Stratigraphic chart of the eastern part of the Nuclear Sector, incorporating architectural stratigraphy, radiocarbon dates, ceramic and figurine debris, offerings, and stratigraphic columns. C14 date ranges are marked in the narrow columns between wider columns dedicated to the ceramic evidence. Each point of evidence for occupation during a particular phase (ware, mode, offerings, excavation evidence, etc.) adds an additional shade of gray for that phase block in the chart (compiled by the author).

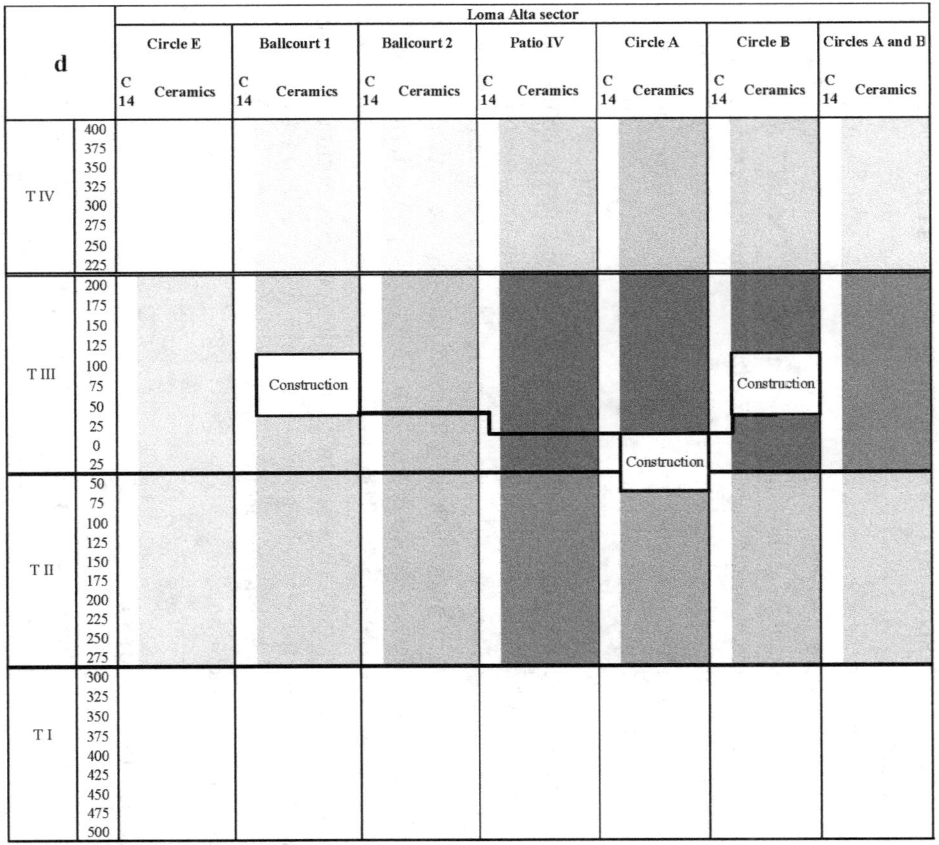

Figure 2.4d. Stratigraphic chart of the Loma Alta Sector, incorporating architectural stratigraphy, radiocarbon dates, ceramic and figurine debris, offerings, and stratigraphic columns. C14 date ranges are marked in the narrow columns between wider columns dedicated to the ceramic evidence. Each point of evidence for occupation during a particular phase (ware, mode, offerings, excavation evidence, etc.) adds an additional shade of gray for that phase block in the chart (compiled by the author).

continuous band at about the same elevation stretching from La Joyita A in the northwest, through Circle 4, Circle 2 and Ballcourt 1, and Circle 7 in the southeast, as well as ER-2. No remains from that phase have been identified within the Loma Alta Sector.

Tequila II marks the beginnings of major construction at the site, with a community perhaps expanding out of La Joyita A, which shows occupation from the very beginning of the phase. Construction followed in Circles 6, 7, 1, and probably 5. Ballcourt 1 was also constructed. The Gran Plaza and La Joyita B were probably built as well. In Loma Alta, the large residential group Patio IV and the small Circle B formed the nucleus for later settlement.

Tequila III is the greatest period of occupation at the site until the Late Postclassic. There are new offerings in Circle 6, occupation in La Joyita B, construction of Circles 2 and 3, and some activity in Circles 5, 10, and 8. All those architectural groups from Loma Alta that have been studied to date were built by this time, and Patio IV continued in use. In Tequila IV, the primary evidence is in Circle 8, although there may also have been activity in La Joyita B, Circle 7, and Loma Alta Circle B.

In summary, the data suggest a process in which a modest residential occupation preceded the construction of ceremonial architecture in the nuclear sector. The first evidence in the Loma Alta sector is also residential, but it is later and begins with a very large residence in the Patio IV group. The occupations in both nuclear and Loma Alta sectors peaked in Tequila III, and lack much evidence of the final Tequila IV phase, indicating that there is some problem with Tequila IV, or that the Los Guachimontones site declined precipitously while the Teuchitlán culture continued in other areas. Although there will be additional analyses in the future to consider, we can use this sequence to propose new models for the foundation of the site of Los Guachimontones.

Acknowledgments

I wish to thank Verenice Heredia and Josh Englehardt for their invitation to participate in the first inaugural Simposio sobre la Arqueología del Occidente en Mesoamérica, held at the Colegio de Michoacán in September 2015. Thanks are due to Catherine Johns, Naomi Ripp, Jones LeFae, Valerie Simard, Patricia Alonzo, Kim Sumano Ortega, David Arturo Muñiz García, Nichole Abbott, and Tony DeLuca for their work on the Los Guachimontones collections. Juan José Cortés Guzmán has worked tirelessly to put the

PAT collections into searchable format and deserves special recognition. My thanks also go to Verenice Heredia for her invitation to be involved in working with the PAT materials. Financial support came from the University of Colorado–Denver, and permission came from the Instituto Nacional de Antropología e Historia.

Note

1. Note that this date was originally reported as 1710±40 BP, but the Beta Analytic datasheets say it should be 1960±50 BP. The former was the age determination prior to applying the C13 correction, and the calibration was mistakenly run on the wrong figure in Beekman and Weigand 2008. This was a critical date, as it extended site occupation later than other sources indicated.

References

Aronson, Meredith Alexandra

1993 *Technological Change: West Mexican Mortuary Ceramics*. PhD dissertation, Department of Material Science and Engineering, University of Arizona, Tucson.

Beekman, Christopher S.

1996a *The Long-Term Evolution of a Political Boundary: Archaeological Research in Jalisco, Mexico*. PhD dissertation, Department of Anthropology, Vanderbilt University, Nashville, TN; University Microfilms, Ann Arbor, MI.

1996b El Complejo El Grillo del Centro de Jalisco: Una Revisión de su Cronología y Significado. In *Las Cuencas del Occidente de México: Época Prehispánica*, coordinated by Eduardo Williams and Phil C. Weigand, pp. 247–291. Colegio de Michoacán, Zamora, México.

2001 *Informe Final de la Temporada 2000, Proyecto Arqueológico Regional Valles de Tequila*. Report submitted to the Instituto Nacional de Antropología e Historia, Mexico City.

2006 The Chronological Context of the Central Jalisco Shaft Tombs. In Mortuary Practices and Shaft Tombs in Western Mexico: An *Homenaje* to Phil Weigand, a Special Section of *Ancient Mesoamerica* 17(2): 239–249.

2008 Corporate Power Strategies in the Late Formative to Early Classic Tequila Valleys of Central Jalisco. *Latin American Antiquity* 19(4): 414–434.

2016a Conflicting Political Strategies in Late Formative to Early Classic Central Jalisco. In *Political Strategies in Pre-Columbian Mesoamerica*, edited by Sarah Kurnick and Joanne Baron, pp. 95–117. University Press of Colorado, Boulder.

2016b Built Space as Political Fields: Community vs. Lineage Strategies in the Tequila Valleys. In *Alternative Pathways to Complexity: Households, Markets, World-Systems, and Political Economy: Essays Honoring the Work of Richard E. Blanton*, edited by Lane Fargher and Verenice Y. Heredia Espinoza, pp. 59–78. University Press of Colorado, Boulder.

In press La Secuencia Cronológica Temprana en Los Guachimontones. In *Nuevos Enfoques en la Arqueología de la Región de Tequila: La Primera Reunión de la Cátedra Phil C. Weigand*, edited by Joshua D. Englehardt and Verenice Y. Heredia Espinoza. El Colegio de Michoacán, Zamora.

Beekman, Christopher S., and Robert B. Pickering (editors)

2016 *Shaft Tombs and Figures in West Mexican Society: A Reassessment.* Gilcrease Ancient Americas Series, Gilcrease Museum, Tulsa, OK.

Beekman, Christopher S., and Phil C. Weigand

2000 *La Cerámica Arqueológica de la Tradición Teuchitlán, Jalisco.* El Colegio de Michoacán and Secretaría de Cultura del Estado de Jalisco, Zamora and Guadalajara.

2008 Conclusiones, Cronología, y un Intento a Síntesis. In *Tradición Teuchitlán*, edited by Phil C. Weigand, Christopher Beekman, and Rodrigo Esparza, pp. 303–337. El Colegio de Michoacán, Zamora, México.

2010 La Secuencia Cronológica de la Tradición Teuchitlán. In *El sistema fluvial Lerma-Santiago durante el Formativo terminal y el Clásico temprano. Memoria del Segundo Seminario-Taller sobre Problemáticas Regionales*, edited by Laura Solar Valverde, pp. 237–261. Instituto Nacional de Antropología e Historia, Mexico City.

Blanco, Ericka, Sean M. Smith, Jorge Herrejon, and Phil C. Weigand

2010 La Tradición Alfarera en Guachimontones y Loma Alta: Tipos, Formas y Usos. *Ecumene* 1(1): 88–119.

Blanton, Richard E., Gary M. Feinman, Stephen A. Kowalewski, and Peter N. Peregrine.

1996 A Dual-Processual Theory for the Evolution of Mesoamerican Civilization. *Current Anthropology* 37:1–14.

Breton, Adela C.

1903 Some Mexican Portrait Clay Figures. *Man* 3: 130–133.

Cach Avendaño, Erik Orlando

2008 La Exploración Arqueológica del Edificio Seis de Los Guachimontones y sus Implicaciones Socioculturales. In *La Tradición Teuchitlán,* edited by Phil C. Weigand, Christopher Beekman, and Rodrigo Esparza, pp. 89–122. El Colegio de Michoacán, Zamora.

Davenport, Bryce

2011 Analysis of Rancho El Guaje Materials. Unpublished manuscript in possession of the author.

DeLuca, Anthony

2014 Excavation Drawings and Maps. *Proyecto Arqueológico Teuchitlán. Informe de Laboratorio 2014*, pp. 17–20. Manuscript on file, Proyecto Arqueológico Teuchitlán, Jalisco.

2017 *Architectural Energetics and the Construction of Circle 2, Los Guachimontones, Jalisco.* MA thesis, Department of Anthropology, University of Colorado–Denver.

Galván Villegas, Luis Javier

1979 Informe del Sitio Arqueológico que va afectarse con las Obras del Fraccionamiento Bugambilias, Zapopan, Jalisco. Unpublished manuscript in possession of the author.

1983 Rescate en el Fraccionamiento Bugambilias, Zapopan, Jal. Unpublished manuscript in possession of the author.

1991 *Las Tumbas de Tiro del Valle de Atemajac.* Instituto Nacional de Antropología e Historia, Mexico City.

Glassow, Michael A.

1967 The Ceramics of Huistla, a West Mexican Site in the Municipality of Etzatlán, Jalisco. *American Antiquity* 32: 64–83.

Heredia Espinoza, Verenice Y., and Christopher S. Beekman.

2018 *Informe del Proyecto Arqueológico Ex–Laguna de Magdalena.* Informe al Consejo de Arqueología, Instituto Nacional de Antropología e Historia, Mexico City.

Hollon, Kristie

2015 *Analysis of Social Space: Analyzing Lived Experiences in the Los Guachimontones Central Ceremonial Area.* Unpublished MA thesis, Department of Anthropology, University of Colorado–Denver.

Johns, Catherine

2014 *Ceramic Activity Analysis of Navajas Circle 5 and the Need for Practice Theory in Unusual Monumental Architecture.* Unpublished MA thesis, Department of Anthropology, University of Colorado–Denver.

LeFae, Jones

2015 Offerings in Burials and Caches. In *Proyecto Arqueológico Teuchitlán. Informe de Laboratorio 2015*, pp. 6–14. Manuscript on file, Proyecto Arqueológico Teuchitlán, Jalisco.

2017 *Craft Specialization in the Burials and Offerings at the site of Los Guachimontones, Jalisco, from 300 B.C.–A.D. 1600.* MA thesis, Department of Anthropology, University of Colorado–Denver.

López Mestas C., Lorenza

2011 *Ritualidad, Prestigio y Poder en el Centro de Jalisco durante el Preclásico Tardío y Clásico Temprano. Un Acercamiento a la Cosmovisión e Ideología en el Occidente del México Prehispánico.* Unpublished doctoral dissertation, Department of Social Sciences, Centro de Investigaciones y Estudios Superiores en Antropología Social, Guadalajara.

Montejano, Marisol

2007 *El Complejo El Grillo en la Porción Oriental del Valle de Atemajac, Jalisco. Desarrollo Cultural en Tonala, durante el Clásico Tardío/Epiclásico (450–900 D.C.).* Tesis de Licenciatura, Universidad Autónoma de Guadalajara.

Nance, C. Roger, Jan De Leeuw, Phil C. Weigand, Kathy Prado, and David Verity

2013 *Sequencing Potsherds through Correspondence Analysis: Ceramics from the Long-Glassow Collection.* University of New Mexico Press, Albuquerque.

Ramos de la Vega, Jorge, and M. Lorenza López Mestas Camberos

1996 Datos Preliminares sobre el Descubrimiento de una Tumba de Tiro en el Sitio de Huitzilapa, Jalisco. *Ancient Mesoamerica* 7: 121–134.

Smith Márquez, Sean Montgomery

2008 Comparación entre Círculos Monumentales y No Monumentales. Aproximación a una Tipología. In *La Tradición Teuchitlán*, edited by Phil C. Weigand, Christopher S. Beekman, and Rodrigo Esparza López, pp. 123–142. El Colegio de Michoacán, Zamora.

Weigand, Phil C.
1979 The Formative-Classic and Classic-Postclassic Transitions in the Teuchitlán-Etzatlán Zone of Jalisco. In *Los Procesos de Cambio: XV Mesa Redonda,* vol. 1, pp. 413–423. Sociedad Mexicana de Antropología, Mexico City.
1985 Evidence for Complex Societies during the Western Mesoamerican Classic Period. In *The Archaeology of West and Northwest México*, edited by Michael Foster and Phil C. Weigand, pp. 47–91. Westview Press, Boulder, CO.
1993 *Evolución de una Civilización Prehispánica*. El Colegio de Michoacán, Zamora.
1999 *Primer Informe del Proyecto Arqueológico Los Guachimontones de Teuchitlán.* Report submitted to the Instituto Nacional de Antropología e Historia, Mexico City.
2008 La Tradición Teuchitlán del Occidente de México: Excavaciones en los Guachimontones de Teuchitlán, Jalisco. In *Tradición Teuchitlán*, edited by Christopher S. Beekman and Rodrigo Esparza, pp. 29–62. El Colegio de Michoacán, Zamora.

Weigand, Phil C., and Acelia García de Weigand
2000 *Segundo Informe del Proyecto Arqueológico Los Guachimontones de Teuchitlán, Jalisco. Temporada 2000.* Report submitted to the Instituto Nacional de Antropología e Historia, Mexico City.

3

Revisiting the Archaeology of the Huetamo Area, Southeastern Michoacán, Mexico

JOSÉ LUIS PUNZO DÍAZ, DIEGO RANGEL, ERIKA IBARRA, JESÚS ZARCO, AND MIJAELY CASTAÑÓN

In 2014, the Mexican government initiated a plan for the construction of an irrigation dam in the community of Chigüero, a few miles north of the city of Huetamo, the most important city in the Tierra Caliente area within the Balsas River Basin in southeastern Michoacán (Figure 3.1). Because of that, in 2015 and 2016 a group of archeologists from the Centro INAH-Michoacán[1] started an archaeological salvage project in the area that will be affected by the construction of the dam, as well as the surrounding agricultural irrigation zone.

The Middle Balsas region contains one of the densest concentrations of archaeological sites yet recorded in western Mexico, but unfortunately very little is known of its archaeological history. As we expected from the outset of the salvage project, the area was full of archaeological remains. We have been able to record 59 sites through intense surveys of the affected zones, and we performed archaeological excavations at 10 of these sites, primarily those that would be most directly impacted by the dam construction project.

In this chapter, we present the preliminary results of the archaeological investigations carried out to date, highlighting the contributions made by this work in terms of advancing scholarly understanding regarding the pre-Hispanic history of this region. First, however, we wish to present a brief review of the limited previous archaeological research in the area in order to comprehend why so little is known about the history of a region with such a dense record of archaeological sites, and to provide a contextual background for our own work in the area.

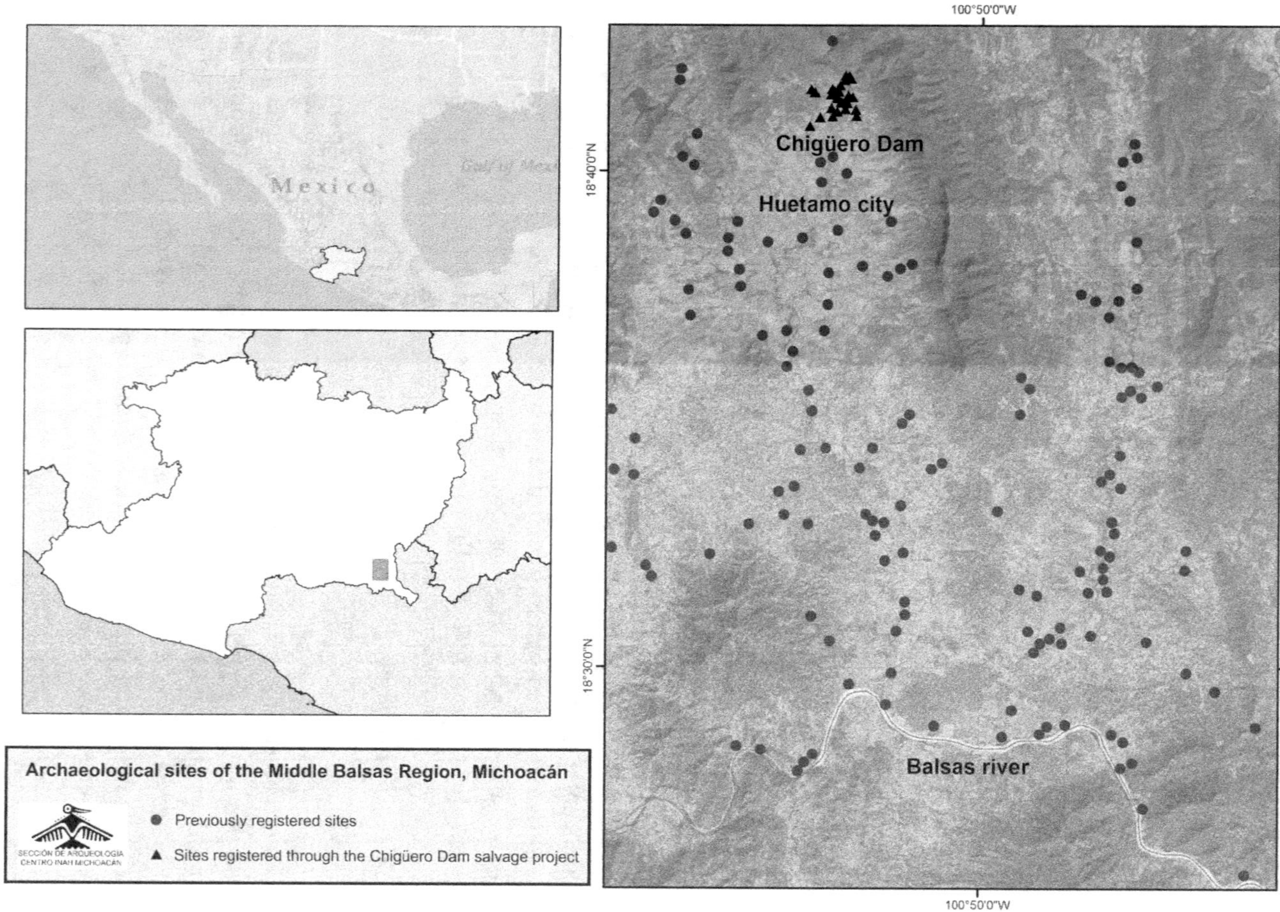

Figure 3.1. Location of the study area and the sites registered in the region of southeastern Michoacán (map by PAPACSUM-INAH).

Prior Research in the Middle Balsas Region

In 1939, Donald Brand (1943) carried out an archaeological reconnaissance in the state of Michoacán, including many areas that had not been previously surveyed, as was the case for the Middle Balsas region. One of Brand's students, Douglas Osborne, performed an important survey of the area surrounding the city of Huetamo, at the border of the states of Michoacán and Guerrero, publishing his work in 1943. Osborne recorded many of the most important regional sites, such as Mexiquito, Characuaro, and La Laguna, among others, and presented a variety of archaeological remains recovered through his survey. In addition, he mapped many of these sites, and also proposed the first areal ceramic typology. Osborne's research was the first systematic investigation that recognized the archaeological significance of the Middle Balsas area. At the same time, Pedro Armillas (1944) visited the area and also reported on the large site of Mexiquito, an extensive settlement on the south bank of the Balsas River in southeastern Michoacán. Robert Lister—another of Donald Brand's students—investigated the area through two field seasons in 1939 and 1941. His research served to refine Osborne's ceramic typology, and Lister further proposed the first architectonic typology, including the different styles of earth mounds—locally known as *yácatas*—and ballcourts (Lister 1947).

Since this initial flourish, archaeological research in the area was essentially abandoned for more than 40 years. Two significant exceptions were the large-scale salvage projects coordinated, respectively, by José Luis Lorenzo (1964) and Jaime Litvak (1968) during the 1960s, in response to the construction of two large dams on the Balsas River: Infiernillo and La Villita, 70 miles downriver from the Huetamo area. Both projects recorded hundreds of sites and performed excavation in almost two dozen of them. These works provided a wealth of information on the archaeological materials in the affected areas, and resulting publications detailed shellworking technologies (Suárez 1971, 1974, 1977), textiles (Mastache 1971), lithics and metal (Maldonado 1980), and settlement patterns and burials (González 1979).

Nonetheless, it was not until the end of the 1980s when regional investigations in southeastern Michoacán were taken up again, through INAH's Atlas Arqueológico Nacional project under the coordination of the archaeologist Mirna Medina. She and her team undertook extensive surveys in the municipalities of Huetamo and San Lucas, recording 557 archaeological sites, 47 raw material sources and deposits, and 28 rock art sites (Medina

1989: 495–498). Unfortunately, that project was focused primarily on recording sites, and very little extensive research or excavation was carried out on the sites identified.

In the late 1990s, Brigitte Faugère and Véronique Darras from the Centro de Estudios Mesoamericanos y Centro Americanos (CEMCA) of the French government carried out limited investigations focused on three rock shelters along the Huarimio Creek near Chigüero (Faugère-Kalfon and Darras 2002). They reported on the rock art they encountered and performed small-scale excavations in these shelters. Thus, despite a dense concentration of hundreds of sites and a variety of archaeological materials in the region—known to archaeologists for over 75 years—the pre-Hispanic history of the Huetamo area remains relatively unknown, and severely understudied (but compare Silverstein's [2000, 2001] studies on the border areas between Guerrero, Estado de Mexico, and Michoacán). It is against this backdrop that we began the Chigüero Dam archaeological salvage project in 2015. Through this project, we have obtained new data that aid in creating a more nuanced understanding of the regional history and its pre-Hispanic population, and in identifying salient characteristics of the social groups that inhabited the area over at least the past 1,500 years.

Preliminary Results: Sites and Regional Chronology

Based on radiocarbon dating of materials recovered through excavations of 10 sites in the vicinity of the Chigüero Dam construction project, as well as archaeomagnetic techniques that measure magnetic paleointensity in ceramics and stucco floors, we can now propose four pre-Hispanic occupational phases in the region, in addition to a fifth, postcontact phase that is also attested to in colonial documents of the sixteenth century (Punzo in press). In Table 3.1, we include all of the radiocarbon and archeomagnetic dates available for the area. This preliminary chronology is presented in Figure 3.2. Each phase is briefly discussed below.

Ancón Phase (AD 0–200)

The materials excavated and dated to this phase correspond to the little mound of El Ancón (Figure 3.3) and other small sites within the survey area. This settlement was constructed at the middle of the first century and was built directly on the alluvial terrace just beside the Chigüero stream. The site's location opens on to the floor of a valley surrounded by low hills

Table 3.1. Radiocarbon and archaeomagnetic dates in the Huetamo area

Sample/site	Technique	Material	Date AD	Date AD	Source	CI
Quesería	Radiocarbon	Charcoal	10–230		Meanwell (2007)	95%
LPI-6_2	Archaeomagnetic	Ceramic	88–114	531–955	Punzo (In press)	65%
Ancón	Radiocarbon	Charcoal	56–211		Punzo (In press)	95%
LPI RT104	Radiocarbon	Charcoal	235–385		Punzo (In press)	95%
LPI RT144	Radiocarbon	Charcoal	385–475	485–535	Punzo (In press)	95%
Itzímbaro	Radiocarbon	Charcoal	440–640		Meanwell (2007)	95%
LPI RT172	Archaeomagnetic	Stucco	528–579		Punzo (In press)	
Piaiticuro	Radiocarbon	Charcoal	556–644		Punzo (In press)	95%
LPI-6_1	Archaeomagnetic	Ceramic	568–926		Punzo (In press)	65%
LPI–RT197	Archaeomagnetic	Stucco	615–699		Punzo (In press)	
Quesería	Radiocarbon	Charcoal	620–770		Meanwell (2007)	95%
LPI RT168	Radiocarbon	Charcoal	650–690	750–760	Punzo (In press)	95%
LPI RT152	Radiocarbon	Charcoal	650–690	750–760	Punzo (In press)	95%
Mexiquito	Radiocarbon	Charcoal	660–880		Meanwell (2007)	95%
Itzímbaro	Radiocarbon	Charcoal	660–890		Meanwell (2007)	95%
La Casita	Radiocarbon	Charcoal	672–861		Punzo (In press)	95%
Tamarindos	Radiocarbon	Charcoal	1296–1404		Punzo (In press)	95%

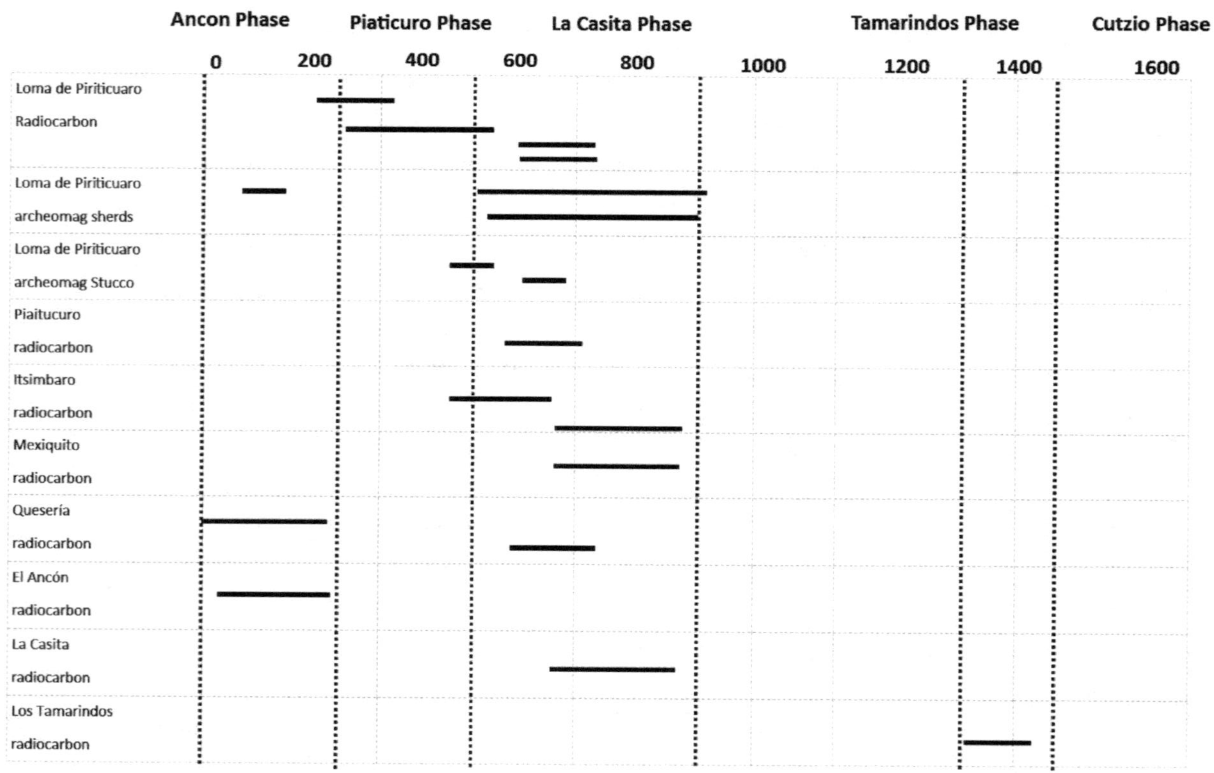

Figure 3.2. Chronology proposed for the Middle Balsas Region, Michoacán (compiled by PAPAC-SUM-INAH).

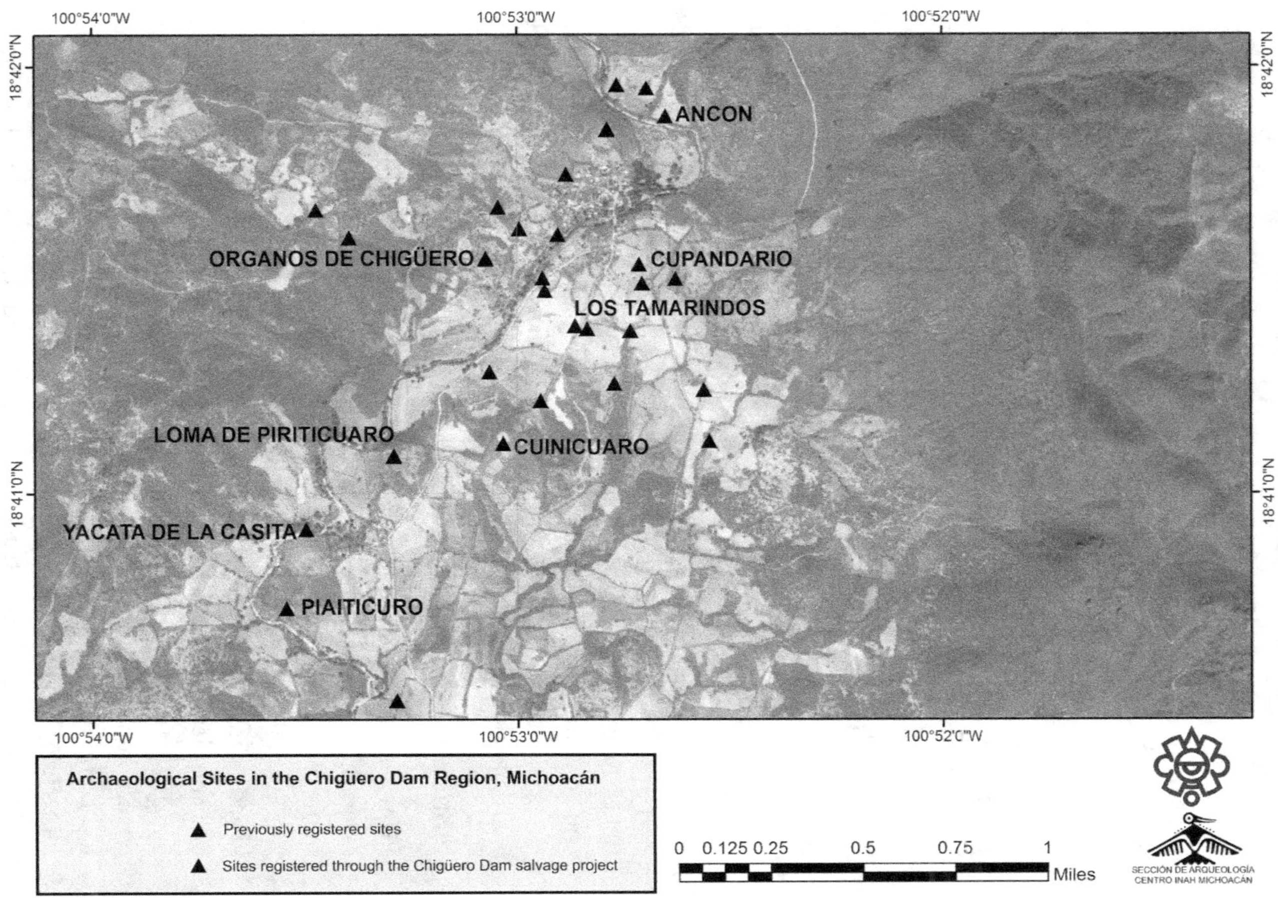

Figure 3.3. Relevant sites in the Chigüero Dam area (map by PAPACSUM-INAH).

at the base of the sierra. The mound was situated close to several water sources, not only adjacent to the stream, but also near several freshwater springs. Probably as a foundational offering to the single structure identified at the site, the body of an infant accompanied by a small vessel was deposited in the base of the structure. Over this offering, earthworks raised the mound to a height of 2 meters, and the mound was covered with large- and medium-sized stones joined by mortar. Wall foundations were also encountered on the upper part of the structure. These were arranged in rows of double or single stones, over which adobe and wattle-and-daub walls were erected. Both the walls and the floors enclosed by them were covered by stucco. We propose that the roof would have been made of some lightweight material, to cool the structure and maintain a stable temperature inside the rooms (Punzo et al. 2015).

At the El Ancón site, three burials were identified, all under the room's floor. The burials were very poor in terms of preservation and the grave goods with which the bodies were deposited, thus limiting our ability to extrapolate mortuary practices of the site's early inhabitants. The infant was the first burial, the remains placed directly on the gravel stones of the alluvial terrace. Accompanying the infant was a small bowl placed face up on one side of the body. The second burial was found 80 centimeters below the upper part of the structure. Unfortunately, due to the poor conservation of the osteological remains, the sex and age of this individual could not be determined. However, we did recover a collar made of shell, lithics, and ceramics beads. The third burial was located directly underneath the stucco floor at a depth of 40 centimeters. This burial was accompanied by a completely broken vessel placed face down on the right side of the body.

Piritícuaro Phase (AD 200–500)

Pirítícuaro-phase sites, like El Ancón, also appear to have been situated adjacent to streams, offering access to fresh water, good agricultural fields in the surrounding areas, and the opportunity to exploit riverine resources (for example, fishing). During this phase, sites appear to have been established primarily on hilltops, thus allowing their inhabitants a wide viewshed and clear sightlines between sites like Loma de Pirítícuaro or Organos de Chigüero. At many sites from this phase we encountered evidence for terrace construction and retaining walls—unsurprising given that hilltop construction requires additional work and labor to adapt the natural terrain, leveling the hilltop to create flat and open spaces on which to build. Within these constructed areas we found stucco floors, both inside

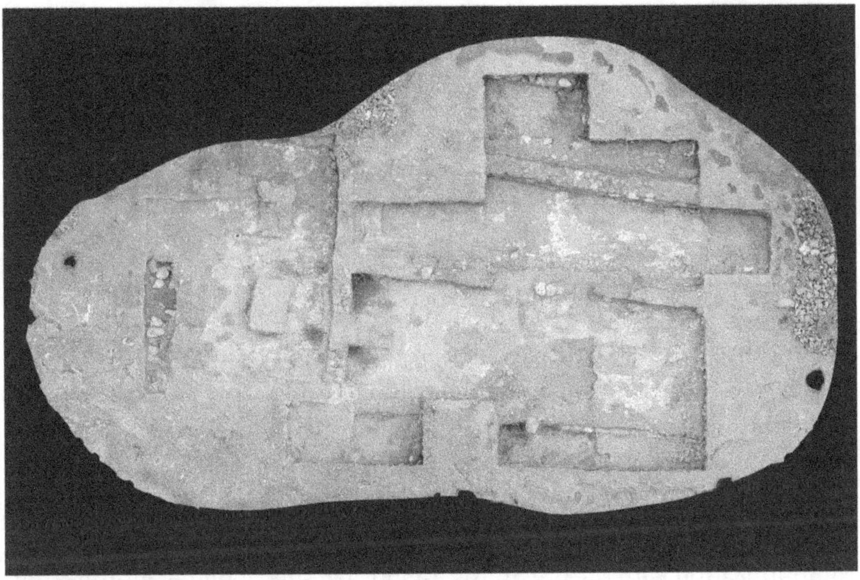

Figure 3.4. Orthophotograph of the main excavations at the Loma de Piritícuaro site (photograph by PAPACSUM-INAH).

individual rooms and in outside patios. Adobe walls and columns were also covered with multiple layers of stucco plaster.

The constructed spaces at the Loma de Piritícuaro site (Figure 3.4) demonstrate that its builders had a firm mastery of earthen architecture construction techniques. The sheer size of massive, overlapping adobe brick columns of 1.70 by 1.10 meters at the site suggest that these columns supported large beams that sustained a heavier roof structure, relative to the preceding phase. Walls were similarly constructed of adobe, to a thickness of 12–15 centimeters. Over time, constructed spaces were adapted, maintained, and repurposed, as suggested by the releveling of some spaces and closure of some rooms. These later modifications also evidence the use of varied construction techniques. For example, large adobes appear to have been employed to relevel spaces, and rooms were cloistered with earth and stone walls. It is also noteworthy that the quality of the stucco appears to have declined in later construction stages toward the end of this phase (Punzo et al. 2015). It is probably in this phase that we start to identify a developing complexity among the inhabitants of the area.

The habitational spaces in the Loma de Piritícuaro site were organized in a series of rooms arranged around a small sunken patio, probably surrounded by a portal allowing the people to use those spaces both for ritual and quotidian activities. On one of the floors inside a room within one such residential unit, we discovered a series of small holes that could have been used in a localized version of the pre-Hispanic game *patolli* (called *K'uilichi Chanaku* in central Michoacán). In the future, we plan to realize chemical analyses of the outside floors to determine the activities that may have occurred in habitational areas at the site (for example, food preparation, butchery, and so on).

La Casita Phase (AD 500–900)

During this phase we observe continuity in the occupation of the sites founded from the previous phase, including Loma de Piritícuaro, as well as the foundation of new, relatively smaller sites, like La Casita or Piaíticuro among others. Many of these new sites were established along the banks of rivers and streams below the hills on which the Piritícuaro-phase settlements were constructed. La Casita–phase sites thus maintained visual communication with the previously established sites at higher altitudes, but occupied the fertile farmland in the valleys below. The Yácata de Piaíticuro site is exemplary of this phenomenon. This phase appears to have witnessed a growth in regional population, evident not only through the establishment of new settlements, but also by the fact that previously constructed larger sites continued to increase in size, with new episodes of construction that incorporated an expanded repertoire of architectural features, particularly evident at the Loma de Piritícuaro site.

The majority of the newly established sites were constructed on the elongated low hills of valley floors. Housing platforms were erected on top of these topographic forms, some of which measure as large as 130 meters long by 30 meters wide. In the construction of the platforms, builders appear to have taken advantage of the natural shape of the hills themselves, making small adjustments such as erecting retaining walls or terracing on the low hillsides to extend the area available, in which both residential and ritual spaces were constructed. This dynamic is evident at the sites of Yácata de La Casita (Figure 3.5) and Piaíticuro. Earthen architecture construction techniques were retained form previous phases, as were the use of stucco on walls and floors. One new, potentially significant development that emerged in the La Casita phase is the use of earth cobs and rammed earth in the construction of walls, in addition to the square adobe bricks evident in previous phases (Punzo et al. 2015).

We also identified a number of burials at structures dating to the La Casita phase. Usually, these interments were located directly under the

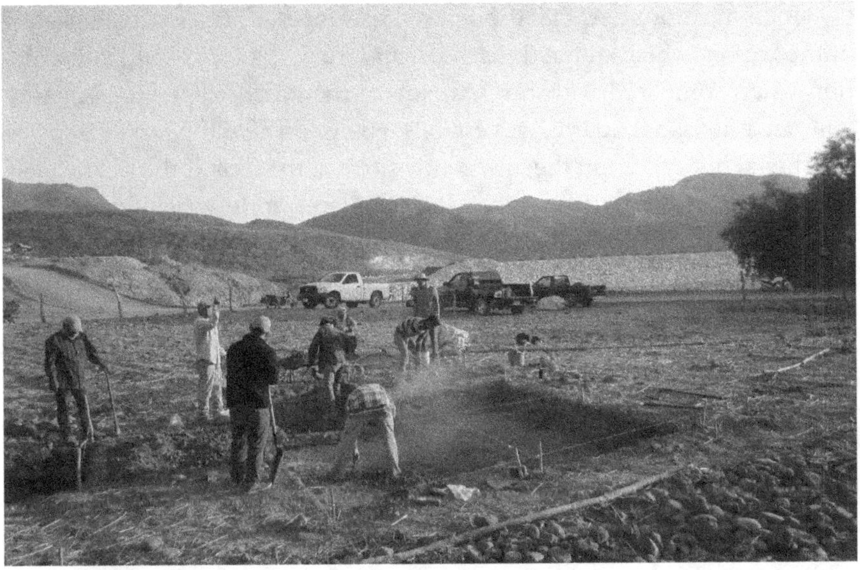

Figure 3.5. Excavations at the La Casita site with the construction of the dam at the background (PAPACSUM-INAH).

habitational floors, and excavation revealed how inhabitants broke through the stucco floor and dug a cavity down to bedrock to create the burial chamber. La Casita–phase burials are notable for the absence of offerings or burial furniture, and, in contrast to previous phases, the bodies of the deceased usually appear to have been placed in specific positions and orientations, sometimes in a prone, extended position with an east-west orientation, and in other instances in a flexed supine position with the arms crossed in a northeast-southwest orientation.

Tamarindos Phase (AD 1250–1400)

There appears to have been something of a hiatus in the regional settlement pattern between AD 900 and 1250. Unfortunately, we have not identified any sites in the research area that date to this temporal context. We speculate that this phenomenon may be related to an economic and social reorganization due to the breakdown of regional social networks that connected different areas and settlements within western Mesoamerica. Such a "collapse" is evident in other parts of West Mexico at this time, such as central Michoacán where those changes appear at the outset of the Tarascan period. So the partial abandonment of many sites in the Middle Balsas region may be a localized manifestation of this dynamic.

Beginning around AD 1250, however, new types of sites were established in the region, which differed markedly in terms of location and construction when compared with the settlements identified in previous phases. One such site, Cupandario, was established not on a hilltop, valley floor, or riverbank, but rather on the low, gentle slope at the foot of the sierra itself. Although such areas were not conducive to large-scale agriculture, sites of this phase were strategically situated near freshwater springs, a very important resource in this hot and dry region. Cupandario builders modified the mountain slope on which the site was established by creating a large platform divided into five flat levels at roughly 80 centimeter intervals. A series of rooms was constructed on each of these levels. The foundations for the walls of these rooms were made with single or double stone rows. It is possible that the walls were erected using stone-earth construction techniques, but we also encountered evidence for wattle-and-daub construction, such as the remains of earthen fragments that would have covered wood scaffolding, as well as organic, lightweight roofing materials—since a light roof structure is all that can be supported by this type of wall.

In terms of burials during this phase, we encountered evidence for two forms of mortuary practices. On the one hand, some individuals were interred below house floors directly on the bedrock without any offerings or burial furniture, as in the Cupandario site. Other individuals, however, appear to have been cremated after death. Their remains, as well as some metal, shell, and stone objects, were placed into small vessels, sealed with mud, and covered with another vessel. These containers were then placed in holes made in the bedrock and then buried. In our excavations at the Los Tamarindos site, we encountered 42 such urns in a space that appears to have been used exclusively for the deposition of these funerary vessels, since no habitational structures were associated with this feature (Figure 3.6). In some of these funerary vessels at Los Tamarindos, microexcavation and noninvasive CT scans of some vessels enabled us to find copper bells and rings that were identical to similar objects recovered in distinct contexts at the Cupandario site, suggesting a temporal correspondence (Punzo et al. in press).

Cutzio Phase (AD 1400–1525)

Historical documents, such as the *Relación de Michoacán* (Alcalá 2008 [1540]), indicate that during the time when Tarascan kings Hiripan, Tanganxoan, and Hiquingare ruled in the central Michoacán cities of Ihuatzio, Tzintzuntzan, and Pátzcuaro, respectively, a large-scale effort was begun

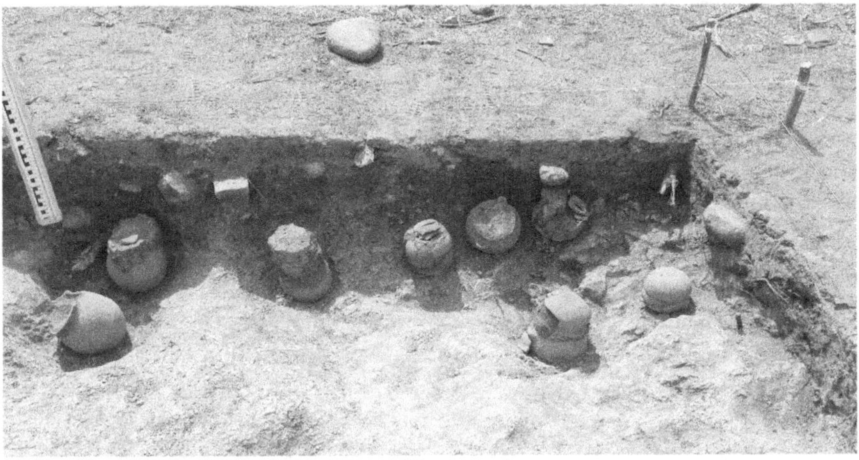

Figure 3.6. Funerary vessels from the Los Tamarindos site (photograph by PAPACSUM-INAH).

to conquer the territory immediately to the south of the Tarascan empire. This Tarascan expansion was first initiated in the Turicato region, at the entrance of the Tierra Caliente of southern Michoacán (the region to the south of the Trans-Mexican volcanic belt), and eventually sought to incorporate the territories along the Balsas River, including the ancient towns of Cutzio, Purechucho, Guayameo, and Sirándaro. This territorial expansion eventually resulted in the confrontation between the Tarascan and Mexica-led triple alliance in the Cutzamala River Area, 20 miles east of our study region.

When Tzitzipandacure, the son of Tanganxoan, was the *calzonci* (ruler) of the major site of Tzintzuntzan in the Pátzcuaro basin from AD 1465 to 1490; the most important settlement in his expansionist policy was located in the southeastern part of his territory at Cutzio. In the center of the modern town of Cutzeo,[2] we have identified a large *yácata*, which was likely the epicenter of the historical Tarascan settlement—unfortunately, this feature is under the urban footprint of the present town. Nonetheless, materials recovered in the vicinity of Cutzeo suggest that the pre-Hispanic town was most likely a multicomponent site with a long occupational history dating to at least the Classic or Epiclassic periods. Similarly, historical sources detail that a large group of Matlatzincas from the Toluca Valley requested permission from Tzitzipandacure to settle in Tarascan territory in the middle of the fifteenth century AD. The ruler granted that request, consenting to the establishment of the town of Huetamo, near Cutzio. In the center of the modern town of Huetamo, sherds and other archaeological materials indicative of this episode are frequently recovered in local excavations but without any systematic study as yet.

The Spanish arrived in the region in 1525. After contact, the indigenous population was centralized in the towns, mainly in Cutzio and Huetamo (Roskamp 2003; Warren 1977). On the top of the Cutzio *yácata* the Spaniards constructed a church, and the town became an important regional center in the colony. Nevertheless, the introduction of cattle ranching, European farming techniques, and new crops, and especially the beginning of intense mining activities, resulted in a profound change in the pre-Hispanic way of life in the Middle Balsas region.

Preliminary Analysis of Recovered Archaeological Materials

Ceramics

As mentioned above, there are few studies of archaeological ceramics in the region of Tierra Caliente and even fewer of Huetamo local pottery. The first Huetamo ceramic typology was made by Osborn in 1943; he studied surface materials collected by Brand in 1939 at Huetamo and neighboring areas in Michoacán and the Cutzamala area of Guerrero. His typology has eight wares: Huetamo Red Coarse, Huetamo Fine Red, Cutzeo Polished Black, La Laguna Buff, Cutzamala Incised Black, La Huizachal Incised Red, La Huizachal Orange, and Mexiquito Red-on-White (Osborne 1943). The same year, Goggin published his ceramic studies and typology of the Tepalcatepec River Basin. Ceramic types defined by him for this area are: Apatzingán Coarse, Apatzingán Red, Apatzingán Grooved Brown, Red wares, Red-on-Buff, Red-on-White, Chandio White-and-Red, Smudged ware, Chandio Black, and Delicias Polychrome (Goggin 1943). In 1947 Robert Lister published his ceramic typology for the Middle Balsas River Basin. The proposed types are: Balsas Red Coarse, Balsas Fine Red, Cútzeo Polished Black, La Huisachal Orange, Chandio White-on-Red, Zimátepec Black-on-White, La Huisachal Incised Red, and Totolapa Red-on-Tan (Lister 1947). Within the neighboring Middle and Lower Balsas regions, the ceramic studies reported were those made by Florence Müller (1979), Paul Schmidt (1990), Eliseo Padilla (2009), Jennifer Meanwell (2007), and Maria de Lourdes López-Camacho and Salvador Pulido Méndez (2010). These previous studies served as the baseline against which we identified

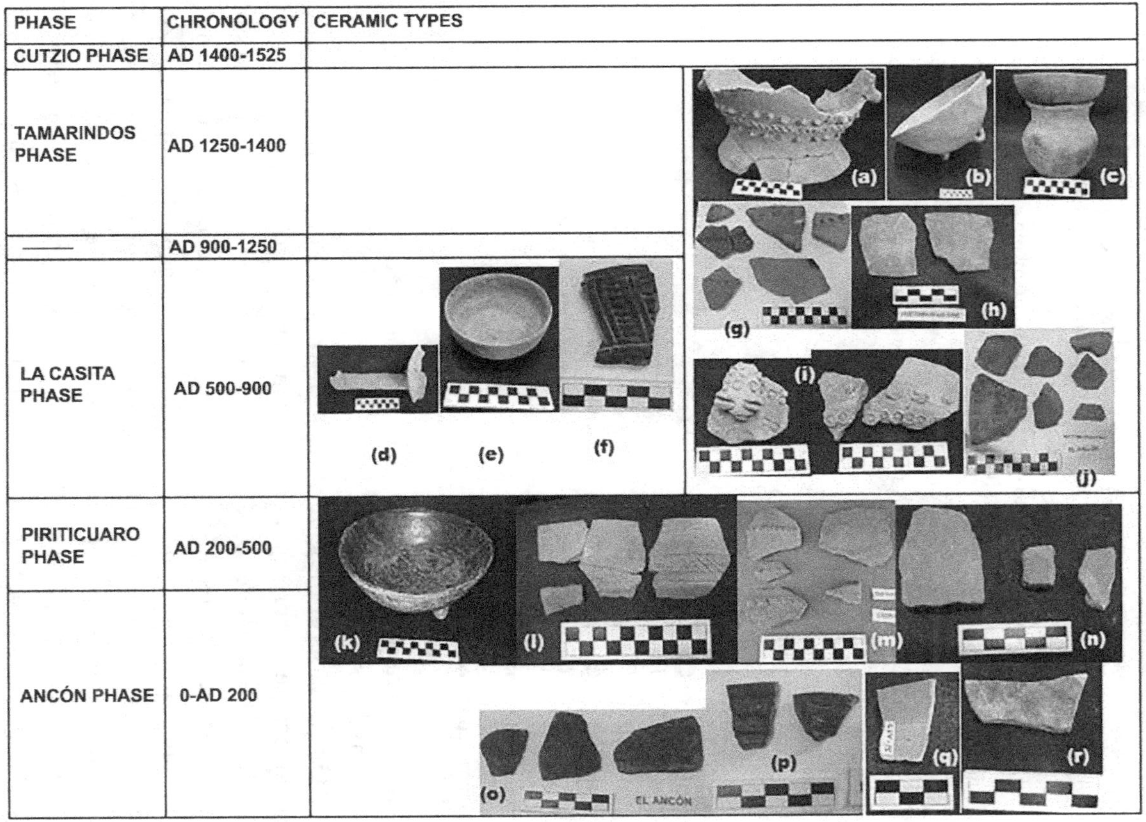

Figure 3.7. Proposed ceramic chronology of the Middle Balsas region of southeastern Michoacán: (a) Chihuero con aplicaciones, (b) Huizachal Anaranjado Pulido, (c) Huizachal Café pulido, (d) Huizachal Anaranjado Alisado, (e) Huetamo rojo/naranja, (f) Cutzio negro inciso, (g) banda decorativa con impresión de dedos, (h) Huetamo rojo, (i) Huizachal con decoración de círculos incisos, (j) Huetamo café, (k) Cutzio negro y café bruñido, (l) Ancón esgrafiado, (m) banda decorativa con impresión de uñas, (n) Cutzio rojo, (o) Cutzio rojo sobre negro, (p) Cutzamala negro inciso, (q) Mexiquito rojo sobre blanco, (r) Zimatepec negro sobre blanco (PAPACSUM-INAH).

Table 3.2. Paste types identified in ceramic sample

Paste Type	Description
Paste 1	*Color*: 5YR 5/6 yellowish red to 2.5YR 4/6 red, color change depending on the degree of exposure to fire *Texture and inclusions*: fine and compact paste with fine size inclusions and some sand *Fracture*: straight
Paste 2	*Color*: 2.5YR 4/6 red *Texture*: medium paste with very fine inclusions sometimes not visible *Fracture*: even
Paste 3	*Color*: 7.5YR 6/6–5/6, 7.5YR 4/4 brown, 5YR 6/6 reddish brown, 5YR 5/6 yellowish red, 7.5YR 6/4 light brown, color change depending on the degree of exposure to fire *Texture*: medium to coarse paste with many big white inclusions and sometimes red and brown inclusions (chamotte) *Fracture*: even
Paste 4	*Color*: 2.5YR 4/6 red *Texture*: fine compact and tough paste with very fine and medium-size inclusions *Fracture*: straight
Paste 5	*Color*: 5YR 6/4 light reddish brown *Texture*: medium to coarse paste with many big white and black inclusions (sand and rock) *Fracture*: brittle
Paste 6	*Color*: 2.5YR 6/6 a 6/8 light red *Texture*: fine paste with sand inclusions *Fracture*: brittle
Paste 7	*Color*: 7.5YR 4/3, 5/4 brown to 2.5YR 3/1 very dark gray *Texture*: fine to medium paste, very compact and very thin (4–9mm thick) with fine white inclusions *Fracture*: straight

Source: Compiled by author.

and comparatively analyzed the surface and excavation ceramic materials collected during the project.

On the basis of these analyses, we were able to propose a new ceramic sequence (see Figures 3.2 and 3.7), using a taxonomic classification by type-variety. The attributes considered in our ceramic analyses include surface finish, paste features, surface color, vessel form, decorative technique, and manufacturing technique, all of which were organized temporally according to the chronological contexts of the sites in which the materials were recovered. Seven types of paste were identified (Table 3.2). Two of these types (paste 1 and paste 2) were identified in ceramic materials recovered from almost all sites and were made out of local clays. Polished and smoothed brown and orange ceramic types (made out of type 2, 3, and 4 pastes), along with unslipped red and gimcrack slipped red ceramics (paste type 1) appear to have been used relatively consistently over time throughout the study area. The vessels themselves, however, do present morphological changes at distinct sites and in diverse temporal contexts. One exception are larger vessels that were possibly used as water containers, which appear throughout the sequence and in diverse contexts. Nonetheless, these vessels also presented variations in decorative motifs. Although analyses are ongoing, in general, from the studies so far conducted, we note significant differences between Ancón, La Casita, and Tamarindos ceramics.[3] This variation is discussed below. Differences in the wares identified at specific sites assisted in establishing site temporality as well as the distinct occupational phases described above.[4]

Late Formative/Early Classic period, Ancón phase

In the Ancón phase (AD 0–250), brown monochrome ceramics with polished and smoothed surfaces are ubiquitous, as are polished and burnished black and reddish ceramics. The dominant pastes are thin pastes with medium and medium-fine inclusions, compact, tan, brown to reddish sand-temper paste, although reddish pastes of brittle fracture are predominant (types 1, 4, and 7). The most common shapes are thin-walled tripod bowls and *apaxtles*. In this phase, red burnished, thin paste bowls are common (paste types 1 and 7), some of which display negative decoration. Also noteworthy is the presence of incised red burnished, brown burnished, and orange smoothed ceramics (paste types 1, 4, and 7). Diverse decorative techniques are evident. Stuccoes, fugitive white paint-on-slip decorations, and *sgraffito* ceramics were present. Common motifs are geometric, including sets of points and parallel, perpendicular, and reticular lines, interspersed with circles of 1 centimeter in diameter. Some bichrome ceramics are also present, often with red painting or negative decoration, such as red linear motifs on black slips or red bands on tan or orange surfaces. Most sherds present good slip quality, unlike those of Epiclassic sites (La Casita phase) and Postclassic sites (Tamarindos phase). During the Ancón phase, the most commonly used manufacturing technique was modeling, although occasionally molding was employed for both figurines and in other ceramic forms.

Epiclassic Period, La Casita Phase

Ceramics of the Epiclassic La Casita–phase sites (AD 550–900), exhibit relatively less variability in types, vessel shape, and decoration. The surface finish and paste of these ceramics also tends to be coarse. In the La Casita phase, two pastes were most common: the first is buff to orange color, sand-tempered, medium texture, compact with a high percentage of medium- large–size inclusions (type 3); the second is reddish to reddish-brown, sand-tempered, medium texture, with a high percentage of whitish medium to large-size inclusions (type 2). These two pastes were used for the elaboration of Huizachal Anaranjado, Huetamo Rojo and "*Chigüero con aplicaciones*" ceramic types. A significant number of thin paste ceramic objects are also evident (paste type 7); these present a diversity of surface finishes. In general, fewer decorative techniques are evident on ceramics recovered from La Casita–phase sites. Most vessels are monochrome (brown or black smoothed and polished), although some bichrome ceramics were also found, such as red-on-tan and black-on-red, as well as Mexiquito Rojo ware on Orange with negative decoration.

In this phase, *sgraffito* ceramics disappear almost completely and are replaced by incised pottery (brown and black polished, red and orange smoothed and polished) with motifs of medium and thick lines; decorative bands of fingerprints are also found on some sherds. Most of the recovered sherds have no slip and they have a very coarse surface. However, it seems that the majority of sherds present polished slip remains. We may speculate that local potters used a poor-quality slip on the vessels. A low-quality slip, combined with regional climatic conditions, soil acidity, and erosion would result in the deterioration of the slip over time. The most common surface colors are red, brown, tan, orange, and clay. It is potentially significant that incised, red-slipped ceramics are relatively less frequent in La Casita–phase ceramics as compared to other phases. Vessel forms elaborated with thin pastes include small bowls, plates, and grater bowls, whereas thicker pastes (paste types 2 and 3), were generally used for the elaboration of larger objects like compound jars, large grater bowls, and incense burners. In this phase, decorated jars with studded applications are also very distinctive. Modeling and rolling are the most common manufacturing techniques evident on ceramics that date to this phase.

Postclassic Period, Los Tamarindos Phase

It is perhaps noteworthy that typical Tarascan ceramics were not encountered in Los Tamarindos–phase assemblages. Only Chigüero con aplicaciones ware has some similarity with the Yarahuato Cream ware, defined by Pollard (1993, 2001) in the Pátzcuaro Basin. That said, the application technique appears to differ substantially from Tarascan wares, particularly as regards the size and distribution of decorative elements. Huetamo's Postclassic period pottery bears a strong resemblance to the local ceramics and pastes evident in the Epiclassic La Casita phase. It is thus possible that ceramic production and consumption were locally focused in both periods, and that Los Tamarindos ceramics are an extension of earlier La Casita traditions, despite the hiatus between these occupational phases.

Thin paste ceramics (paste type 7) appear to have disappeared completely in this phase. Most Los Tamarindos domestic ceramics continue to present reddish or tan sand-tempered pastes with medium to large inclusions (paste type 2), continuing the trend noted in the preceding La Casita phase. However, thinner and more compact pastes (types 4 and 5) also appeared at this time. Significantly, these were the most predominant in Los Tamarindos–phase strata. Surface finishes tend to be smoothed and polished brown, orange, or tan monochromes. Coloration, however, is quite heterogeneous, as these three colors can be present in just one earthenware vessel. Further, sherds dating to this phase tend to be relatively coarse, and the slips have the same low quality noted in the preceding La Casita–phase assemblages. In general, the presence of polished brown ware increases and red slip ceramics (Huetamo Red Fine) decreases, although earthenwares with reddish surfaces are also abundant. The predominant shapes are grater bowls, tripod bowls, incense burners, and small, handled jars.

In terms of decoration, few polychrome wares were present in Los Tamarindos assemblages. The most common decoration was made of red slip lines (simple designs of bands and lines) over a tan or clay-colored slip. Fingerprint banding also continued, and Chigüero con aplicaciones ware remain present. Three sherds of fine paste reddish ceramic were also found—these appeared to have been painted with black lines, small triangles, and very fine parallel lines forming triangles or rhomboid patterns over a cream slip. These fine paste ceramics may correspond to the Zimatepec Black ware described by Lister (1947). Modeling and rolling continued to predominate as the manufacturing techniques in the Los Tamarindos phase.

In sum, we can differentiate at least two ceramic traditions in the Huetamo area: one that was predominant during the Late Formative and Early Classic period Ancón phase, which produced relatively more refined ceramics with greater variability, and a second that corresponded to the Epiclassic and Postclassic period La Casita and Los Tamarindos phases, whose products were coarser and less variable. We speculate that during these latter phases, the production and consumption of ceramic materials was more locally focused.

Lithic Technology

The pre-Hispanic groups that inhabited the Middle Balsas region of Michoacán—like those in neighboring regions—exploited locally available lithic raw materials for the production of tools and implements to help them perform their daily tasks. Nonetheless, previous studies of lithics in the Huetamo region are virtually nonexistent. In fact, Maldonado's (1976, 1978) investigations of the projectile points and the prismatic blades recovered through the Infiernillo Dam salvage project stands as the only research to date. As such, Maldonado's work served as both a starting point and baseline for our own analyses. In the course of our research and excavations, we recovered a variety of different types of lithic materials and artifacts, ranging from tools developed to perform specific functions, to production debitage indicative of the working of several raw lithic materials, including andesite, flint, and obsidian.

Recovered lithic materials were analyzed utilizing a technological approach in order to identify the manufacturing sequence of each artifact. Based on this analysis, lithic artifacts and debitage were grouped in analytical categories that shared similar characteristics, thus allowing for the identification of localized lithic industries that spanned hundreds of years. In the study region, two such industries appear to have been locally developed—these primarily involved the working of andesite and flint. Obsidian, on the other hand, was not a local resource, and as such we can surmise that it figured prominently in trade relations between the inhabitants of the region and groups further afield that enjoyed access to this raw material. Is important to mention that the obsidian-sourcing studies are currently under way.

In terms of morphology and function, it would appear that the vast majority of stone artifacts collected were designed to cut and scrape. Beginning in the Ancón phase, we observe that many such cutters and scrapers were constructed from blades or flakes of andesite. This raw material is abundant in the region, and easily located on the surface throughout the study area. The manufacturing process began with the procurement of raw materials, either in the form of nodules or small blocks of andesite, which were either fractured by direct percussion at one end or broken in half to form a plane of percussion (in several cases, more than two planes of percussion are evident). After extracting decortical flakes by direct percussion, large percussion flakes and blades were removed to take advantage of the natural edge on one or both sides, which could be used as a cutting or scraping edge. Such a function is inferred from use-wear analysis of the artifacts' functional edges, which display patterns formed through the use of specific artifacts on other raw materials. During the analysis of the andesite artifacts, we also identified chopping tools, which were made from medium nodules by direct percussion on one end to form a sharp and winding edge. On the other hand, the development of scrapers involved direct percussion to a block to remove a large percussion flake, whose functional edge was retouched to obtain a convex edge.

The production of andesite tools is evident at all the sites since the Ancón phase, although a marked increase in lithic use and production is noted during the Piritícuaro and La Casita phases. At sites like Piaíticuro, Loma de Piritícuaro, or La Casita, we noticed an increase in the presence of lithic artifacts of these specific types of tools. Further, certain technological advances were evident during these periods. For example, we encountered artifacts with unifacial notches on one edge and rejuvenation flakes, indicating that the artifacts could be used in cutting harder materials such as wood. In the Tamarindos phase, the presence of andesite artifacts decreases in relation to previous phases, likely due to a concomitant increase in the production of prismatic obsidian blades during this phase.

Analysis of flint artifacts also suggests that this was a locally developed industry. The use of this raw material appears first during the Piritícuaro phase and extends into the Tamarindos phase. Artifacts and debitage of this raw material were found during excavations and surface collections at the sites of the Piritícuaro phase, like Loma de Piritícuaro; La Casita–phase sites, like Piaíticuro and Organos de Chigüero; and Tamarindos-phase sites, like Cupandario. This suggests the presence of a long-standing technological tradition associated with this raw material in the region. Specialized artisanal knowledge is also suggested through the presence of complex shapes such as bifacial arrowheads and knives in La Casita–phase assemblages. In this sense, although flint was exploited since the Piritícuaro phase—and completed knives and scrapers dating to this period are evident—it would appear that the La Casita phase witnessed the apogee of flint exploitation and technology. Lithic artifacts from these temporal contexts were of greater morphological complexity (for example, including bifacial projectile points and knives) and were produced in higher quantity and quality.

Furthermore, at the La Casita site we encountered the greatest amount of flint debitage associated with the production process. This process began with obtaining a medium block, from which decortical flakes were extracted by direct percussion on one plane, with the intention that subsequent extractions could result in the removal of large percussion blades by the same technique. Once a blade or flake of the desired size was obtained, the craftsman could then prepare a preform, applying all intrusive retouches by pressure on both sides to thin the piece, and then making marginal retouches on the edge of both sides through pressure flaking to achieve the form desired. As was the case with andesite artifacts, flintworking appears to have decreased in the Tamarindos phase with the advent of locally produced prismatic blades of obsidian, thus reflecting a transformation in the production and consumption of raw materials at this time.

That said, from a general point of view, obsidian artifacts were present in the study area since the early Ancón phase, which suggests that, throughout the areal occupational sequence, groups in the Middle Balsas region developed trade relations with foreign groups to obtain this and other raw materials not locally available. Analysis of obsidian artifacts revealed the presence of two primary types of obsidian: translucent green obsidian and gray veined obsidian. Likewise, two obsidian working industries are evident, along the lines that Clark (1988: 11) has suggested; one artisanal, likely involving specialized craftsmen and characterized by fine quality and relatively more complex shapes, and another unspecialized industry, available to any individual but that was only capable of producing relatively simpler shapes (Clark 1988: 11). This latter, unspecialized industry is present from the Ancón through Tamarindos phases. In this industry, we observe decortical and percussion flakes of various sizes that were extracted by direct percussion of veined gray obsidian casual cores, and whose form was most likely determined by function (although this may not necessarily be the case, insofar as percussion flakes are removed for all production processes; compare Clark 1988: 16). In temporally discrete sites such as Cuinicuaro, Loma de Piritícuaro, and Tamarindos, casual cores formed by the exploitation of obsidian blocks were found, from which blades and flakes were removed by direct percussion (to take advantage of the raw edge [for cutting] formed through reduction). Therefore, we suggest the existence of an unspecialized regional obsidian industry that spanned the Late Formative, Classic, and Postclassic periods.

A specialized industry is evident in the production of bifacial tools like knives and arrowheads, in addition to the extraction of prismatic blades in their specific stages and the development of "eccentric" materials whose function is unknown. The origins of specialized obsidian production are found in the gray obsidian tools first observed in the Pirtícuaro phase and extend to the Tamarindos phase. During these phases, bifacial knives were excavated at sites such as Loma de Pirtícuaro, La Casita, and Cupandario. Such artifacts were produced by pressure flaking both sides of a prismatic blade to thin the artifact, detaching microflakes from both sides to produce straight edges. At the Tamarindos-phase site of Cupandario, more than 12 types of projectile points were made from prismatic blades, and only second- and third-series blades were used in the production process (see Figure 3.8). This fact suggests a technological innovation by Tamarindos-phase artisans, since points of this type were not encountered in other contexts.

The production, use, and consumption of prismatic blades made with gray obsidian are evident during all phases. Taking into account the amount of materials derived in the extraction and production process, we can say that during the Ancón phase, production of obsidian artifacts is relatively low, and primarily involved the extraction and use of percussion flakes. In the following Pirtícuaro and La Casita phases, relatively more specialization is apparent, involving a three-stage production process (the manufacture of a platform on the core, the removal of decortical and percussion flakes by direct percussion to form the prismatic core, and the extraction of prismatic blades through pressure flaking [Clark 1988]). But without a doubt the Tamarindos phase witnessed a significant increase in specialized obsidian production, particularly in terms of prismatic blade manufacturing. At this time, we note the highest concentration of exhausted polyhedral cores, and decortical and percussion flakes resulting from core preparation, as well as flakes derived from the rejuvenation of prepared platforms through pitting or abrasion.

The case of green obsidian is distinct because it is present in all phases, but only in the form of prismatic blades. Artifacts and debitage recovered suggest that objects made of green obsidian were locally produced,

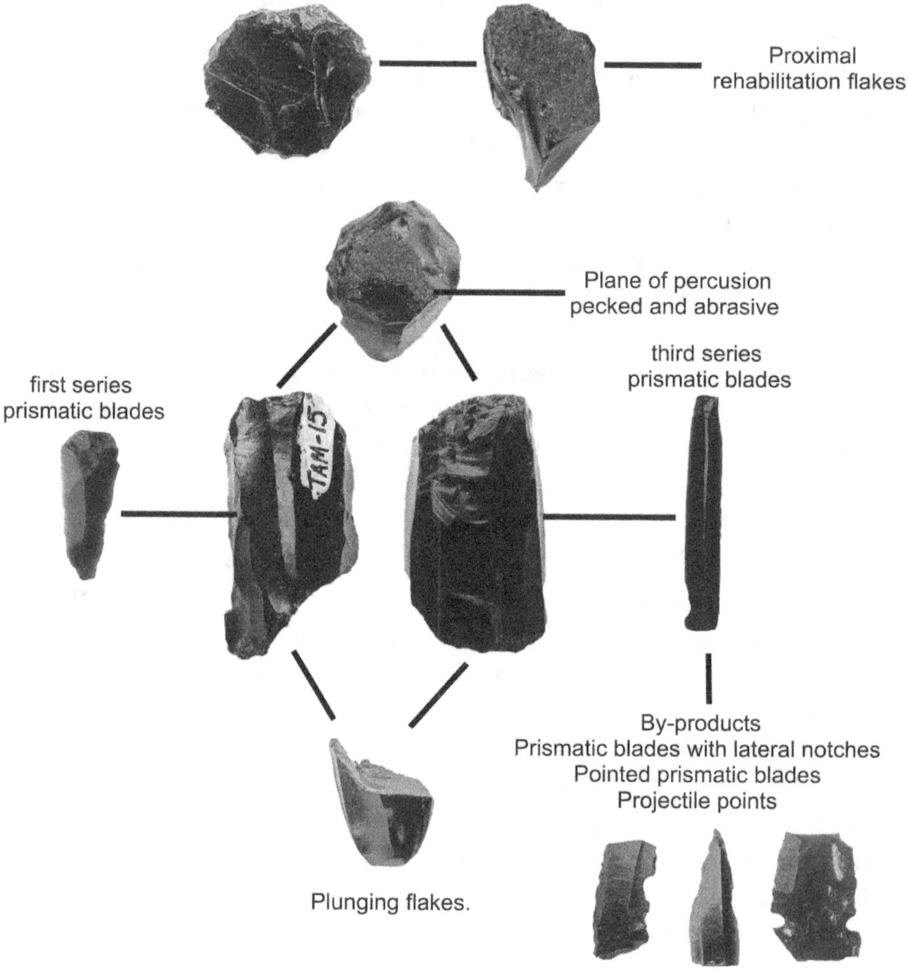

Figure 3.8. General diagram of Tamarindos-phase prismatic blade technology (image by PAPACSUM-INAH).

although such materials were only found only in the second and third stages of extraction, in the form of elongated, straight, and trapezoidal shapes in cross section (Clark 1988, 1997). Many of these artifacts, however, present functional use-wear on one or both edges indicative of cutting activities. The presence of objects fabricated with green obsidian in these states of reduction suggests that it was a commodity that was exchanged throughout the occupational history of the region. Nevertheless, the relatively greater presence of green obsidian during the Ancón, Pirtícuaro, and

La Casita phases may suggest a greater amount of interaction with central Mexico in the Late Formative, Classic, and Epiclassic periods. In the Postclassic period, particularly in the Tamarindos phase, the production and use of green prismatic blades is reduced, likely due to the control of obsidian sources in central Mexico by the Mexica triple alliance—in conflict with the Tarascan empire—as well as the increase in the exploitation of the gray obsidian quarry at Zináparo-Ucareo by the Tarascans at this time. As such, at sites like Tamarindos and Cupandario, the presence of this gray obsidian is higher, both relative to green obsidian and to previous phases.

Final Remarks

This paper presents only a brief summary of the preliminary research carried out in the Chigüero Dam area. Nonetheless, these preliminary results allow for a more nuanced understanding of a profound and complex regional history. To recapitulate, the first occupational phase at the El Ancón site reveals a relatively smaller and less complex settlement, but one that nonetheless evidences the beginnings of several significant cultural components (for example, ceramics, lithic industries, architecture, and construction techniques) that are present in later phases. The second phase, Piritícuaro, corresponding to the Early Classic period, reflects an increase in social and technological complexity, both in terms of architectural techniques, the use of different raw materials, more sophisticated ceramic production, population growth, and the expansion of interregional trade—a trend that continued and indeed was augmented during the apogee of the following Epiclassic La Casita phase. Sadly, an Early Postclassic period hiatus limits our comprehension of this crucial transitional period. Nevertheless, the Late Postclassic Tamarindos phase reveals significant changes, in terms of architecture and construction techniques, relatively less sophisticated ceramics, the development of a specialized local lithics industry, new burial practices, and a decrease in interregional exchange. These changes during the Postclassic continued through the period of Tarascan expansion and into the terminal Postclassic and Early Colonial periods, where they were observed by the Spanish chroniclers during the sixteenth century. Of course, this contribution represents only a first step in constructing a deeper understanding of over 1,500 years of pre-Hispanic history in the area. As such, a great deal of work remains to be done in the future in order to address the multiple lacunae in our knowledge of the Middle Balsas region of southeastern Michoacán, its relation to neighboring areas in western Mexico, and Mesoamerica as a whole.

Notes

1. Centro INAH-Michoacan is the state office of the Instituto Nacional de Antropología e Historia, a part of the Mexican federal government in charge of archaeological archaeological is correct research and conservation.

2. Cutzeo is one mile north of the center of Huetamo; the two towns may be considered a conurbation.

3. We do not discuss Piritícuaro-phase ceramics in this chapter as analyses are ongoing, and more detailed studies are needed to refine our understanding of the ceramics of this phase.

4. A more complete description of the sequence and common attributes can be consulted in the report submitted to the INAH Consejo de Arqueología (Punzo et al. 2016).

References

Alcalá, Fray Jerónimo

2008 [1540] *La Relación de Michoacán*. El Colegio de Michoacán, Zamora.

Armillas, Pedro

1944 Mexiquito, Gran Ciudad Arqueológica en la Cuenca del Río de las Balsas. *El México Antiguo* 6(7–8): 254–262.

Brand, Donald

1943 Recent Archaeologic and Geographic Investigations in the Basin of the Rio Balsas, Guerrero and Michoacan. In *27ª Congreso Internacional de Americanistas*, pp. 147. ICA, Mexico City.

Clark, John

1988 *The Lithic Artifacts of la Libertad, Chiapas, Mexico: An Economic Perspective*. Papers of the New World Archaeological Foundation 52. Brigham Young University, Provo, UT.

1997 Prismatic Blademaking, Craftsmanship and Production: An Analysis of Obsidian Refuse from Ojo de Agua, Chiapa, Mexico. *Ancient Mesoamerica* 8: 137–159.

Faugère-Kalfon, Brigitte, and Véronique Darras

2002 Las obras rupestres de Huarimio, Tierra Caliente de Michoacán. *Arqueología* 28: 21–48.

Goggin, John

1943 An Archaeological Survey of the Rio Tepalcatepec Basin, Michoacán. *American Antiquity* 9(1): 44–58.

González, Norberto

1979 *Patrón de asentamiento prehispánico en la parte central del Bajo Balsas: un ensayo metodológico*. Colección Científica 73. Instituto Nacional de Antropología e Historia, Mexico City.

Lister, Robert

1947 Archaeology of the Middle Rio Balsas Basin, Mexico. *American Antiquity* 13(1): 67–78.

Litvak King, Jaime

1968 Excavaciones de rescate en la presa de La Villita. *Boletín del INAH* 31(28): 30.

Lorenzo, José Luis

1964 Primer informe sobre los trabajos arqueológicos de rescate efectuados en el vaso de la presa de "El Infiernillo," Guerrero y Michoacán. *Boletín del INAH* 17: 24–31.

López-Camacho, María de Lourdes, and Salvador Pulido Méndez

2010 Las cerámicas tempranas en el área del delta del Balsas. *Arqueología* 43: 85–98.

Maldonado, Rubén

1976 Puntas de proyectil del Infiernillo. *Boletín del Departamento de Monumentos Prehispánicos* 1: 1–22.

1978 Navajas de obsidiana del Infiernillo. *Boletín de la Escuela de Ciencias Antropológicas de la Universidad de Yucatán* 29: 2–36.

1980 *Ofrendas asociadas a entierros del Infiernillo en el Balsas, México*. Colección Científica 91. Instituto Nacional de Antropología e Historia, Mexico City.

Mastache, Alba Guadalupe

1971 *Técnicas prehispánicas del tejido*. Colección Investigaciones 20. Instituto Nacional de Antropología e Historia, Mexico City.

Meanwell, Jennifer

2007 *El Proyecto del Balsas Medio: Una Investigación Sobre Funcionalidad y Cronología de la Cerámica*. FAMSI, Crystal River, FL.

Medina, Mirna

1989 Informe de los resultados del proyecto Atlas Arqueológico Nacional para el estado de Michoacán. In *Memoria del Proyecto Atlas Arqueológico Nacional, 1985–1988*, vol. 5, pp. 460–550. Manuscript on file in the Archivo Técnico del Departamento de Registro Público de Monumentos y Zonas Arqueológicas-INAH, Mexico City.

Müller, Florence

1979 *Estudio tipológico provisional de la cerámica del Balsas Medio*. Colección Científica 78. Instituto Nacional de Antropología e Historia, Mexico City.

Osborne, Douglas

1943 An Archaeologic Reconnaissance in South-Eastern Michoacán. *American Antiquity* 9(1): 59–73.

Padilla, Eliseo

2009 La Cerámica Blanco Granular de Guerrero: Implicaciones de su Distribución Temporal y Espacial. Unpublished MA thesis, Instituto de Investigaciones Antropológicas, Universidad Nacional Autónoma de México.

Pollard, Helen

1993 *Taríacuri's Legacy: The Prehispanic Tarascan State*. University of Oklahoma Press, Norman.

2001 *Proyecto desarrollo del Estado Tarasco: Los Señoríos Urichu, Xaracuaro, y Pareo. Cuenca de Pátzcuaro, Michoacán, 1990–1998*. Report submitted to the Instituto Nacional de Antropología e Historia, Mexico City.

Punzo Díaz, José Luis

In press Una propuesta cronológica prehispánica para la región de Huetamo, en el Sureste de Michoacán, México. *Arqueología Iberoamericana.*

Punzo Díaz, José Luis, Alfonso Gastélum, and Ingris Peláez

In press Conservación y estudio arqueológico no invasivo mediante la reconstrucción virtual 3D de 8 urnas cinerarias de la tierra caliente michoacana. *Intervenciones.*

Punzo Díaz, José Luis, Diego Rangel, Erika Ibarra, and Jesús Zarco

2015 Primeros datos sobre el uso de adobe y cal en época prehispánica en la región michoacana del río Balsas medio, México. In *Tierra, Sociedad, Comunidad: 15° Seminario Iberoamericano de Arquitectura y Construcción con Tierra* (SIACOT). SIACOT-Universidad de Cuenca, Cuenca, Ecuador.

Punzo Díaz, José Luis, Diego Rangel, Erika Ibarra, Jesús Zarco, and Mijaely Castañón

2016 *Informe Final del Proyecto Salvamento Arqueológico en el Emplazamiento de la Presa Chigüero, en el Municipio de Huetamo, Michoacán.* Report submitted to the Instituto Nacional de Antropología e Historia, Mexico City.

Roskamp, Hans

2003 Los Códices de Cutzio y Huetamo: Encomienda y Tributo en la Tierra Caliente de Michoacán, Siglo XVI. El Colegio de Michoacán y El Colegio Mexiquense, Zamora.

Schmidt, Paul

1990 *Arqueología de Xochipala, Guerrero.* Universidad Nacional Autónoma de México, Mexico City.

Silverstein, Jay E.

2000 A Study of the Late Postclassic Aztec-Tarascan Frontier in Northern Guerrero, Mexico: The Oztuma-Cutzamala Project. Unpublished PhD dissertation, Department of Anthropology, Pennsylvania State University, University Park.

2001 Aztec Imperialism at Oztuma, Guerrero: Aztec-Chontal Relations during the Late Postclassic and Early Colonial Periods. *Ancient Mesoamerica* 12(1): 1–30.

Suárez, Lourdes

1971 *Técnicas de manufactura y tipología del material de concha encontrado en la presa Presidente Adolfo López Mateos (Michoacán, Guerrero).* Unpublished MA thesis, Escuela Nacional de Antropología e Historia, Mexico City.

1974 *Técnicas prehispánicas en los objetos de concha.* Colección Científica 14. Instituto Nacional de Antropología e Historia, Mexico City.

1977 *Tipología de los objetos prehispánicos de concha.* Colección Científica 54. Instituto Nacional de Antropología e Historia, Mexico City.

Warren, Benedict

1977 *La Conquista de Michoacán 1521–1530.* Estudios Michoacanos. FIMAX, Morelia.

4

The Early Postclassic Aztatlán Colonization of the Pacific Coast of Jalisco

JOSEPH B. MOUNTJOY, FABIO GERMÁN CUPUL-MAGAÑA,
RAFAEL GARCÍA DE QUEVEDO-MACHAIN,
AND MARTHA LORENZA LÓPEZ MESTAS CAMBEROS

The focus of this chapter is a recently discovered archaeological site, Arroyo Piedras Azules, located on the northern Pacific coast of Jalisco, Mexico. Materials discovered by local people and our follow-up excavations in 2015, 2017, and 2018 are providing considerable information about the purpose, nature, and date of the colonization of this area of West Mexico by people associated with the Aztatlán archaeological culture of the Early Postclassic period, and about the nature of the expansion of the Aztatlán phenomenon in West Mexico.

In order to be able to discuss the importance of the Aztatlán settlement at Arroyo Piedras Azules, we first provide some context about what the Aztatlán archaeological culture is thought to represent, when it dates, its geographic extension in West Mexico, and some ideas as to its place in the broader context of pre-Hispanic Mesoamerican culture.

The Context

Susan Toby Evans, in her encyclopedic summary of the archaeology of ancient Mexico and Central America (2008: 410), has provided a recent description of what the Aztatlán archaeological culture is thought to represent:

> This term [Aztatlán Complex] was applied in the Early Colonial period to the cultural complex found along the Pacific Coast of West Mexico and the Northwestern Frontier. Archaeological research has indicated that the sites in the Aztatlán complex emerged in the early

> Postclassic. The pottery that characterizes these sites is so similar to Mixtec codex style that some researchers once speculated that Aztatlán was the result of a colonization effort by Mixteca-Puebla peoples. These coastal and uplands regions were clearly part of the period's extensive trade networks, but Aztatlán sites were not just trading centers. They were the focal communities in river valleys along the coast, supported by local farming and by production of metal objects and obsidian as well as pottery.

As to when the Aztatlán phenomenon dates, we have currently available 59 radiocarbon dates associated with remains identifiable as pertaining to the Aztatlán archaeological culture that can be used to estimate the beginning and end dates of the Aztatlán phenomenon, and therefore also its duration. The chronological ordering of these dates indicates the beginning of Aztatlán at about AD 900, and that the hegemony Aztatlán represents broke up about AD 1300, although Aztatlán material seems to persist in some locations in northern Nayarit and southern Sinaloa into the late 1400s.

This gives a total span of about 400 years for the duration of the Aztatlán hegemony, and current evidence indicates there were two developmental phases to this process, divided at about AD 1200. The first phase is characterized by relatively simple Mesoamerican-type iconography, with some forms and decoration of pottery similar to that of the Gran Nicoya area in Costa Rica and Nicaragua. The second phase has much more elaborately decorated Mixteca/Puebla codex-style iconography that may be related to the development and expansion of the Mixtec kingdom at Tututepec in southern coastal Oaxaca.

As for the geographic distribution of Aztatlán remains in West Mexico, based on the location of major Aztatlán habitation sites, cemeteries, and ceremonial centers, the area covered (Figure 4.1) is considerable, about the same size as the extension of the Maya sites on the Yucatan Peninsula during the Classic period.

The northernmost major site of Aztatlán occupation thus found is at Mochicahui in the municipality of Fuerte in northernmost Sinaloa (Manzanilla and Talavera 1988). Southward along the Pacific coast, the farthest south that a significant amount of Aztatlán material has been found is at the Higuera Blanca site near the mouth of the San Nicolás River (Mountjoy and Schöndube 2014). There are reported to be Aztatlán occupations farther south along the Jalisco coast, but the existence of such sites has not yet been substantiated by archaeologists.

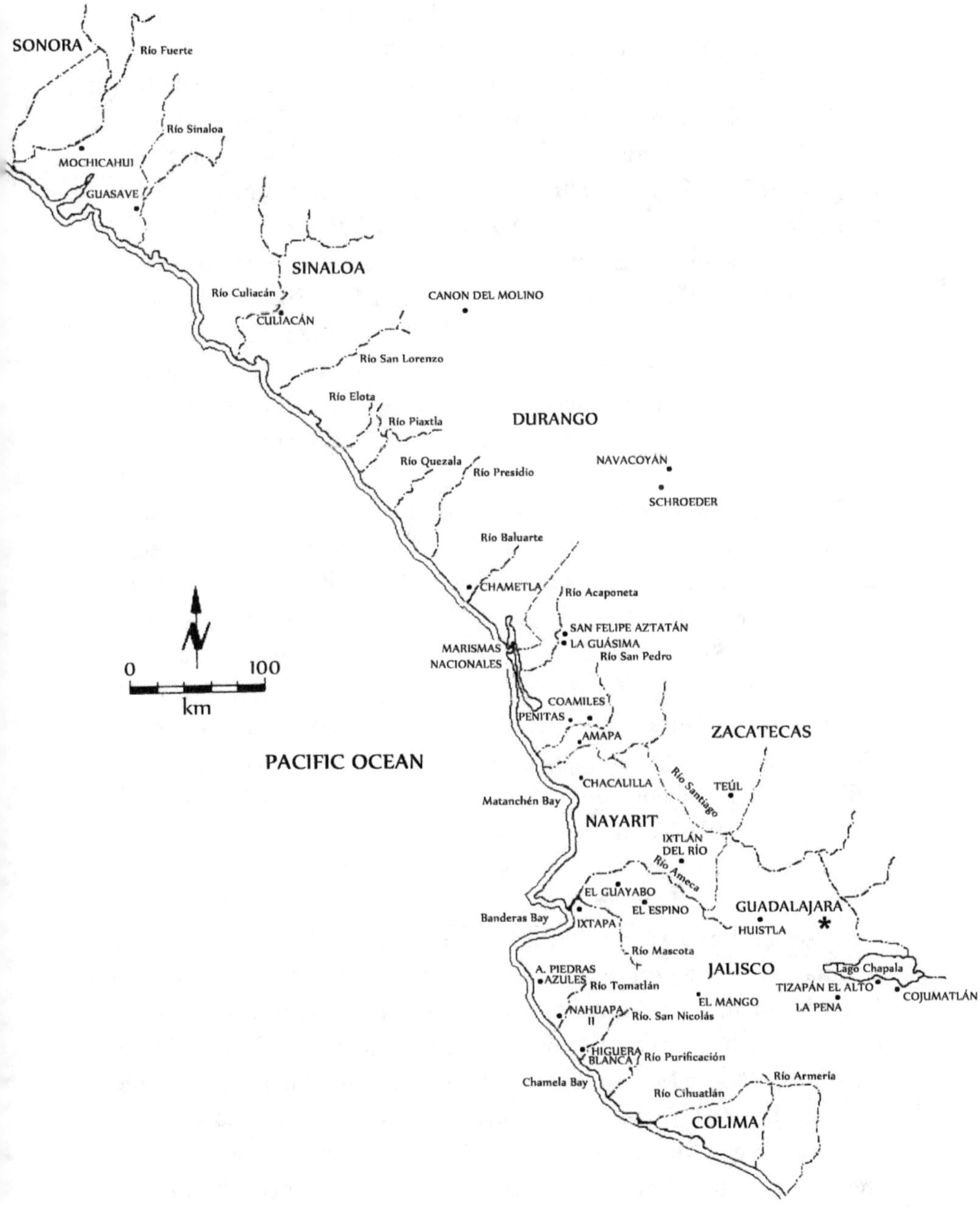

Figure 4.1. The location of selected important Aztatlán sites in western Mexico (map by Joseph B. Mountjoy).

The northeastern part of Aztatlán territory extends up the western side of the Central Mexican highland plateau where three sites in the state of Durango show major Aztatlán occupations. These sites are, north to south, El Cañón del Molino, Navacoyán, and Schroeder (Ganot and Peschard 1995). However, Aztatlán material has been reported as far north in Durango as the site of Zape, although the nature of Aztatlán presence there and at several other sites in Durango is not well documented.

El Teúl in southern Zacatecas marks the eastern border of the Aztatlán distribution. At this site, Peter Jiménez and Laura Solar have conducted several field seasons of excavations. Their excavations have revealed an Aztatlán-period ballcourt with associated stone sculptures of ballplayers, temple mounds (one of which is associated with a large serpent carving in the round), and clear evidence of an industry of metalworking that includes a smelting pit.

The southeastern distribution of Aztatlán sites is concentrated in the area around Lake Chapala in Michoacán and Jalisco. This includes the sites of Cojumatlán, Michoacán (Lister 1949), and Tizapán el Alto, Jalisco (Meighan and Foote 1968) as well as the site of La Peña in the Sayula Lake Basin of Jalisco (Liot et al. 2006). There is some evidence of Aztatlán occupation in the municipality of Villa Purificación (Mountjoy 2008), and there is a not-yet-well-understood relationship between the Postclassic polychrome pottery of the Autlán Valley of southeastern Jalisco (Kelly 1945b) and the Aztatlán phenomenon. Both areas are near the border with Colima.

Much has been published about a possible connection between Aztatlán and a Mixteca-Puebla iconographic tradition (Ekholm 1942) or with a more generalized "International Style" Mesoamerican iconographic tradition (Smith and Heath-Smith 1980; Pohl 2012). It is apparent that the Aztatlán phenomenon in West Mexico was in some ways part of a much larger Mesoamerican Early Postclassic Mesoamerican phenomenon. However, exactly how the Aztatlán archaeological culture of West Mexico relates to this larger Mesoamerican development—politically, economically, or as to religious beliefs and practices—is not well understood. As to the religious aspect, Mathiowetz (this volume), stresses an Aztatlán relation to the Mesoamerican Xochipalli cult, and to representations of the fire serpent Xiuhcóatl on Aztatlán pottery. However, several other Mesoamerican gods or goddesses, including Mictlantecuhtli, Xipe Tótec, and Tlazoltéotl, are also represented in Aztatlán materials.

There were close contacts between western Mexico and the rest of Mesoamerica in the Early Postclassic. At this time, quite a few cultural ideas and even specific items were coming into far western Mexico from areas to the east and southeast, including Mazapán figurines and Plumbate pottery, and it is possible that cotton garments, tobacco, cacao, *Spondylus limbatus* (*calcifer*) shells, and pearls were being exported from western Mexico to other Mesoamerican areas at this time.

The Arroyo Piedras Azules Site

The Arroyo Piedras Azules site (CC-22) is located in the municipality of Cabo Corrientes on the eastern edge of the village of Maito, and 1.5 kilometers from the Pacific coast of Jalisco (Figure 4.1). The Pacific coast at Maito contains a major estuary and a sandy beach that extends some 6 kilometers northward up to the mouth of the Tecolotlán River. South of Maito, there are about 3 kilometers of rocky shoreline that includes the Punta de Tehuamixtle. Southward of the Tehuamixtle Bay, the coastline changes again to sandy beach and estuaries that extend for some 6 kilometers farther south.

The Arroyo Piedras Azules site (N-20° 15.711' by W-105° 34.503'; UTM 13Q 0439944 × 2240563; Figure 4.2) is a broad, peninsula-like piece of land with an elevation of about 29 meters above sea level. It extends out from the foothills to the southeast, and the habitation area is elevated approximately 9 to 15 meters above the coastal plain that borders the peninsula to the northwest. The Piedras Azules arroyo is located a short distance to the north of the site.

The pre-Hispanic habitation area extends over a little more than 3 hectares, although the Aztatlán remains extend over about half of that. This land appears never to have been cultivated or subjected to any major earth-moving activities in historic times, but the Late Postclassic occupation damaged the earlier remains of the Early Postclassic (Aztatlán) occupation, except for the deep Aztatlán trash deposits that had accumulated down the sides and at the base of the hill. There are no mounds or other obvious signs of construction on the surface of the site. We registered 40 fragments of granite *metate* grinding basins and companion *mano* handstones on the surface of the site, plus one granite stela 87 centimeters high, 43 centimeters wide, and 15 centimeters thick. However, these all appear to pertain to the Late Postclassic occupation of the site.

The Arroyo Piedras Azules site was discovered by some local young men who noticed artifacts that had been uncovered when an access road was cut to the Villa del Mar *ejido* (communal) property owned by Ramón de la Cruz. In 2014, these men excavated an irregular explorative pit that

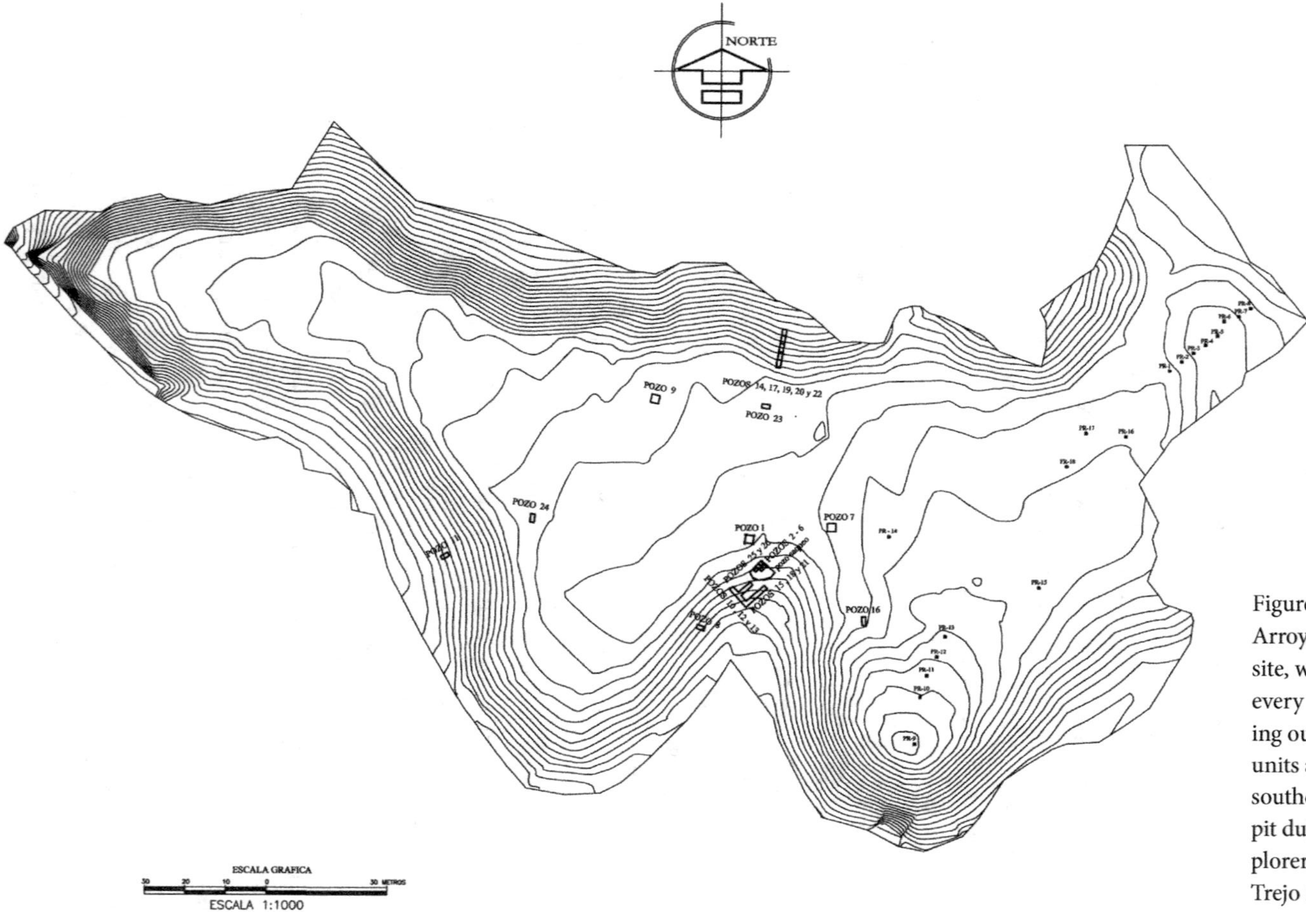

Figure 4.2. Map of the Arroyo Piedras Azules site, with contour lines every 50 cm, showing our 48 excavation units as well as the southern edge of the pit dug by the local explorers (map by Félix Trejo Hernández).

eventually reached approximately 4 (N–S) by 5 meters (E–W) by 2 meters in depth (Figure 4.2). They screened the excavated deposit, saving not only the decorated pottery, figurines, metal objects, and so on, but also samples of shells and bones. One of the men was pursuing the idea of using the material to set up a small archaeology museum in the village of Maito.

About this time Michelle Havlik, one of Mountjoy's archaeology students at the University of Guadalajara, Puerto Vallarta campus, took photographs of some of the polychrome pottery and called the discovery to Mountjoy's attention. Mountjoy visited the site in April of 2015 and with the help of the original explorers and the permission of the owner of the property, Ramón de la Cruz, he conducted a preliminary study of the site to assess its research potential.

Here we report on some of what the original explorers discovered and eventually donated to the municipality, but we also include some of the results of Mountjoy's excavations at the site in 2015, 2017, and 2018. These excavations included 48 pits that ranged from 50 centimeters to 2 meters on a side (Figure 4.2) and focused on the remains of the Aztatlán occupation of the site, especially as preserved in domestic trash deposits that in some places reached a depth of 2 meters. These investigations indicate that the whole site area extends over approximately 2.9 hectares, including the top and the sides of the hill. The top, flattish part of the hill that is useful for habitation, however, is about 1.48 hectares in area, and the portion of that area in which Aztatlán remains were found is only about .92 hectares.

When Mountjoy inspected and photographed the local collection, he was especially impressed by the quantity of high-quality and beautifully decorated polychrome pottery (Figure 4.3). Some of the decorative motifs he recognized to be present in abundance in the Punta de Mita area of far southern coastal Nayarit, on the northern side of Banderas Bay, as well as on the central coast of Nayarit, notably, Iguanas Polychrome found at Amapa (Meighan 1976). Other polychrome pottery was the same as the Mangos Polychrome found at Amapa (Meighan 1976), and some resembled polychrome pottery found in the area of Culiacán, Sinaloa, by Kelly (1945a).

Three facial profile images on polychrome sherds found by the local men were recognized as similar to profiles that had been found on Aztatlán pottery at Mochicahui and Guasave in Sinaloa (Manzanilla and Talavera 1998; Ekholm 1942); Amapa in Nayarit (Meighan 1976); and El Rincón del Guayabo de Emilia in Jalisco (Mountjoy and Schöndube 2014). In addition, some of the sherds found at the Arroyo Piedras Azules site had decoration

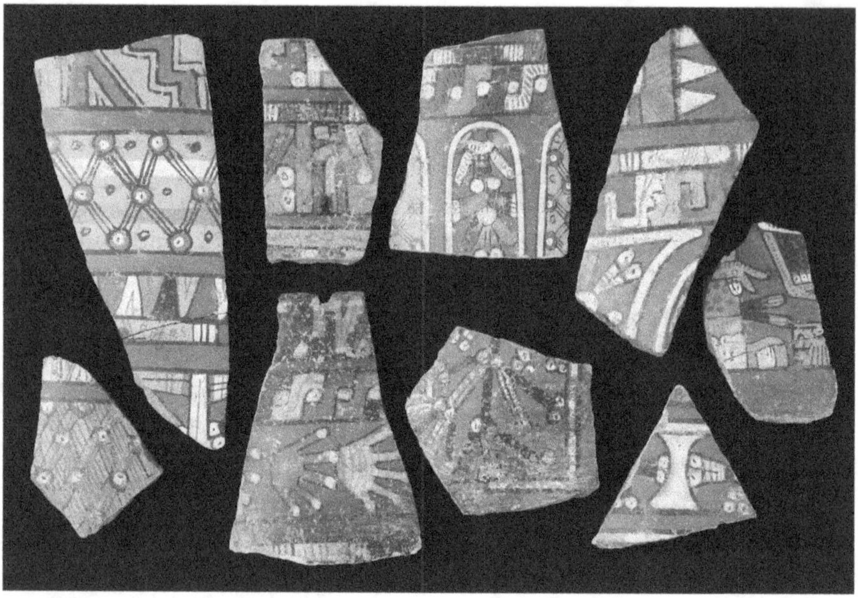

Figure 4.3. Sherds of polychrome pottery from the Arroyo Piedras Azules site (local collection) (photograph by Joseph B. Mountjoy).

likely based on cotton textile designs of elite clothing (see Mathiowetz, this volume).

Our 2017 and 2018 excavations increased considerably the sample of both facial and textile designs on this Iguanas-like polychrome pottery. Facial profiles increased by 18 examples and they include depictions of nose bars, ear ornaments, facial painting, and headdress elements (Figure 4.4a) that link these profiles to the Borgia codex attributed to somewhere in an area that extends from central Puebla to northern Oaxaca (Byland 1993: xiv). We recovered one small polychrome plate (Figure 4.4b) with what appears to be a person depicted in the guise of Mixcóatl, the patron god of the North and of the hunt, as he is shown on Plate 25 of the Borgia Codex (Byland 1993: xxi and plate 53). This same plate depicts two sets of four burial bundles that may represent this person's deceased ancestors. The person is also shown as someone playing the ballgame (Figure 4.4c).

Other extremely common iconography on this pottery includes what we believe to represent bone punches or awls of the type depicted in the Nuttall Codex attributed to the Mixtec of Oaxaca. Such a punch is shown in the Nuttall Codex (Nuttall 1975: 52) being used in the ceremonial piercing of 8 Deer's nose in order to insert a nose bar. At Arroyo Piedras Azules, these bone awl motifs are commonly associated with what appear to be multiple

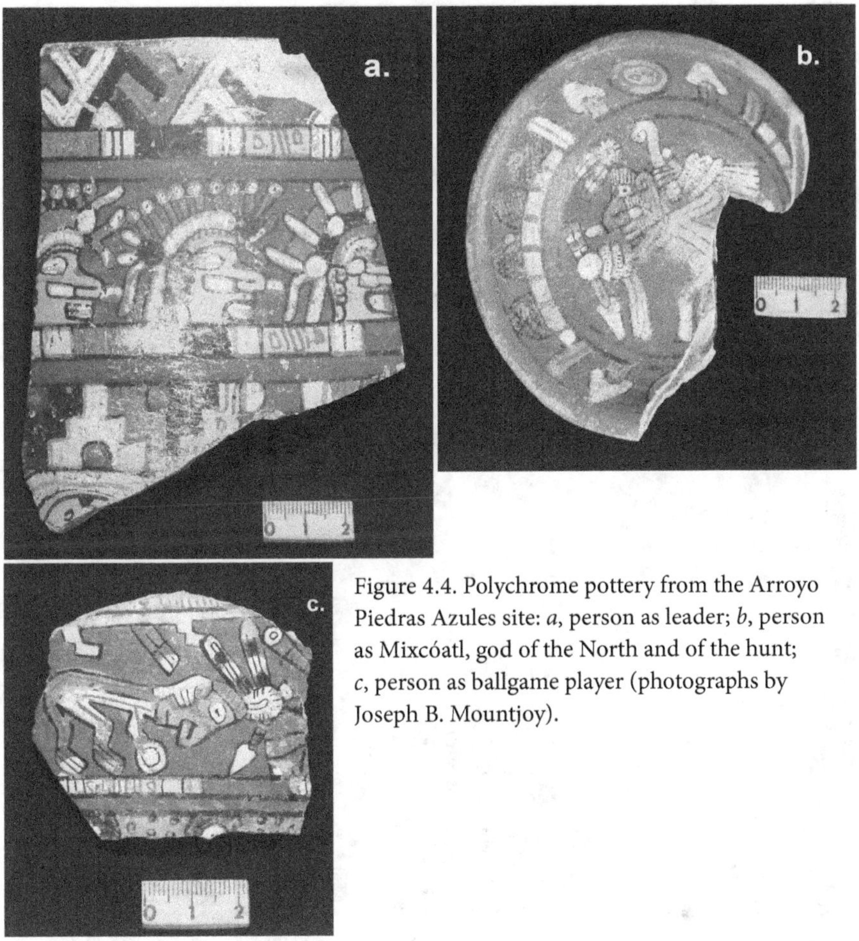

Figure 4.4. Polychrome pottery from the Arroyo Piedras Azules site: *a*, person as leader; *b*, person as Mixcóatl, god of the North and of the hunt; *c*, person as ballgame player (photographs by Joseph B. Mountjoy).

hand motifs (Figure 4.5a). However, another polychrome bowl from the site (Figure 4.5b) depicts panels of these punches as part of a decorative sequence that includes other panels with textile designs (Figure 4.5b). Decorative panels similar to these occur on some Cholula Polychrome vessels (Solís et al. 2006: 91, fig. 4). The fourth panel from the left appears to show the last punch in the row inserted into the side of a nose. In other examples, the punches seem to end at a design representing the hole in the nose.

The stratigraphic sequence at the Arroyo Piedras Azules site shows that this codex type of pottery becomes abundant after the Aztatlán settlement was well established and the inhabitants had begun intensive exploitation of the local shellfish resources.

The more common Aztatlán decorated pottery types found at Arroyo Piedras Azules were the typical Red-on-Buff (Figure 4.6a) and Red-and-

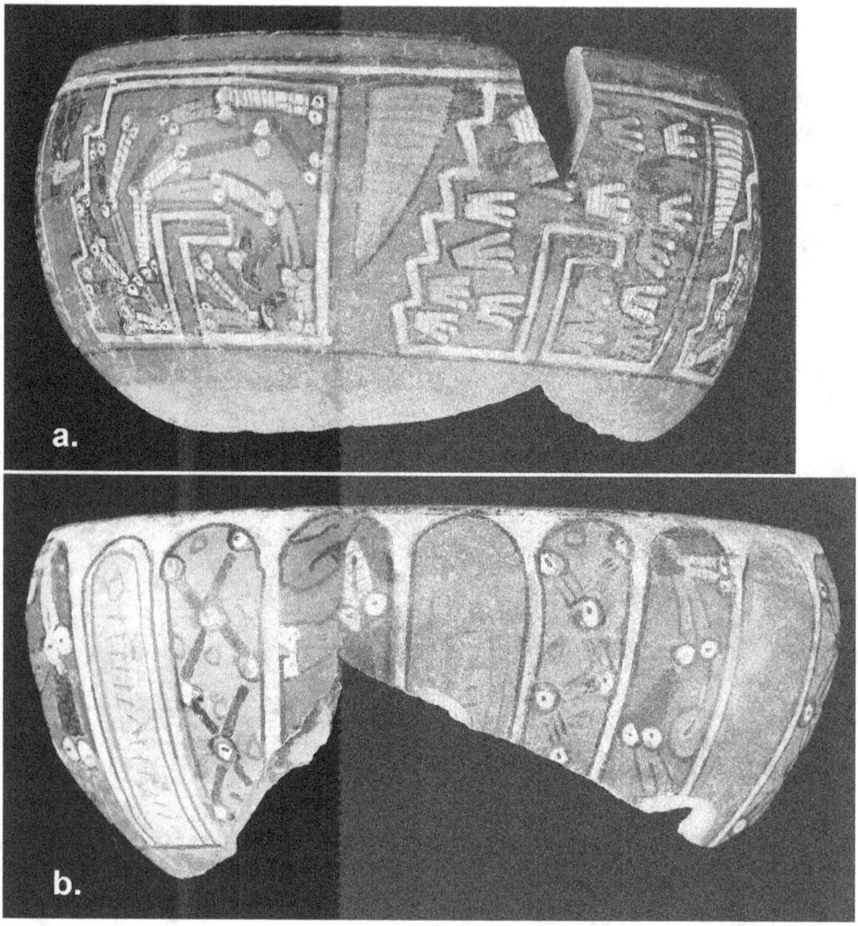

Figure 4.5. Polychrome pottery from Arroyo Piedras Azules: *a*, decorated with bone punch/awl and hand designs; *b*, decorated with bone punch/awl and textile designs (local collection) (photographs by Joseph B. Mountjoy).

White-on-Buff wares. The former is commonly of bowl form and often has a *molcajete* interior grinding surface and either annular base, or hollow, rattle supports. The latter decorative type is typically, but not wholly, associated with large jars. At least the Red-on-Buff ware occurs over the entire range of Aztatlán presence in Sinaloa, Nayarit, Jalisco, Durango, Zacatecas, and Michoacán. These types of decorated pottery at Arroyo Piedras Azules are associated with two, rarer, types of polychrome pottery: one incised that corresponds to Cerritos Polychrome at Amapa in Nayarit (Meighan 1976: 435, plate 129, f–l), and the other not incised that appears identical to Mangos Polychrome at Amapa (Meighan 1976: 490, plate 3, p–u).

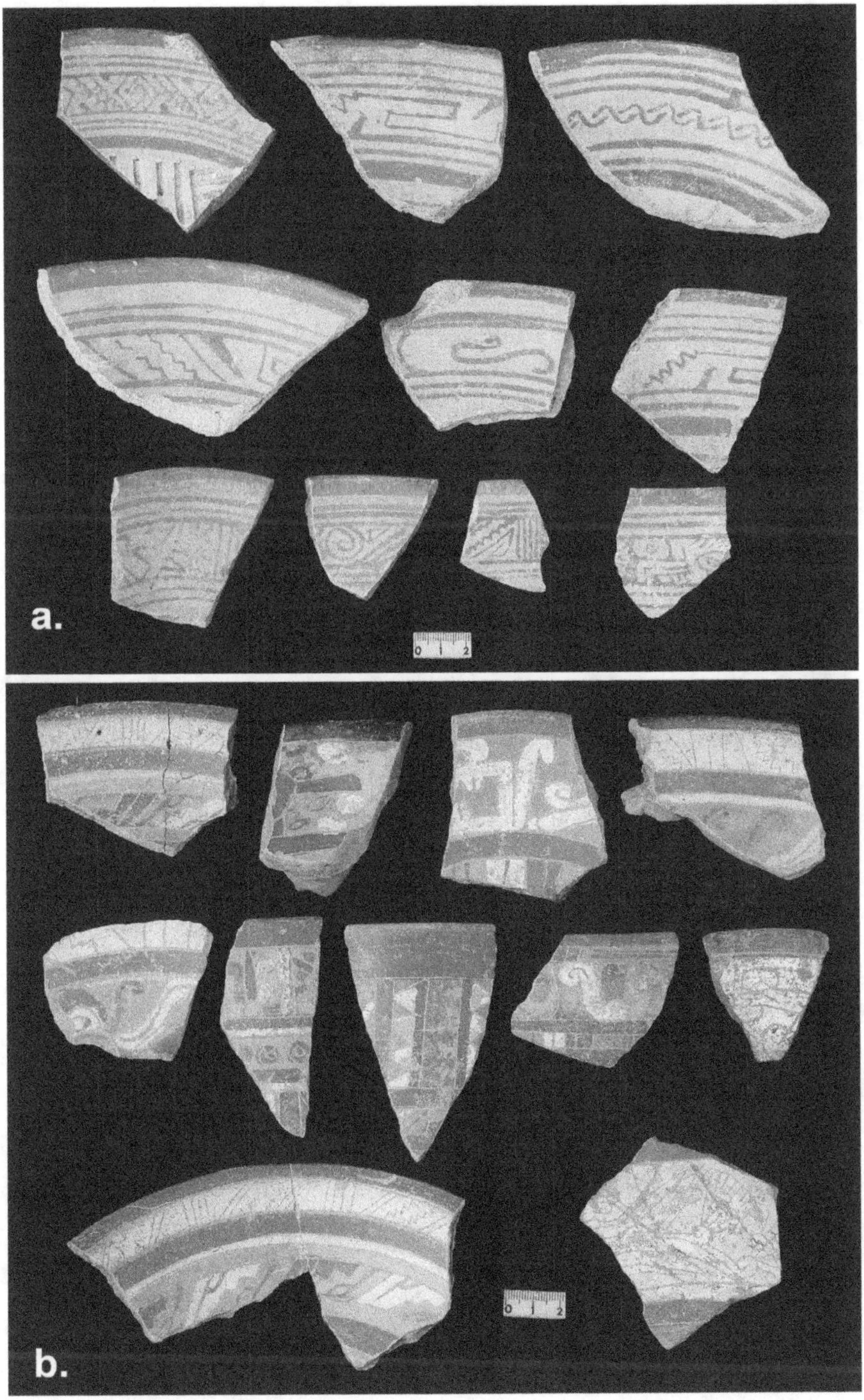

Figure 4.6. *a*, Red and Buff; *b*, Polychrome Incised Aztatlán pottery (local collection) (photographs by Joseph B. Mountjoy).

Decorative elements on the simple and incised forms of red-and-white-on-buff pottery bowls as well as on some of the incised polychrome pottery bowls include at least nine examples from the site of a skull with a crossed-bones motif (Figure 4.6b [upper row, sherd 3]). This motif was carved, on an Aztec (Mexica) stone monument depicting a *tzompantli* skull rack, and this motif, as well as that of solitary skulls, can be found decorating some Cholula pottery where it likewise signifies human sacrifice (Matos Moctezuma 2012: 35; Solís et al. 2006: 120–121).

Curiously, the local excavators did not find any Mazapán mold-made figurines, but in our excavations, we did recover two fragments of Mazapán figurines from Early Postclassic deposits. The main figurines the local collectors recovered (Figure 4.7a) were of what appears to be a crude, locally produced type of figurine not known elsewhere along the Jalisco coast. We found several of these figurines in the Early Postclassic (Aztatlán) deposits and suggest that they might be crude local attempts to produce Mazapán-like figurines.

Local collectors also recovered a few pottery stamps from the site, and we recovered several pottery stamps from our excavations in Aztatlán deposits. Two of these stamps were of the cylindrical type. Several more stamps were of the handled type, with either a square or a round stamping surface. One of these stamps was carved from pumice stone. All the motifs of these Aztatlán stamps are geometric except for a square one that has an elaborate Xiuhcóatl (fire serpent) head depicted in profile (Figure 4.7b).

Local collectors also recovered many fragments of small pottery bells. Such bells are abundant in Late Postclassic excavated contexts in the nearby Tomatlán Valley (Mountjoy 1982). At Arroyo Piedras Azules, we recovered pottery bells from both Aztatlán and Late Postclassic contexts. The Aztatlán bells are usually larger than those from the Late Postclassic, they lack a slit on the end above the stringing hole, and one of them is decorated with a crude humanlike face that has both modeled and incised features.

Of additional special note in the local collection were four copper fishhooks, as well as several copper ringlets found by the local excavators associated with human bones deposited in a large pottery urn. In 2017, we excavated another of the burial urns. It was determined to pertain to the Late Postclassic occupation of the site. No burial offerings accompanied the disarticulated skeletal remains of the young male that we recovered from this urn.

However, we did recover some items of copper from Aztatlán trash contexts at the site. These included copper ringlets, fishhooks, bells, and one

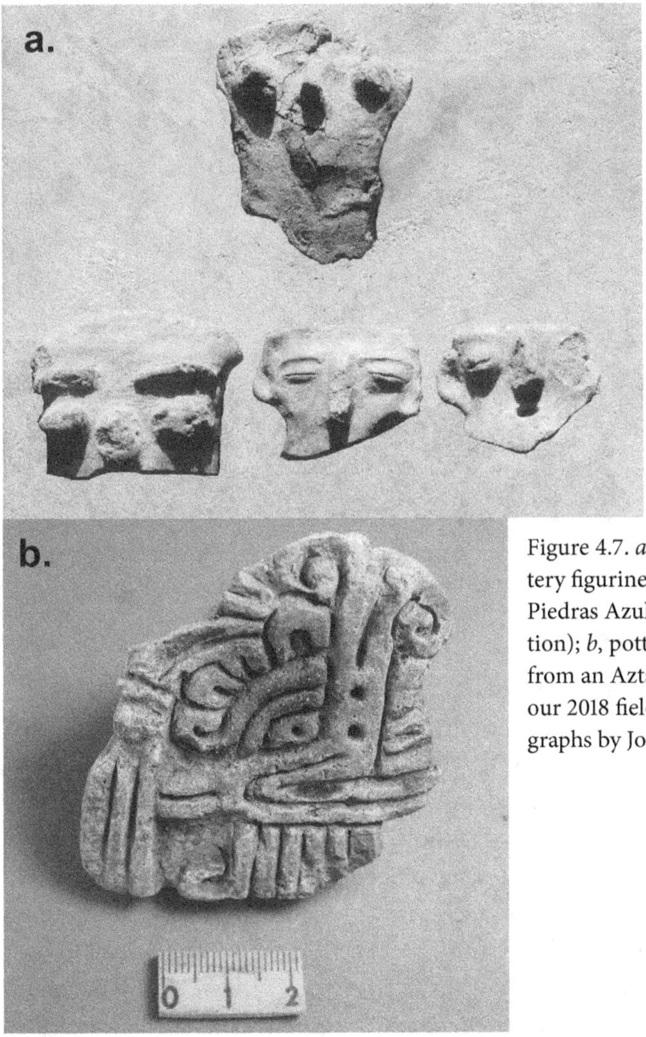

Figure 4.7. *a*, Fragments of pottery figurines from the Arroyo Piedras Azules site (local collection); *b*, pottery stamp recovered from an Aztatlán stratum during our 2018 field season (photographs by Joseph B. Mountjoy).

small celt (Figure 4.8). The ringlets of the Aztatlán occupation are about the same size as those found by locals in the Late Postclassic urn, but the Aztatlán ones are appreciably more delicate.

Also of Early Postclassic date at the site are abundant prismatic blades of obsidian. Most of them are golden-brown in color and therefore are likely from the La Joya source near Magdalena in the central highlands of Jalisco. However, other prismatic blades, as well as most Early Postclassic obsidian projectile points, were made from gray obsidian. One major source of gray obsidian is located near the major Aztatlán site of Ixtlán del Río in Nayarit.

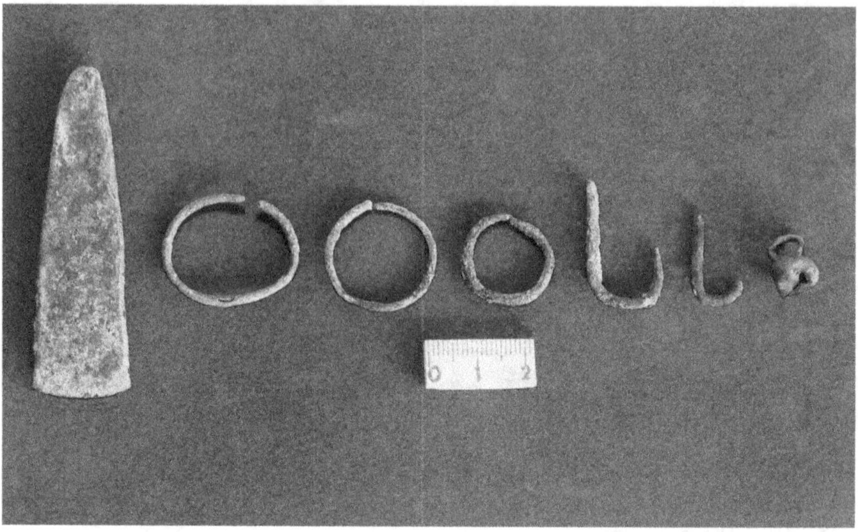

Figure 4.8. Copper celt, ringlets, fishhooks, and a bell recovered from Aztatlán deposits at the Arroyo Piedras Azules site during the 2017 and 2018 field seasons (photograph by Joseph B. Mountjoy).

Other items of stone recovered by the local explorers include two cream-colored stone cylinders that may have been inserted through the perforated septum of the nose (Figure 4.9). We recovered a small cylinder made of shell from an Early Postclassic deposit at the site.

We found both shell fishhooks and shell bracelets, as well as the pumice stone reamers that appear to have been used to make them, in Early Postclassic (Aztatlán) deposits at the Arroyo Piedras Azules site. The shell fishhooks were usually of *Pinctada mazatlanica* (mother-of-pearl) shell, and the bracelets made mainly from *Glycymeris gigantea* shells, although *Megapitaria aurantiaca* and *Peryglypta multicostata* shells were also sometimes used.

Also found in both the local collection and Early Postclassic deposits we excavated were a few shell beads and *Olivella* shell "tinklers." Our excavations revealed the use of both *Spondylus limbatus* (*calcifer*) and mother-of-pearl (*Pinctada mazatlanica*) shell in the fabrication of a few items of shell jewelry during the Early Postclassic. In addition, from the same Aztatlán deposits we recovered small drills fashioned from prismatic obsidian blades, objects that were probably used to drill the holes in shell beads.

Other items in the local collection were tubular beads of bird bone, as well as two polished flat slivers of bone and a bone comb, all items believed to have been used in weaving cloth (Figure 4.10a; see Mathiowetz,

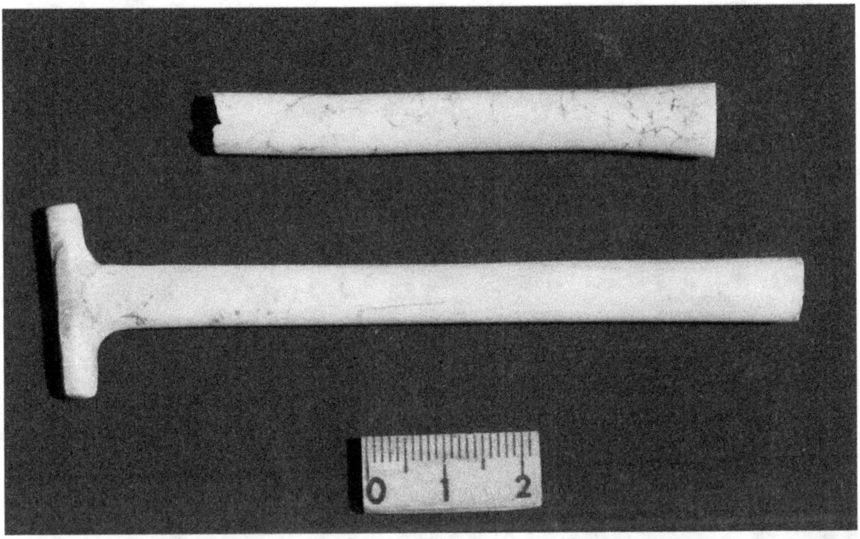

Figure 4.9. Cylindrical stone objects, possibly for inserting in a pierced nose septum (local collection) (photograph by Joseph B. Mountjoy).

this volume). We recovered similar tubular bone beads and polished bone slivers from Early Postclassic (Aztatlán) deposits, as well as pointed bone implements that Mathiowetz (this volume) calls "picks" and that he relates to the weaving of cotton textiles. Similar bone tools, including combs, were found at Amapa in Nayarit (Meighan 1976: 124–125; 418) and they were attributed to the Aztatlán occupation at that site.

Other, more common items in the local collection include pottery spindle whorls that attest to a local industry of cotton thread spinning. We recovered quite a few spindle whorls from both Aztatlán and Late Postclassic deposits at the site. Most of the Aztatlán spindle whorls are medium-sized and undecorated, although a few have simple, crude incised-line decorations.

Two pottery sherds with multiple notches carved along one edge that seem to have been used for the same purpose as the bone combs (Figure 4.10b) were noted in the local collection. One of these is definitely of Aztatlán decorated polychrome pottery. We recovered three other possible pottery sherd combs from our excavations in Aztatlán contexts. In addition, it seems noteworthy that we found cotton growing wild about a kilometer from the Arroyo Piedras Azules site.

Mountjoy's 2015 investigation included a 1-meter-square test pit (Pit #1) dug into the flat, upper part of the site just north of the local explorers'

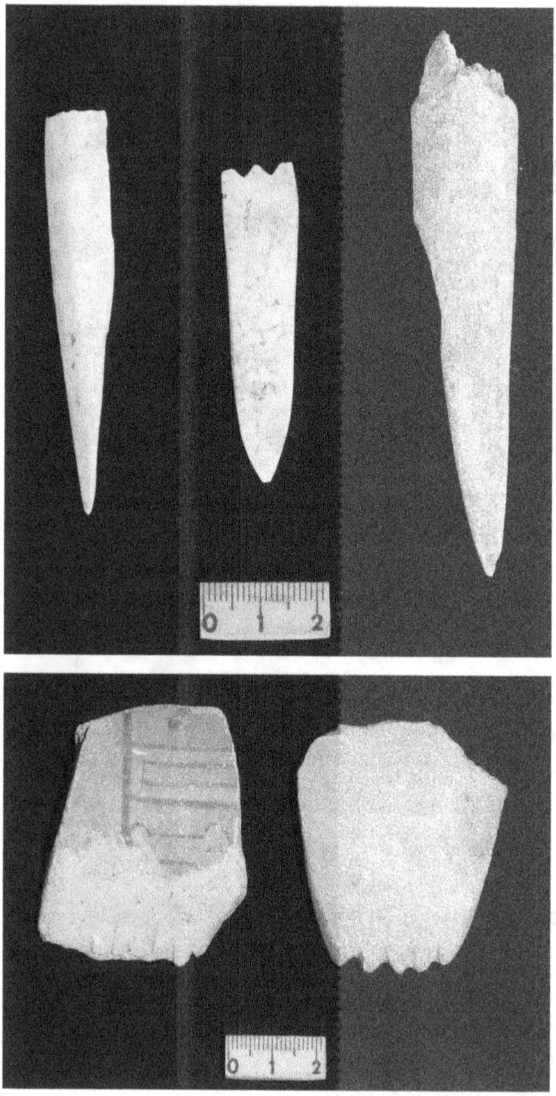

Figure 4.10. Weaving tools made of: *top*, bone; *bottom*, pottery sherds (local collection) (photographs by Joseph B. Mountjoy).

pit, and a strata cut (Pit #2) on the northwestern edge of the explorers' pit (Figure 4.2). The test pit (#1) was dug in 10 centimeter levels and revealed from 20 to 30 centimeters of dark earth with lots of fragmented marine shells. Two decorated sherds recovered from this stratum were of Nahuapa Red-and-Black-on-Buff, a well-known Late Postclassic type of pottery that was produced in the Tomatlán Valley some 50 kilometers to the southeast of the Arroyo Piedras Azules site, a type of pottery representative of the post-Aztatlán habitation at the Nahuapa site (Mountjoy 1982).

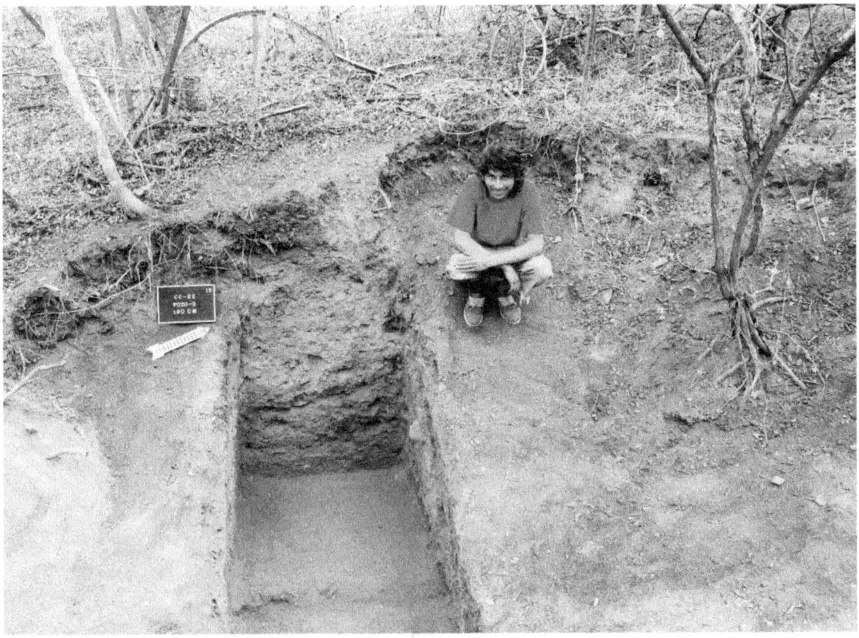

Figure 4.11. Isiah Valdovino at the Pit #2 strata cut, Arroyo Piedras Azules site (photograph by Joseph B. Mountjoy).

From 30 to 80 centimeters in Pit #1, the soil was loose, gray, and had only Aztatlán decorated pottery, associated with abundant marine shells and some bones. In the southeastern corner of the pit at a depth of 40 to 47 centimeters, we found a cluster of four stones, one of which was a *mano* hand-grinding stone. In the center of the excavation, we discovered an area of burned earth, and fragments of burned clay flooring were also recovered within this level. The cultural deposit reached a maximum depth of 75 centimeters where it rested upon decayed granite bedrock.

Pit #2 (2015) was a 1-meter-square cut into the northwestern side of the local explorers' pit (Figures 4.11 and 4.12). In this case, we tried to excavate levels that coincided with the visible layering of the soil deposit, but only started collecting material below the 50 centimeter level in an attempt to avoid sampling any Late Postclassic, non-Aztatlán material.

Due to time constraints and the great abundance of material, we only saved pottery rim sherds, bases, and decorated pieces, plus any other important objects (we recovered one copper fishhook). We collected all the bones, but only saved a sample of the different species of shellfish. This strategy produced 14 cloth collection bags full of material, with a total sample of 762 pottery sherds that included 307 decorated sherds classifiable

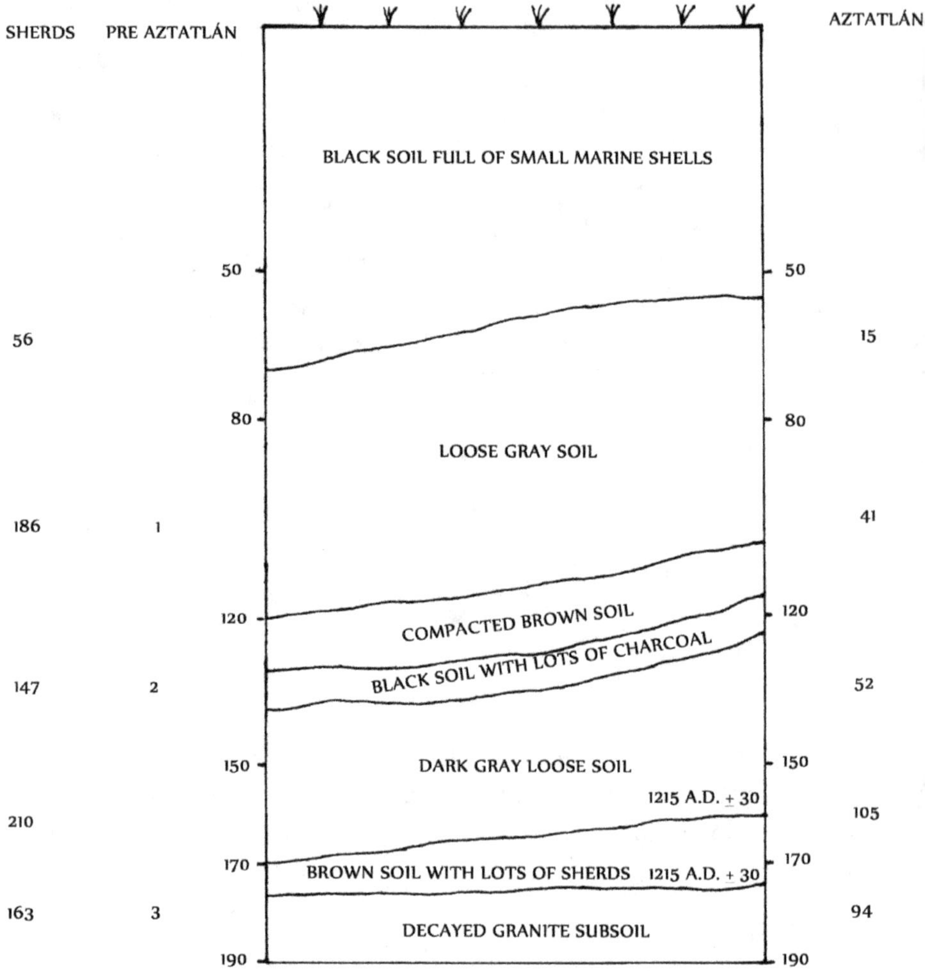

Figure 4.12. Northern face of Pit #2 stratigraphic cut at the Arroyo Piedras Azules site, showing the visible stratification, the excavation units by depths, counts of decorated pottery sherds (both Aztatlán and pre-Aztatlán), and the two radiocarbon dates obtained (figure by Joseph B. Mountjoy).

as Aztatlán, 6 sherds classifiable as pre-Aztatlán, and no sherds classifiable as post-Aztatlán (Figure 4.12). Of the Aztatlán decorated pottery, 61 percent was of Aztatlán Red on Buff; 33 percent, Aztatlán Red and White on Buff Incised; 3 percent, Aztatlán Black on Buff; and 3 percent, Aztatlán Polychrome (Figure 4.13). All of these pottery types are associated with the Cerritos phase (Aztatlán) occupation at the Amapa site in central coastal Nayarit (Meighan 1976).

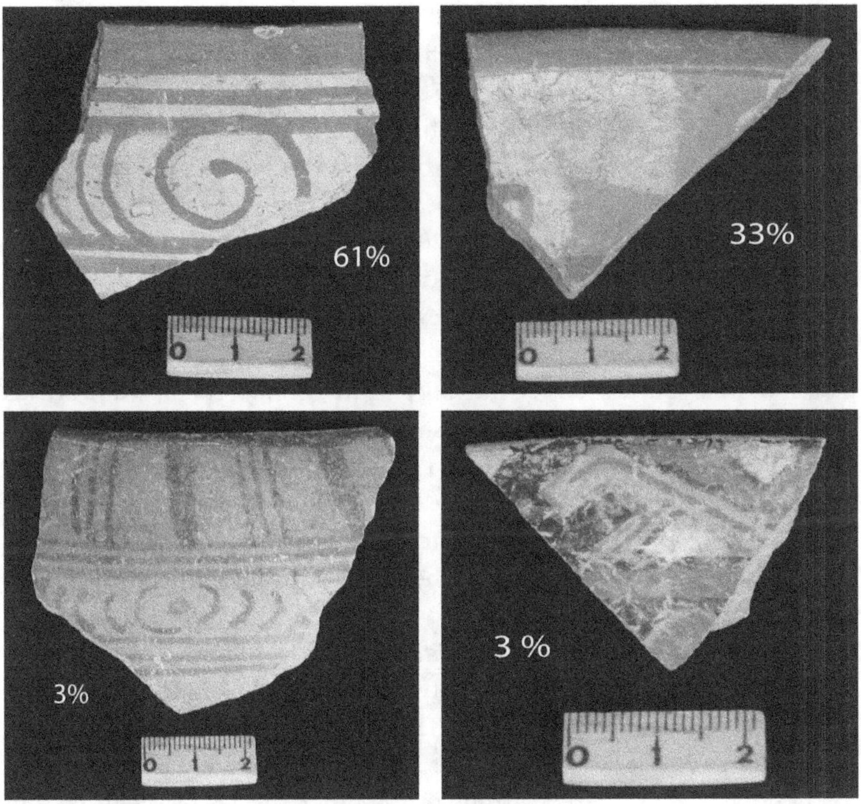

Figure 4.13. The four types of decorated Aztatlán pottery (*left to right, top to bottom*: Red on Buff, Red and White on Buff Incised, Black on Buff and Polychrome) recovered from the excavation of stratigraphic Pit #2 at the Arroyo Piedras Azules site, arranged according to descending percentage of representation (photographs by Joseph B. Mountjoy).

From each of the lowest two excavated levels, we recovered large samples of charcoal (Figure 4.12). This charcoal probably came from the initial slash-and-burn clearing of the area by the Aztatlán people. Both charcoal samples gave identical dates of AD 1215 + 30 (Beta 419370 and 419371), calibrated to 95 percent certainty. So these two dates establish rather well the time of initial habitation of this site by people of the Aztatlán tradition, and, by extension, help to better date the Cerritos phase at Amapa.

Excavations in 2017 adjacent to and east of the 2015 strata cut were able to discern a stratum of discontinuity between the Early Postclassic and Late Postclassic occupations of the site, indicating a period of abandonment of the site between the two phases of the Postclassic.

The shell and bone refuse from Pit #1 located in the habitation area was fragmented, so only a general identification of genera was possible for the bones from the Aztatlán levels. These bones included fish, mammal, and reptile remains. For Pit #2, however, the bones were preserved well enough to enable the specific identification of Pacific Ridley (*Lepidochelys olivacea*) sea turtle; Spotted Wood (*Rhinoclemmys rubida*) terrestrial turtle; and green iguana (*Iguana iguana*). The 2017 and 2018 excavations in Aztatlán deposits added the presence of Mexican cottontail rabbit (*Sylvilagus cunicularius*); dog (*Canis lupus familiaris*); white-tailed deer (*Odoicoileus virginianus*); armadillo (*Dasypus novemcinctus*); river crocodile (*Crocodylus acutus*); and three species of fish (*Ardenna pacífica; Caranx caninus*; and *Scarus perrico*). Considerable numbers of fish vertebrae that are not identifiable as to species have come from the Early Postclassic (Aztatlán) deposits.

Identifiable shells from these same levels in Pit #2 were overwhelmingly of small *Polymesoda mexicana* (inflated marsh) clams, but 13 other species of bivalves were also present. Since the Pit #2 excavations, three additional species of bivalves have been added to the list, a list that includes *Peryglypta multicostata* and *Megapitaria squalida* clams, *Choromytilus palliopunctatus* mussels, and very large shells of two species of oyster (*Crassostrea corteziensis* and *Striostrea prismatica*). The 2017 and 2018 excavations recovered a large number of thin granite slab axes that we believe were used to chop oysters and perhaps some of the other species of shellfish from the rocks along the coast. We have also recovered pieces of several small, pointed objects of zinc-like stone that may have been used to pry open the oysters.

In addition, there are 26 species of gastropods represented in the Aztatlán food refuse, which includes the shells of three large sea snails: *Triplofusus princeps*; *Lobatus galeatus*; and *Hexaplex erythrostomus*, as well as the giant limpet *Scutellastra mexicana*. These gastropods include many shells of *Plicopurpura columellaris*, used by some indigenous peoples to dye cloth. Also found in the same deposits were abundant segments of *Polyplacophora* or chitons *(Chiton articulatus)* and quantities of crab claws *Cardisoma crassum* and *Uca* spp. from both rocky intertidal and estuarine areas.

As for carbonized vegetable remains, we recovered nine very small, carbonized maize cobs from Early Postclassic deposits. According to the authority on pre-Hispanic maize in western Mexico, Bruce Benz (personal communication, 2017), this small size is typical for maize at AD 1200 in western Mexico. However, coupled with the relative scarcity of grinding stones in the Early Postclassic trash deposits as compared to those of the

Late Postclassic at the site, this seems to indicate that maize was not a major component of the diet of the Aztatlán people at the Arroyo Piedras Azules location.

The shell and bone refuse indicates the utilization of a broad spectrum of marine, coastal estuary, lagoon, and scrub-forest habitats. However, it seems unlikely that these coastal food resources were the primary reason a small group of relatively well-to-do Aztatlán people settled at the Arroyo Piedras Azules site. These people had ready access to fine polychrome pottery for use in their daily lives, as well as prismatic blades of high-quality obsidian and both decorative and utilitarian items of metal. Yet they had come from somewhere apparently on the coastal plain of Nayarit in order to establish a specialized settlement at this spot on the northern coast of Jalisco.

We suspect the main draw for these people was access to the abundant *Spondylus limbatus* (*calcifer*) and mother-of-pearl *Pinctada mazatlanica* shellfish that could be (and still can be) found along the 3 kilometers of rocky coastline in the area of the Punta de Tehuamixtle. These were shells held in high esteem by pre-Hispanic Mesoamerican peoples. They could be used for producing jewelry at major Aztatlán centers located far to the north of the Arroyo Piedras Azules site where there were no rocky shoreline habitats for such shells.

Several disc-shaped beads made from *Spondylus limbatus* (*calcifer*) shell were recovered from the site by local explorers, but only two *Spondylus limbatus (calcifer)* shells were recovered from the domestic trash in our 2015 excavations. One of these may be food refuse as the meat from this creature is edible, but the other shell showed evidence of human modification. In 2017, we recovered several pieces of worked *Spondylus limbatus* (*calcifer*) shell from excavated Early Postclassic contexts, including at least one bead. However, no whole *Spondylus limbatus* (*calcifer*) shells have been found among the shells thus far analyzed from the 2018 excavations. Likewise, from Aztatlán deposits we have recovered quite a number of mother-of-pearl (*Pinctada mazatlanica*) fishhooks and some jewelry made from that shell, but no complete shells. Thus, the near absence of complete *Spondylus limbatus* (*calcifer*) and *Pinctada mazatlanica* (mother-of-pearl) shells suggests to us that their importance to the Aztatlán people was not as a source of food, but as an exportable commodity.

Therefore, we suggest that the Early Postclassic (Aztatlán) people established the settlement at Arroyo Piedras Azules primarily for the exploitation of certain highly valued local shells for exportation to major Aztatlán centers far to the north or east where they were worked into jewelry. Excavations at Amapa, Nayarit, recovered shell beads made from *Larkinia grandis* (*Anadara grandis*) *and Spondylus limbatus* (*calcifer*) shells, along with shell refuse produced from manufacturing such beads. In addition, three kits of small stone tools for shell bead manufacture were discovered at Amapa (Meighan 1976: 123).

Furthermore, judging from thread-spinning and weaving implements found at the Arroyo Piedras Azules site, it appears possible that another important activity of the Aztatlán inhabitants was the fabrication of textiles, perhaps from locally grown cotton, an activity that seems to have continued to be important during the Late Postclassic occupation of the site.

Once the Aztatlán settlement had become well established at Arroyo Piedras Azules and the intensive exploitation of local shell resources was fully developed, the settlement began to have strong ties to the Mixteca-Puebla area, perhaps as another destination for exporting the local *Spondylus limbatus* (*calcifer*) and *Pinctada mazatlanica* (mother-of-pearl) shells.

Although our inspection of this part of the coast of Jalisco has been limited, we have identified two other Aztatlán sites in the same general area as Arroyo Piedras Azules. One of these, Potrero Carlos Cárdenas, is located on the north side of the Tecolotlán River, some 6 kilometers to the northwest of Arroyo Piedras Azules. This site is smaller than Arroyo Piedras Azules but it has similar decorated pottery and one rustic granite stela. The other local Aztatlán site is El Pozo. It is larger than Arroyo Piedras Azules and is located on the northern side of the Ipala River some 6 kilometers to the southeast. At this site local people found some objects of copper that include a turtle-shaped finger ring like one recovered at Amapa, Nayarit (Meighan 1976: 409). So Arroyo Piedras Azules was not isolated, but rather a place of specialized activities within a more general system of Aztatlán colonization of this part of the coast of Jalisco.

Conclusions

Based on the data thus far obtained from the Arroyo Piedras Azules site, we offer five conclusions important for understanding the development and the spread of the Aztatlán archaeological culture in West Mexico. First is support for the idea of a relatively late development of Aztatlán along the coast of Jalisco, as compared to its manifestation in the central highlands of Jalisco and the coastal plain of northern Nayarit where according to radiocarbon dates Aztatlán development is 2–300 years earlier than Aztatlán settlements along the Jalisco coast.

A second conclusion has to do with the ability of the Aztatlán people at Arroyo Piedras Azules to adapt to living substantially off natural resources available along a stretch of open Pacific coastline. This adaptation contrasts to that of the riverine, coastal plain, and mountainous habitats where so many major Aztatlán centers are located in Jalisco, Nayarit, and Sinaloa, excepting the *Marismas Nacionales* area at the Nayarit/Sinaloa border.

Although the Aztatlán inhabitants of Arroyo Piedras Azules did have maize, the ears were small and grinding stones are rare in the Early Postclassic deposits. For their subsistence, the Aztatlán inhabitants seem to have relied much more heavily on the natural animal, fish, and shellfish resources acquirable along this part of the Jalisco coast. They consumed a wide variety of shellfish that they obtained from many different coastal environments, including the rocky coastline, sandy bottom, muddy bottom, deep water, and coastal lagoons. Some of these shellfish, if collected live, notably the *Glycymeris gigantea*, *Spondylus limbatus* (*calcifer*), and the *Pinctada mazatlanica*, had to be obtained by diving in deep water along rocky stretches of the coastline.

Third, it is also apparent from the investigations at the Arroyo Piedras Azules site, as well as from explorations at Ixtapa in the Banderas Valley (Mountjoy et al. 2003) and at Nahuapa II in the Tomatlán Valley (Mountjoy 1982), that the expansion of the Aztatlán phenomenon into these areas of coastal Jalisco was by way of actual colonization. This colonization was carried out by people who were closely related to the major Aztatlán centers farther north; for example, Amapa in central coastal Nayarit. They brought with them a new system of religious beliefs and rituals and participated in an economic and political system that exercised strict control on the production and distribution of economically important commodities present in the areas they colonized.

The picture from the investigations at Arroyo Piedras Azules and other Aztatlán sites along the northern coast of Jalisco is definitely not one that supports the idea that the Aztatlán archaeological culture spread through trading networks in mercantile systems organized by *pochteca*-type "mobile merchants," as was proposed by Kelley (2000). This "mobile merchants" model is also not likely to be applicable for Aztatlán expansion into other areas of West Mexico such as northern Sinaloa, western Durango, southern Zacatecas, and some parts of the central Jalisco highlands.

Fourth, major Aztatlán centers in the highlands and coast of Nayarit, at least up as far as the *Marismas Nacionales* on the border with Sinaloa, seem to have specialized in the production of maize, cotton, tobacco, pottery, obsidian blades, and items of metal. At such centers, archaeologists have found great numbers of grinding stones, spindle whorls, extensive obsidian workshops, kilns for making pottery, and facilities for smelting copper. We are inclined to believe that the Aztatlán colonists at Arroyo Piedras Azules participated in this economic network primarily by exporting *Spondylus limbatus* (*calcifer*) and *Pinctada mazatlanica* shells to major Aztatlán centers in Nayarit, where these shells were used to make jewelry. The Arroyo Piedras Azules colonists may also have exported to these centers woven cotton textiles, perhaps including some colored with purple dye obtained locally from the small *Plicopurpura columellaris* gastropod. In return, from the major centers to the north, the Aztatlán colonists at Arroyo Piedras Azules acquired such luxury items as elaborately decorated pottery, prismatic blades of high-quality obsidian, and a variety of metal objects.

Lastly, once the settlement at Arroyo Piedras Azules became firmly established, probably as an important source of Spondylus and mother-of-pearl shells, the stratigraphic sequence at the site reveals the acquisition of large quantities of elaborate new polychrome pottery of a kind also found in southern and central coastal Nayarit. Finally, some of this pottery at Arroyo Piedras Azules has codex-type iconography associated with depictions of an important person, perhaps a culture-hero figure such as 8 Deer of the Mixtecs, which links the Arroyo Piedras Azules site to the Mixteca-Puebla area.

References

Byland, Bruce E.

1993 Introduction and Commentary. In *The Codex Borgia: A Full-Color Restoration of the Ancient Mexican Manuscript*, by Gisele Díaz and Alan Rodgers, pp. xiii–xxxi. Dover, New York.

Ekholm, Gordon F.

1942 *Excavations at Guasave, Sinaloa, Mexico*. Anthropological Papers of the American Museum of Natural History, No. 38. AMNH, New York.

Evans, Susan Toby

2008 *Ancient Mexico & Central America: Archaeology and Culture History*. Thames and Hudson, London.

Ganot R., Jaime, and Alejandro A. Peschard F.

1995 The Archaeological Site of el Cañón del Molino, Durango, México. In *The Gran*

Chichimeca: Essays on the Archaeology and Ethnohistory of Northern Mesoamerica, edited by Jonathan E. Reyman, pp. 146–178. Avebury, Brookfield, IL.

Kelley, J. Charles

2000 The Aztatlán Mercantile System: Mobile Traders and the Northwestward Expansion of Mesoamerican Civilization. In *Greater Mesoamerica: The Archaeology of West and Northwest Mexico*, edited by Michael S. Foster and Shirley Gorenstein, pp. 137–154. University of Utah Press, Salt Lake City.

Kelly, Isabel T.

1945a *Excavations at Culiacán, Sinaloa.* Ibero-Americana, No. 25. University of California Press, Berkeley.

1945b *The Archaeology of the Autlán-Tuxcacuesco Area of Jalisco, Part I.* Ibero-Americana No. 26. University of California, Berkeley.

Liot, Catherine, S. Ramírez, J. Reveles, and O. Schöndube (coordinators)

2006 *Transformaciones socioculturales y tecnológicas en el sitio de La Peña, Cuenca de Sayula, Jalisco.* Editorial Pandora, Guadalajara.

Lister, Robert H.

1949 *Excavations at Cojumatlan, Michoacán, Mexico.* University of New Mexico Publications in Anthropology No. 3. University of New Mexico, Albuquerque.

Manzanilla, Rubén L., and J. A. Talavera

1998 Informe de los trabajos de salvamento e investigación arqueológica en la población de Mochicahui, municipio del Fuerte, estado de Sinaloa, México. Dirección de Antropología Física, Departamento de Salvamento Arqueológico. INAH, Mexico City.

Meighan, Clement W. (editor)

1976 *The Archaeology of Amapa, Nayarit.* Monumenta Arqueológica No. 2. Cotsen Institute of Archaeology, University of California, Los Angeles.

Meighan, Clement W., and Leonard J. Foote

1968 *Excavations at Tizapán el Alto, Jalisco.* Latin American Studies, Vol. 11. University of California, Los Angeles.

Matos Moctezuma, Eduardo

2012 Excavaciones en la Gran Pirámide de Cholula (1931–1970). *Arqueología Mexicana*, Vol. XX, Núm. 115, pp. 31–35.

Mountjoy, Joseph B.

1982 El Proyecto Tomatlán de Salvamento Arqueológico: fondo etnohistórico y arqueológico, desarrollo del Proyecto, estudios de la superficie. INAH Colección Científica, Arqueología, No. 122. INAH, México, D.F.

2008 Arqueología de la Zona Costera de Jalisco y del Municipio de Villa Purificación. In *Miscelánea Histórica de Villa Purificación: Testimonios del 475 Aniversario de su Fundación*, coordinated by Aristarco Regalado Pinedo and Juan Sánchez Vázquez, pp. 21–39. Ayuntamiento Constitucional de Villa Purificación, Jalisco.

Mountjoy, Joseph B., Tammy C. Smith, Ryun Papson, Debbie Guida, John Pleasants, Chris Witmore, and Cheryl Cross

2003 *Arqueología del Municipio de Puerto Vallarta.* Electronic document, www.uncg.edu/arc/Vallarta.

Mountjoy, Joseph B., and Otto Schöndube

2014 Investigaciones Arqueológicas en la Zona Costera de Jalisco. Report submitted to the Consejo de Arqueología del Instituto Nacional de Antropología e Historia, Mexico City.

Nuttall, Zelia (editor)

1975 *The Codex Nuttall: A Picture Manuscript from Ancient Mexico.* Dover, New York.

Pohl, John M. D.

2012 La tradición Aztatlán de Nayarit-Jalisco y el estilo nahua-mixteca de Cholula. *Arqueología Mexicana* 20(115): 60–65.

Smith, Michael E., and Cynthia M. Heath-Smith

1980 Waves of Influence in Postclassic Mesoamerica? A Critique of the Mixteca-Puebla Concept. *Anthropology* 4(2): 15–50.

Solís, Felipe, Verónica Velásquez, and Roberto Velasco

2006 Cerámica policroma de Cholula y de los otros valles de Puebla. *La Gran Pirámide Cholula,* edited by Felipe Solís, Gabriela Uruñuela, Patricia Plunkett, Martin Cruz, and Dionisio Rodríguez, pp. 79–130. CONACULTA/INAH, Mexico City.

PART 2

SPACE

Old Questions, New Methods

5

Architectural Discourse and Sociocultural Structure at Los Guachimontones, Jalisco

KIMBERLY SUMANO ORTEGA AND JOSHUA D. ENGLEHARDT

> The primacy of configuration in the "social logic" of space does not just *happen* to be the case. It originates in the logic of space itself.
>
> (Hillier et al. 1987: 363)

The site of Los Guachimontones in central Jalisco state, Mexico (Figure 5.1), has long been the subject of intensive archaeological research, beginning in the 1970s with Weigand's (1979, 1993, 1996, 2008a, 2008b; see also Mountjoy and Weigand 1975) investigations of the site's circular architectonic configurations. The site evidences a continuous occupation from the Middle Formative period through the Postclassic (Beekman and Weigand 2008), and is perhaps best known for its *guachimontón* architecture. These complexes are comprised of a central circular building—sometimes called an "altar"—surrounded by a patio bounded by numerous rectangular platforms arranged in a circle around the central structure, thus forming a series of concentric circles (Figure 5.2; see also Beekman 2003a: 300, fig. 1; Weigand 2008b: 31). The majority of these complexes were constructed in the Late Formative through Early Classic periods (circa 300 BC–AD 450; phases Tequila II–IV; see Beekman, this volume, Table 2.1), although they appear to have been used throughout the occupational history of the site. Weigand (1993, 2008a, 2008b) argued that the *guachimontón* complexes were public in character and essentially homogeneous in terms of both form and function. Nonetheless, this suggestion has not been closely scrutinized or evaluated with available evidence.

Complicating Weigand's proposal is the fact that the size of the *guachimontones* varies, as does the number of platforms that surround the central altar, both throughout western Mexico and within Los Guachimontones

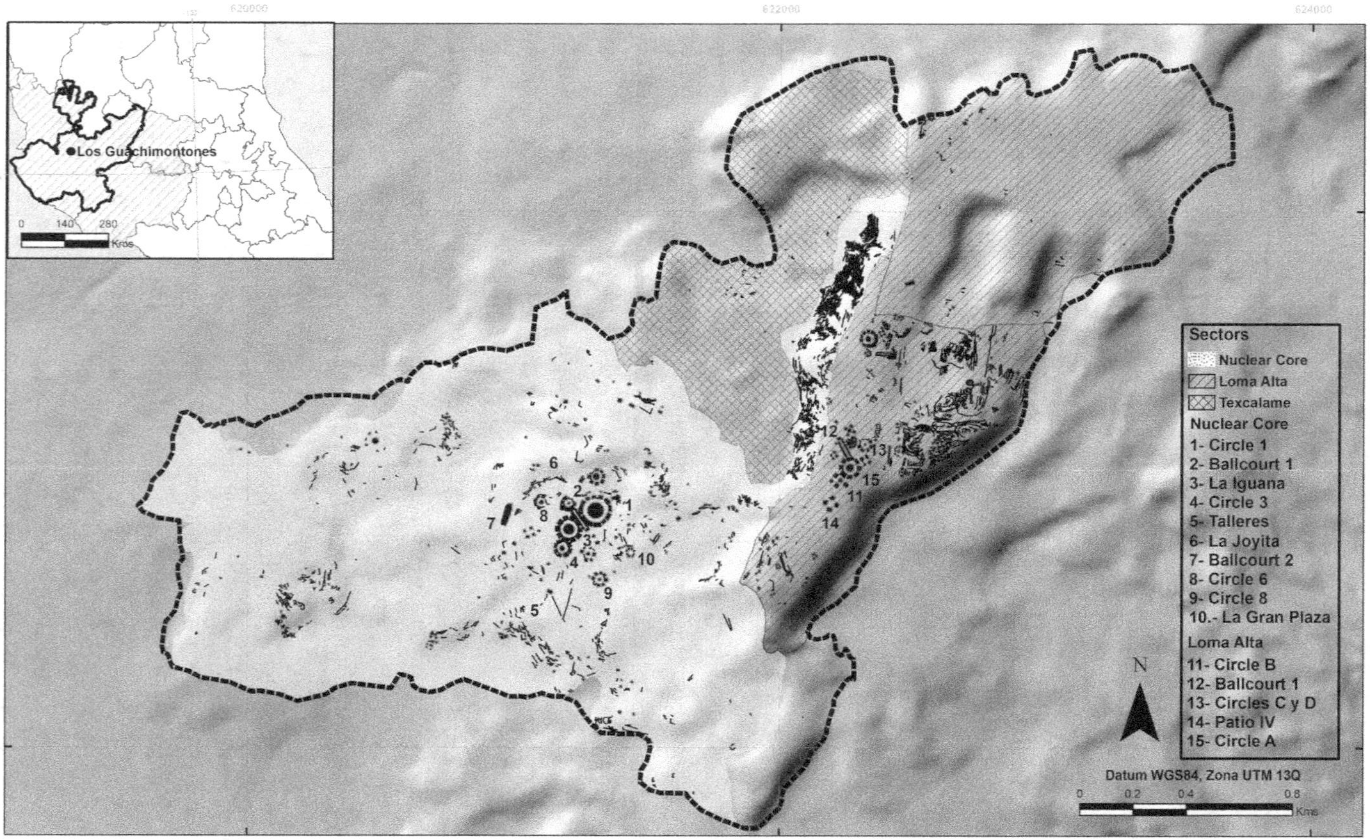

Figure 5.1. Los Guachimontones within the Valles Region of Jalisco State; site map details distinct sectors and architectonic complexes within the site (map by David A. Muñiz García).

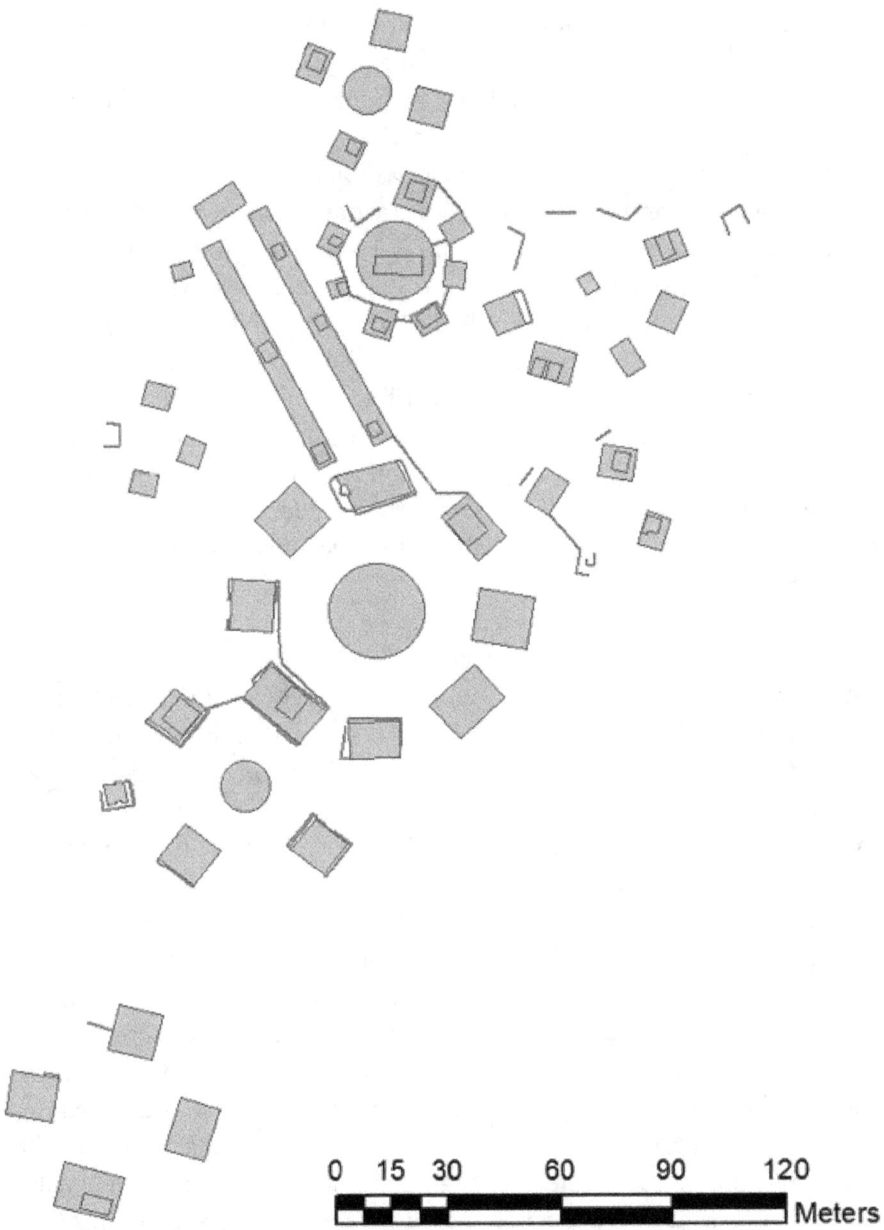

Figure 5.2. Examples of *guachimontón* complexes, from the Loma Alta Sector of the Los Guachimontones site (Heredia et al. 2014).

(see Figure 5.2; compare Smith 2009). A detailed understanding of this intrasite architectonic variability eludes adequate explanation and obscures our comprehension of the internal sociopolitical dynamics of the groups that inhabited the site. To address these lacunae, this chapter compares architectural patterns and spatial organization in two distinct areas of the site—Loma Alta and the nuclear core—that are thought to have been socially differentiated habitational and ceremonial-administrative zones (Weigand and Esparza 2008). Departing from a dual-processual framework (Blanton et al. 1996) and utilizing gamma analysis (Hillier and Hanson 1984; see also Bermejo 2009, 2015), we analyze variability in the spatial syntax, formal characteristics, and distribution of architectural groups in these zones. The principal indicators that we explore are circulation and accessibility, which refer to the degree of connectivity between architectural spaces. We associate these indicators with variable degrees of openness of the individual architectonic complexes that were experienced by inhabitants of the site.

Variability in these aspects may indicate distinct functions of discrete areas within the site, and demonstrate how the sociostructural organization of—and relations between—the groups that occupied Los Guachimontones were negotiated, reflected, and reified in the architectural configurations of the built environment. Through a comparison of Late Formative–Early Classic period architecture in two sectors of the Los Guachimontones site, we demonstrate that discrete physical spaces were utilized in diverse manners as architectural discourse to communicate distinct messages to different social groups, even when the built environment of these sectors presents a high degree of formal homogeneity and contains the same architectonic elements. We seek to add a new analytic axis and alternative framework that provides insight into both architectural variability at Los Guachimontones and the social structures that gave rise to these architectonic configurations. It is not, however, our intention in this chapter to offer a definitive characterization of the sociocultural structure of the Los Guachimontones society, but rather to demonstrate how the analysis of spatial syntax and architectural discourse offers new and intriguing possibilities for approaching this fundamental question. In this sense, our analysis contributes to a more nuanced understanding of the sociocultural dynamics of a site recognized as key to understanding the development of social complexity in western Mesoamerica.

Theoretical Framework: Architecture as a Mirror of Sociocultural Structure

The built environment is an ideal medium to explore the connection between material culture and sociostructural organization, inasmuch as architectural complexes are themselves social objects whose production potentially reifies sociocultural configurations as an embodiment of cultural activities within specific contexts (Bourdieu 1989; Maran 2006; Martin 2001: 168; Mathews and Garber 2004; Rapoport 1988, 1990). Constructed space may thus serve as a form of social communication, defining, negotiating, expressing, and/or questioning particular social relations within an established social order (Englehardt and Nagle 2011: 357; see also Maran 2006; Martin 2001). Since the production of the built environment is a sociocultural process, architectural configurations—and their meanings—are spatially and temporally variable, even within a single site. As a given society changes over time, the architectural spaces that it produces and consumes may be considered as dynamic material expressions of the social behaviors and relations that characterize that society.

Previous studies have demonstrated that dual-processual theory (Blanton et al. 1996; Blanton 1998) is a productive framework in which to elucidate such variability, in both synchronic and diachronic terms (see, for example, Beekman 2000, 2008; Englehardt and Nagle 2011; Small 2009). The dual-processual perspective focuses on variation in political economies, distinguishing between exclusionary, individual-centered configurations versus those that are more corporate or group-oriented (Englehardt and Nagle 2011: 357).[1] Dynamic oscillation between exclusionary and corporate strategies in developing societies is thus mirrored by variation in material culture, which itself represents distinct sociostructural configurations at given points within the history of a particular site (Blanton et al. 1996: 5–6; Marcus 1998).

Insofar as material culture is actively constructed and deployed within historically particular sociocultural contexts, social actors may variously reproduce, reject, or modify dominant structures in the material culture they produce, including architecture. Distinct social, political, or economic relations may thus be expressed in architectural discourse, understood as the material manifestation of the sociocultural structures that underpin the phenomenological experience of the built environment (Bachelard 1958; El-Bizri 2011, 2015; Hillier and Hanson 1984; Norberg-Schulz 1980). In other words, since architecture is an expression of sociostructural configurations, architectural variability—or the lack thereof—may reflect spatial or temporal variations in corporate or network strategies, and the continuous negotiation of the interpersonal and intergroup social relationships produced and implied by these strategies, in discrete spatial, temporal, or sociocultural circumstances (Englehardt and Nagle 2011: 357).

We propose that architectural discourse at Los Guachimontones reflects differing conceptions of sociocultural structure and relations at the site during the Late Formative and Early Classic periods. We evaluate this proposition by examining architectural discourse via variable spatial syntax among architectonic complexes in two discrete sectors of the site. These patterns, in turn, offer a window into the variable corporate and network strategies that both shaped and reflected the sociostructural configurations evident in intrasite architectonic variability. Our goals are to illustrate how architectural configurations can help identify sociocultural strategies and relationships, and to identify patterns and material correlates associated with variable exclusionary and corporate sociopolitical strategies that were subsequently reflected in the built environment (Englehardt and Nagle 2011: 358, n. 19; Small 2009: 207, 217).

Case Study: Los Guachimontones

The site of Los Guachimontones is often presented as the paradigmatic site of the Teuchitlán Tradition (Ohnersorgen and Varien 2008; Weigand 1996, 2008a,b; Weigand and Beekman 1998). It is certainly the largest and most intensively studied of the various sites throughout western Mexico that comprise this tradition, and its nuclear core presents the largest concentration of monumental architecture, in terms of *guachimontón* complexes and ballcourts, of any site in western Mesoamerica. The traditional typological classifications that place the site at the center of the Teuchitlán Tradition, as well as atop the regional settlement hierarchy, are primarily based on the size and sheer number of its architectonic complexes, particularly the 24 *guachimontones* identified to date (see Figure 5.1; Blanco 2009; Heredia Espinoza et al. 2014; Smith 2008, 2009; Weigand 1996, 2008a, 2008b). The monumental architecture at the site, often classified as public and administrative-ceremonial, is in turn frequently associated with domestic units or residential complexes. Los Guachimontones is incorporated within the UNESCO-designated Agave Landscape and Ancient Industrial Facilities of Tequila World Heritage Site.[2]

The conception of Los Guachimontones as the archetypical exemplar of the Teuchitlán Tradition is founded principally on the monumental architecture within the site's nuclear core. There are, however, other sectors within the site that have been intensively studied, among them Loma Alta (see Figure 5.1; Weigand and Esparza 2008). The comparative evaluation of architecture in these sectors may offer insight into intrasite architectural variability, as well as the uses of and meanings assigned to variable architectonic complexes by the inhabitants of the site. We would suggest that the relative sizes and specific configurations of distinct *guachimontón* complexes in discrete areas of the site are not the only aspects that vary. Rather, we propose that a potentially significant aspect of variability lies in the ways in which these structures are situated within and relate to both the physical landscape and other components of the built environment. These aspects, in turn, may assist in describing the sociocultural structure that is reflected in the architecture of Los Guachimontones.

Our comparison considers two sectors of the site: the nuclear core and Loma Alta. The nuclear core is the largest of all the sectors in which the site is divided. It encompasses 10 *guachimontón* complexes, two ballcourts, residential areas, lithic workshops (*talleres*), and a number of terraces, which were likely used for cultivation and associated with the domestic complexes of groups dedicated to agricultural production. At Loma Alta, 500 meters to the northeast, similar architectural complexes are evident, and the material culture and features in the sector are strikingly similar to those of the nuclear core. In this sense, there is a material continuity between the sectors, and little disjunction in terms of artifact distribution is evident between discrete areas of the site. Importantly, as Beekman (this volume) has shown, the two sectors are contemporaneous. Therefore, differences found in the spatial analysis cannot be attributed to chronological variation.

Despite the similarities between Loma Alta and the nuclear core, there are obvious differences between them. The first is the topography of the distinct sectors and the physical environment in which buildings were constructed. While the nuclear core lies within a lower, flatter area more conducive to transit and construction, Loma Alta occupies a zone of higher ground with more pronounced topographic divisions. This fact makes transit throughout the sector relatively more difficult, and would have added considerable costs to both construction—in terms of required labor and sheer force necessary—and the actual use and experience of the space itself. That said, topographic differences between the sectors may themselves have served to limit access or control circulation between distinct sectors of the site, as well as within the built environment in Loma Alta. This suggestion, in turn, may point to differences in the architectural discourse evident in the distinct sectors, which we analyze below.

Methodology: Analysis of Spatial Syntax

The scale of our analysis is at the level of the site, and our analytic units are the two discrete sectors defined above. The architectonic complexes and spatial syntax of these sectors are analyzed separately and then compared utilizing gamma analysis (Bermejo 2009; Grau 2015; Hillier and Hanson 1984; Hillier et al. 1987), focusing primarily on the variables of circulation and accessibility. Our goal in the analysis is to determine whether evident spatial relationships within the built environment are similar in discrete sectors or, alternatively, whether these relationships are the product of architectural variability. Extrapolating on the basis of these spatial relationships, in turn, allows for a consideration of the sociostructural organization that is reflected in them.

Spatial syntax is a descriptive analytic methodology that has frequently been employed in what some have called the "archaeology of architecture" (Bermejo 2009, 2015; Castillo 2015; Fernández 2015; Grau 2015). Such analyses are most often predicated on theoretical models that treat the phenomenology of architecture, spatial logic, and the meaning of the built environment, which allow for a deeper contextual and significant interpretations (see, for example, Bourdieu 1989; Hillier and Hanson 1984; Maran 2006; Norberg-Schulz 1980; Rapoport 1988, 1990; compare Martin 2001; Mathews and Garber 2004). Through the identification of spatial patterns within and between architectonic complexes and their respective compositions, the descriptive mechanism of syntax analysis permits the visualization of social information and content, which can be abstracted and displayed as planimetric graphs (Hillier and Hanson 1984). Here, we follow Bermejo's (2009: 50) conceptions of spatial syntax as a tool for studying "the ways in which the spaces of an architectonic complex are linked and organized, attempting to infer those aspects of social structure that may have influenced its design."

An analogy with textual syntax—and material culture as text—thus follows logically (compare Hodder 1989; Tilley 1991). As a hermeneutic model associated with semiotics, syntactical analysis seeks to identify meaning evident in textual components, on both significant (for example, in terms of icons, indices, or signs) and deeper conceptual levels (Bermejo 2009: 49). In this sense, an architectonic composition can be understood as both an archaeological feature and a purveyor of contextual and specifically significant content; that is, as a text or other element of nonverbal expression (Rapoport 1988, 1990) capable of transmitting messages laden with meaning, as well as reflecting the underlying sociocultural structures that at once constrain and enable that meaning. There exist diverse potential indicators and archaeological correlates by which to examine spatial syntax (see Hillier and Hanson 1984), each of which informs distinct aspects of social structure (Bermejo 2009: 51). In this study, we evaluate the variable spatial syntax of architectural complexes at Los Guachimontones via two distinct mechanisms: gamma analysis of connectivity and the identification of patterns of accessibility and circulation.

Gamma Analysis of Connectivity

Gamma analysis (Hillier and Hanson 1984; Hillier at al. 1987; see also Bermejo 2009: 51) is frequently employed to demonstrate degrees of connectivity or segmentation. Built structures exist within a spatial order that reflects a unitary domain of organizational control. Such essential unity is expressed in two ways: one outward, in terms of what exists within a higher-order spatial unit (the site as a whole, in this case), which defines the limits of the structure; and one inward, in terms of the interior accessibility (or "permeability") of lower-order spatial units for the social groups that use them. Gamma analysis permits the interpretation and graphic representation of these aspects and their internal relationships along the outward-inward continuum of unity within a larger spatial order (Hillier and Hanson 1984: 147). The central premise of gamma analysis is that the built environment transmits sociocultural information through its internal structural configuration(s), which is observable through variation in the basic syntactic parameters that define and organize an architectural complex via the relations between its constituent spaces (Hillier and Hanson 1984: 154; Hillier et al. 1987: 363; compare Rapoport 1988, 1990).

Gamma analysis produces diagrams that are adaptations of axial line graphs (see Hillier and Hanson 1984: figs. 26–28), which visually represent the presence and/or permeability of individual units within a given spatial configuration (Bermejo 2009: 51). Gaps and liminal areas between spaces create permeable relations between them, which can be analyzed by considering the positional location of spaces with respect to larger architectonic compositions (Hillier et al. 1987: 363). Such relations can be visualized through justified graphs, which graphically represent those spatial configurations (Hillier et al. 1987: 363–364, figs. 2c–d; compare Bermejo 2009: 52, fig. 6). Such graphs also illustrate those configurational properties of spatial layouts that appear to articulate cultural ideas and social relations: depth (degrees of distance between spaces) and choice (the "existence or otherwise of alternative routes from one space to another"; Hillier et al. 1987: 364). These properties, in turn, may be quantitatively represented through three measures—integration values, inequality genotypes, and difference factors. Thus, gamma analysis permits a quantitative visualization of accessibility in the form of a diagram of a given spatial configuration.[3]

Gamma analysis aids in understanding relationships between built spaces within a larger architectonic composition, which in turn provides a window into sociocultural principles and relations. Having defined spatial units—in this case the architectural complexes in the two site sectors—the method can be applied by projecting axial lines that further define the minimum spatial and analytic units (Grau 2015: 4). This permits the visualization of potential relationships between spaces, in terms of permeability or accessibility, and thus an appreciation of the logical, organic flow of transit between them. In this way, we can interpret the internal coherence of architectural spaces, clarifying the relationship between elements of the built environment and, by extrapolation, the relations between the groups that experienced these spaces (Bermejo 2009: 51–52; compare Hillier and Hanson 1984; Hillier et al. 1987).

Patterns of Circulation

Like gamma analysis, the interpretation of patterns of circulation is a potentially useful mechanism for the evaluation of spatial relationships within and between constructed spaces. Degrees of permeability and accessibility, derived from gamma analysis, offer suggestions for how people moved within and experienced the built environment. As Palyvou (1987: 195) notes:

> Circulation in an architectural compound involves two major activities: moving in and out of the building and moving within it. The manner in which these activities are dealt with through architecture manifests the gradation between community and privacy, as conceptualised in a specific culture. . . . Two major activities are accommodated and, at the same time, controlled: a) the frequent moving in and out of those who are the primal users of the building; b) the

occasional moving in and out of people indirectly related to the function of the building.[4]

Thus, the ability to access, move freely within, or experience certain spaces may reflect concrete symbolic meanings, such as the status or privileges of specific social groups, as well as the relationships between those groups—internal or external to a given society—that do or do not enjoy access to particular spaces (compare Bourdieu 1989).

Various studies (for example, Castillo 2015; Fernandez 2015; Grau 2015) have suggested that architecture that may formally appear to respond to certain sets of behaviors in actuality exhibits far more variability than may be first evident, thus suggesting significant differences in the uses or formal order of a given spatial configuration. In this sense, interpreting patterns of circulation is particularly apropos in the case of the Los Guachimontones architecture, since it has long been thought—and, sadly, generally accepted—that little variability exists among the architecture at the site (Weigand 1993; compare Smith 2008). Syntactic analyses are capable of elucidating whether or not formally similar spatial arrangements in architecture may in fact reflect variability in terms of the functions, symbolic meanings, and/or social relationships codified within built spaces. In other words, our analyses allow for a deeper evaluation of intrasite architectural variability—or lack thereof. Thus, applying space syntax analyses allows a more nuanced understanding of the internal logic of individual architectonic complexes as well as the higher-order sociostructural and spatial organization in which they exist.

Further, variability in social and structural relationships is reflected in the architectural elements and spaces within and between the two sectors of Los Guachimontones on which we are focusing, thereby permitting a characterization of the sociopolitical organization of the groups that inhabited the site. To this end, we interpret patterns of architectural variability as reflections of points along the dual-processual corporate-network continuum, as detailed above (see Blanton et al. 1996; Blanton 1998). On the one hand, a higher degree of syntactic variability and access control within spaces and among sectors would suggest a more network-oriented society, one less flexible in terms of intergroup structural relationships. On the other hand, less syntactic variability and greater permeability may indicate a more corporate organization, with more open relationships between distinct social groups.

Applying Space Syntax Analysis at Los Guachimontones

In our analysis of space syntax at Los Guachimontones, we employed the open-source programs Depthmap and AGRAPH to locate and represent axial lines within and between architectural spaces in different sectors of the site (Figures 5.3 and 5.4). In turn, these graphs may be interpreted as indicative of syntactical relationships, and also provide clues regarding the directional patterns of circulation within and between defined spatial units. Due to the size of the individual site sectors, the axial line gamma analysis was performed separately for each sector, although we subsequently realized a comparison of similarities and differences in both circulation and use of space between the sectors, as suggested by the axial lines. In Figures 5.3 and 5.4, the axial lines permit an abstracted visualization of the relationships between architectonic complexes (by virtue of a lineal connection or lack thereof) within the spatial units (Manum 2009).[5]

Axial lines also permit an appreciation of nodes of convergence, consisting of the plot of the points where the greatest amount of axial lines meet, which may serve as indicators of social nodes (Bermejo 2009 :51). Maps of convergence are useful tools to identify paths and intersections of axial lines in small but complex systems (Manum 2009: 8)—such as the discrete sectors of Los Guachimontones under investigation. Additionally, the identification of these social nodes potentially indicates circulatory flows between points and within spaces (compare Palyvou 1987: 196–197, figs. 1–2), further suggesting social structures and specific behaviors reflected in variable architectural syntax. These patterns then can be compared between sectors.

An interpretive difficulty arises in attempting to discern which of all the axial lines actually represent specific paths of circulation and connectivity between spaces. To address this issue, we employed a tool native to the Depthmap software called Visibility (Al-Sayed and Turner 2012; compare Hillier et al. 1987: 383–384). This tool calculates viewsheds and lines of sight, using an algorithm that employs a predetermined inclination of 17° to the visual horizon (assumed as standard for the average individual; Al-Sayed and Turner 2012). In basic terms, "visibility" permits an evaluation of the choices (Hillier et al. 1987: 364) with which actors were confronted when navigating the built environment. Depthmap visibility analysis is a two-stage process. First, the program considers the "Agent First Moment," which reveals the range of possible circulatory options with which an individual is confronted on first entering a spatial unit. The second stage, "Point

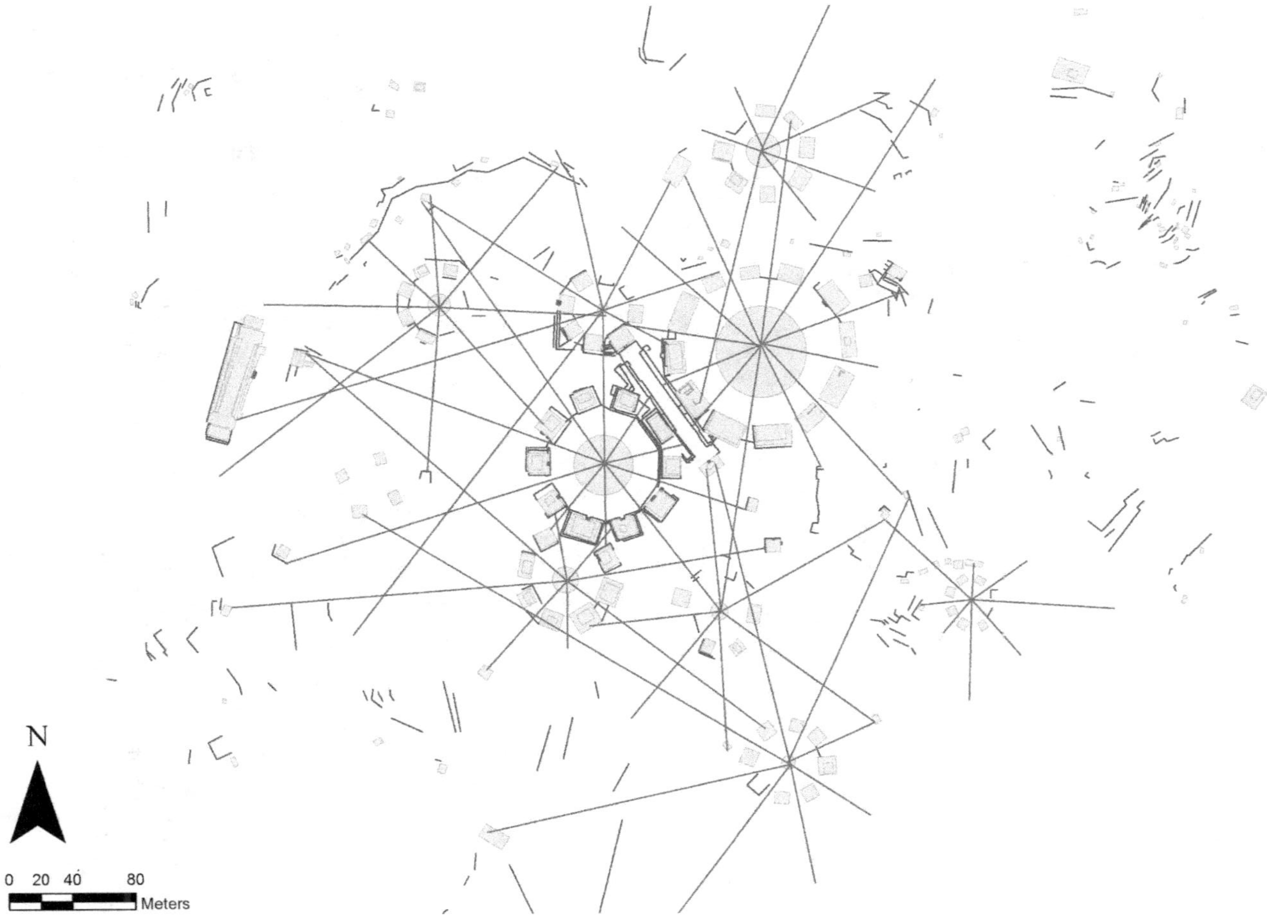

Figure 5.3. Axial line graph for the Nuclear Sector of Los Guachimontones at its peak occupation in the Tequila III phase. Most architectural spaces are identified by circles, whereas the transitional areas—and relations—between them are represented by lines. Note that axial lines do not connect ballcourt interiors to any open or constructed area (cartography courtesy of Verenice Heredia).

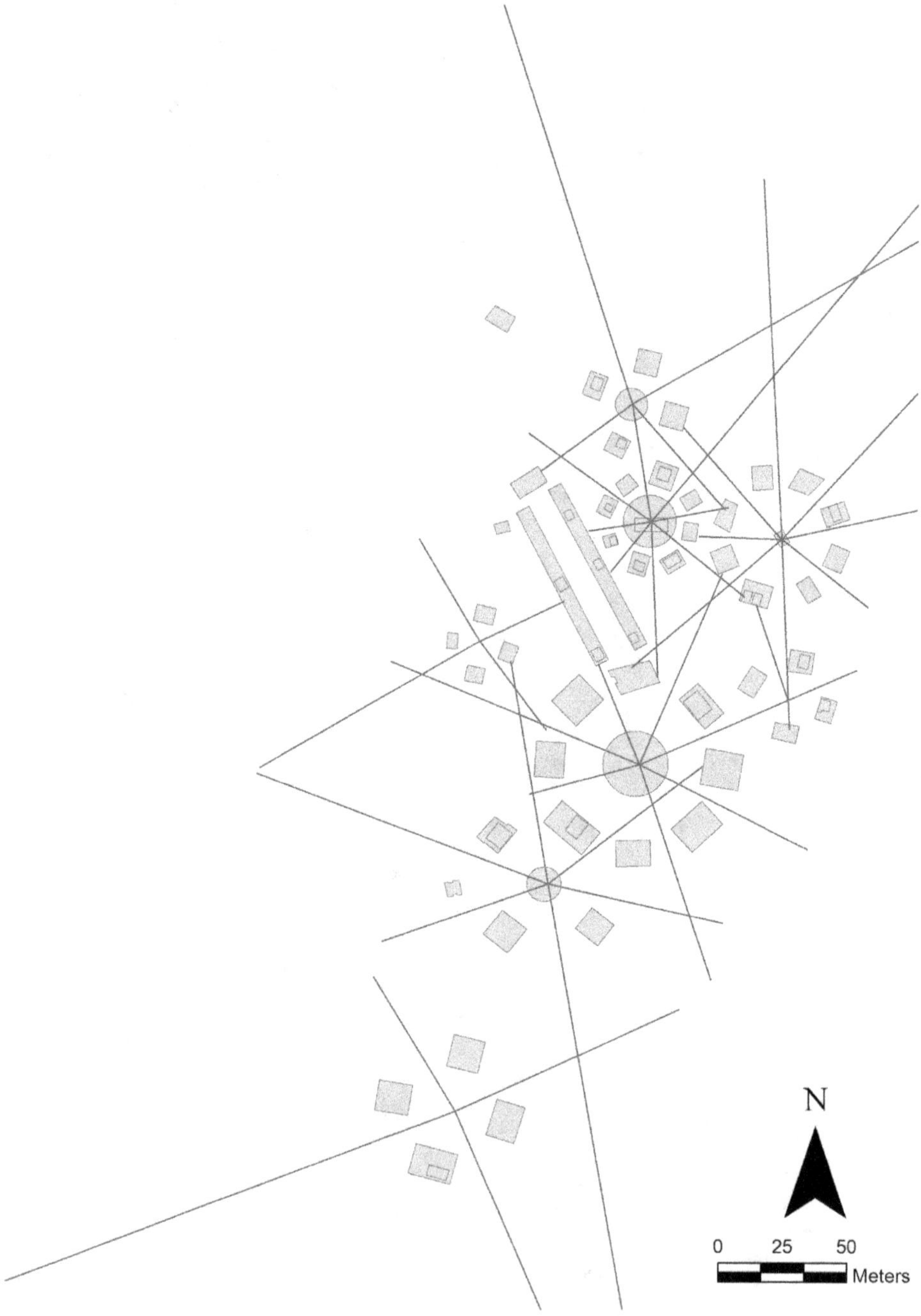

Figure 5.4. Axial line graph for the Loma Alta Sector. As in the Nuclear Sector, axial lines do not appear to connect the ballcourt interior to any open or constructed area (cartography courtesy of Verenice Heredia).

Second Moment," then considers the possible second-level choices for circulation following the initial decision, thus taking into account potential dynamic adaptive behaviors of individual agents as they move through spaces. The Visibility tool thus suggests multiple paths that an individual may have taken when circulating within and among spatial units, including both architecture and topography. We used this tool to produce Figures 5.5 and 5.6.

Finally, to complement the visibility and axial line graphs produced with Depthmap software, we applied gamma analysis utilizing the AGRAPH program (Manum 2009; Manum et al. 2009), which more robustly supports the identification of convergent nodes than Depthmap. Gamma analysis graphed nodes and justified graphs (Figure 5.7) represent the relative depth and accessibility of architectural spaces (compare Hillier et al. 1987: 364). AGRAPH also calculates the quantitative measures of depth, asymmetry, and integration of each node (Tables 5.1 and 5.2). The total depth of a node (TDn) is the total of the shortest distances from the node to the other nodes in the systems (that is, the total of line n in the distance matrix), whereas mean depth refers to the average depth (or average shortest distance) from node n to all the other nodes. Relative asymmetry (RA) describes the integration of a node by a value between (or equal to) 0 and 1, where a low value describes high integration. Finally, the integration value (i), contrary to RA, describes high integration via a relatively higher number (see Manum et al. 2009).[6]

In some cases, gamma analysis is capable of producing graphic visualizations that clearly demonstrate overarching relational structures within the built environment. At Los Guachimontones, however, such relations are perhaps less evident. This may be due to the fact that the architectural types present at the site are in some ways less common than others throughout Mesoamerica, as are, by extension, the nature and degree of the spatial relationships between architectural elements in the site's discrete sectors. For this reason, our principal aim in this analysis is to identify potential points and units that may be indicative of the higher-level spatial order, structured patterns, relationships, and organizational principles—both spatial and sociocultural—reflected in the architectonic complexes at Los Guachimontones (compare Hodder and Orton 1976).

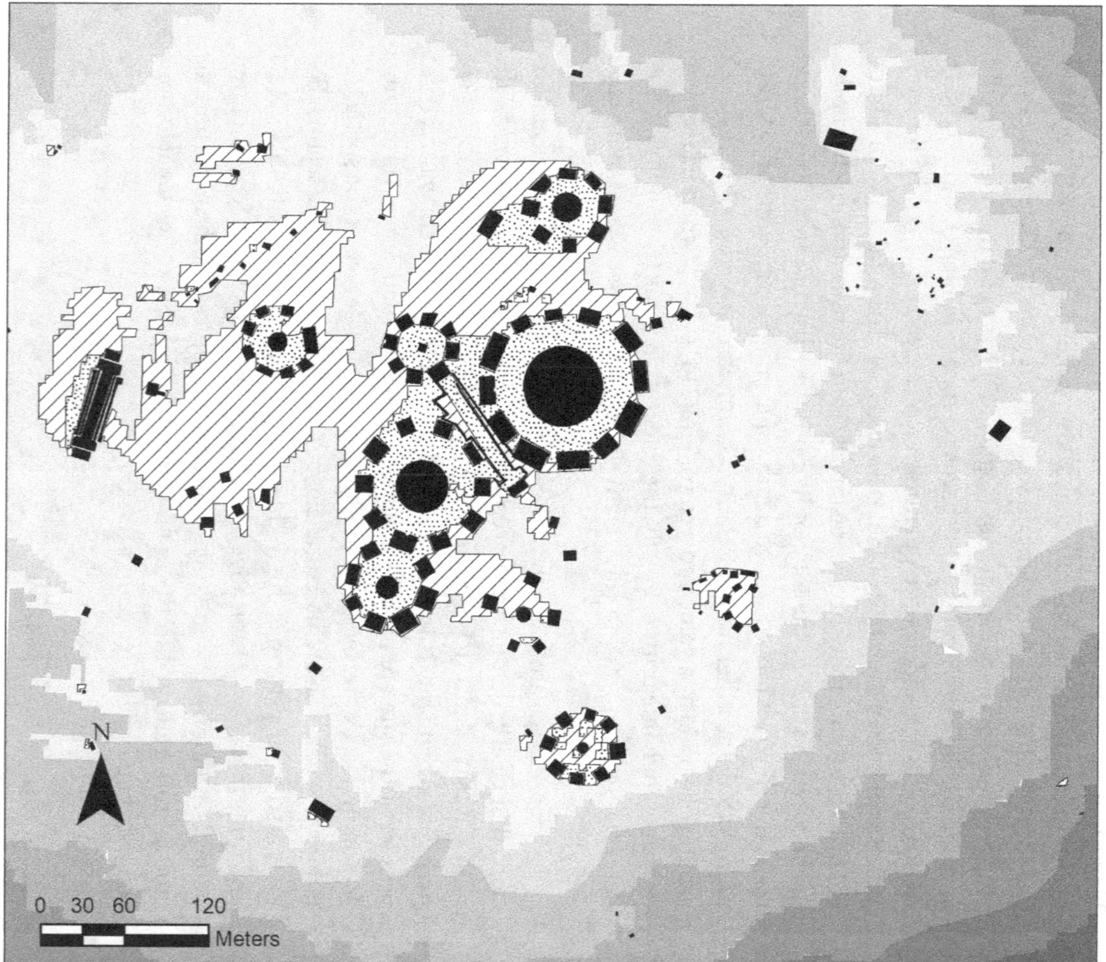

Figure 5.5. Depthmap visibility analysis within the Nuclear Sector at its peak occupation in the Tequila III phase. The stippled pattern indicates higher connectivity values and paths of least resistance at the first moment, whereas the diagonal line pattern details circulatory options at the second moment. Darker shades of gray indicate increasing segregation and lower connectivity values (map modified by David A. Muñiz García with original data courtesy of Verenice Heredia; analysis by the authors).

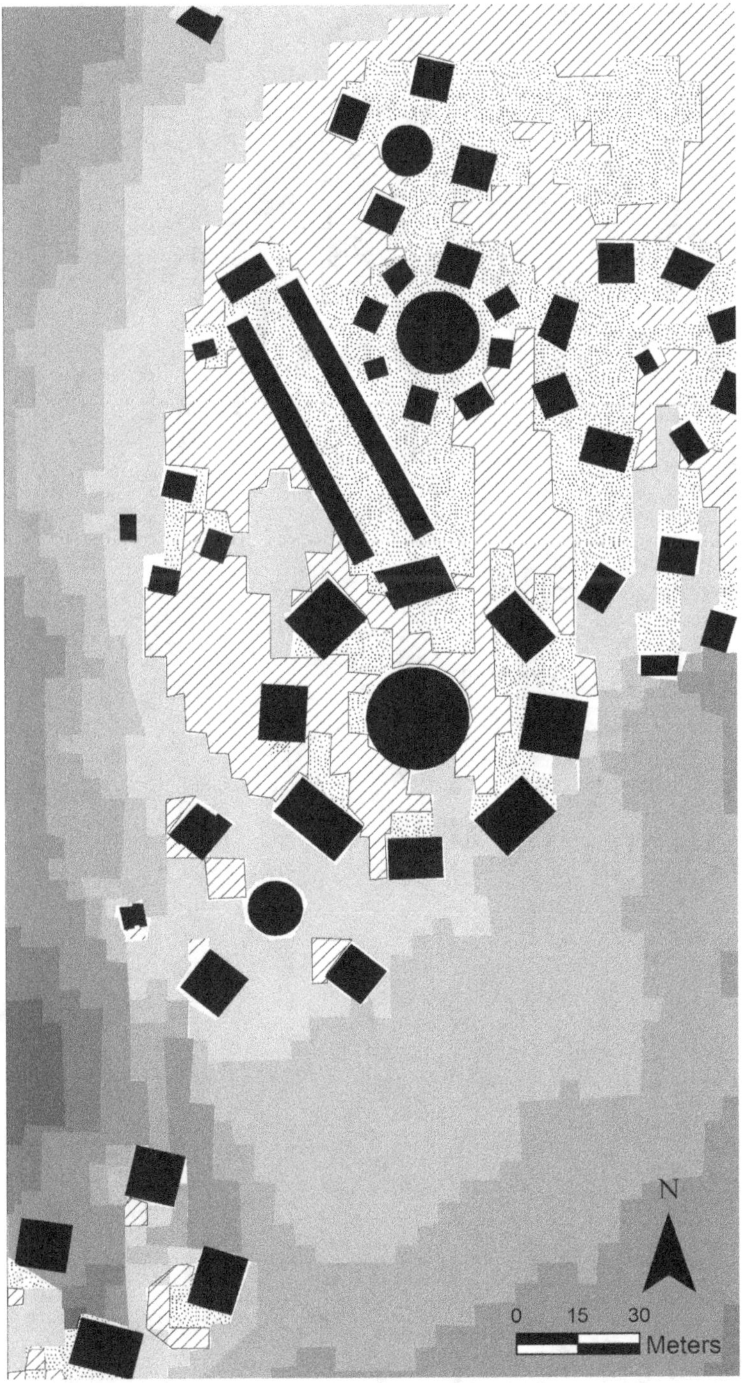

Figure 5.6. Depthmap visibility analysis within the Loma Alta Sector. The stippled pattern indicates higher connectivity values and paths of least resistance at the first moment, whereas the diagonal line pattern details circulatory options at the second moment. Darker shades of gray indicate increasing segregation and lower connectivity values (map modified by David A. Muñiz García with original data courtesy of Verenice Heredia; analysis by the authors).

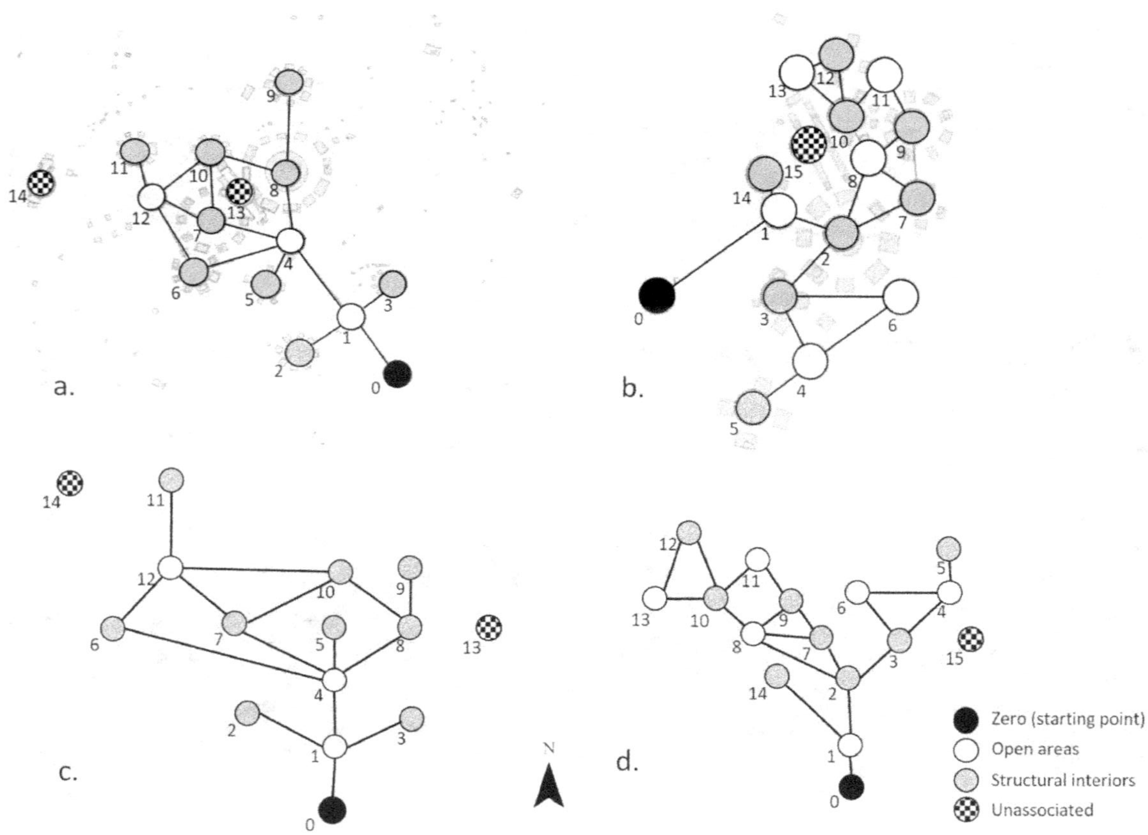

Figure 5.7. Node maps and justified gamma graphs of the Nuclear Sector at its peak occupation in the Tequila III phase (a, c) and Loma Alta (b, d). Convergent nodes are color coded: the zero (starting) point is in black, structural interiors are gray, and open areas, white. The nodes displayed in checkerboard pattern do not display evident links with other spatial units (map and scheme by the authors).

Table 5.1. Integration value calculations for axial nodes in the Nuclear Sector

Node	TDn	MDn	RA	i	CV
0	50	3.57	0.39	2.52	0.33
1	37	2.64	0.25	3.95	2.25
2	28	2.00	0.15	6.50	1.25
3	35	2.50	0.23	4.33	1.08
4	45	3.21	0.34	2.93	1.83
5	58	4.14	0.48	2.06	0.33
6	46	3.28	0.35	2.84	0.66
7	33	2.35	0.20	4.78	0.83
8	30	2.14	0.17	5.68	1.16
9	40	2.85	0.28	3.50	1.08
10	37	2.64	0.25	3.95	1.75
11	47	3.35	0.36	2.75	0.58
12	49	3.50	0.38	2.60	0.75
13	49	3.50	0.38	2.60	0.75
14	50	3.57	0.39	2.52	0.33
Min.	28.00	2.00	0.15	2.06	0.33
Mean	42.26	3.01	0.31	3.57	1.00
Max.	58.00	4.14	0.48	6.50	2.25

Notes: TDn Total depth of a node; MDn Mean depth of a node; RA Relative asymmetry; i Integration value; CV Control Values. Control Values are calculated by assigning each node a total value of 1, thus assuming that the node is equally connected to all other nodes; the CV thus acts as a baseline against which asymmetry and integration values may be evaluated for significance (see Manum et al. 2009).

Table 5.2. Integration value calculations for axial nodes in Loma Alta

Node	TDn	MDn	RA	i	CV
0	66	4.71	0.57	1.75	0.25
1	55	3.92	0.45	2.21	3.20
2	66	4.71	0.57	1.75	0.25
3	66	4.71	0.57	1.75	0.25
4	50	3.57	0.39	2.52	2.41
5	61	4.35	0.51	1.93	0.20
6	56	4.00	0.46	2.16	0.45
7	55	3.92	0.45	2.21	0.78
8	55	3.92	0.45	2.21	1.53
9	66	4.71	0.57	1.75	0.33
10	59	4.21	0.49	2.02	0.91
11	70	5.00	0.61	1.62	0.25
12	59	4.21	0.49	2.02	2.16
13	210	15.00	2.15	0.46	0.00
14	210	15.00	2.15	0.46	0.00
Min.	50.00	3.57	0.39	0.46	0.00
Mean	80.26	5.73	0.72	1.79	0.86
Max.	210.00	15.00	2.15	2.52	3.20

Notes: TDn Total depth of a node; MDn Mean depth of a node; RA Relative asymmetry; i Integration value; CV Control Values. Control Values are calculated by assigning each node a total value of 1, thus assuming that the node is equally connected to all other nodes; the CV thus acts as a baseline against which asymmetry and integration values may be evaluated for significance (see Manum et al. 2009).

Discussion

In formal terms, the architecture of Los Guachimontones appears substantially homogeneous across sectors. Both the nuclear core and Loma Alta evidence similar architectonic elements and spatial complexes: *guachimontones*, ballcourts, quadripartite residential units, and open areas such as plazas and patios. Applying a reading of spatial syntax to the Los Guachimontones built environment, however, reveals differences between the spaces, particularly in terms of their internal relationships, spatial order, and the link between architectural spaces within and between the discrete sectors. The first point we wish to highlight is a distinction regarding access control within each sector. The nuclear core is comprised of much more open, accessible spaces, where architecture exhibits less control over the flow of movement and permits the flexible circulation of sizable groups and free transit—as revealed in the multiple convergences of axial lines in Figure 5.3. Architectural spaces in Loma Alta, on the other hand, are less open and connected (Figure 5.4), and the topography itself serves as a natural barrier to the sector, possibly functioning as the first filter that controls access to the space. In this sense, natural topography is in itself a limiting aspect of Loma Alta, as well an important factor separating it from the more major complex below.

The first significant interpretation suggested by the axial line graphs is that the nuclear core of Los Guachimontones exhibits a greater degree of connectivity between its constituent architectonic spaces—particularly the *guachimontón* complexes—whereas in the Loma Alta sector less convergence among axial lines is evident, suggesting a relatively lower degree of relationships between architectural units. Nonetheless, axial lines appear to suggest a strong relationship between the sectors, which is logical given the site layout, the proximity of the sectors, and the presence of homogenous architectonic elements in both spatial units. What is potentially significant, however, is that analysis revealed greater axiality and convexity *within* each sector than *between* them—possibly due to the natural barriers that restrict flow between the nuclear core and Loma Alta. This suggests that there is very little integration between the distinct sectors, despite homogeneity in architectural elements and some axial line convergence. Rather, with respect to each other, the sectors are quite segmented.[7]

The axial lines also reveal a striking difference between the two sectors. In the nuclear core, several lines emanating from the center of various circles lead to small, apparently isolated structures that are neither close to or evidently associated with the respective circles. Nor do these structures present any evident spatial relations among themselves or with other architectural elements. Nonetheless, the axial connection between such *guachimontón* complexes and seemingly segregated structures does raise the possibility of some spatial relationship between these units and the circle(s) with which they are linked, although such association is unclear at this point. Regardless, this phenomenon is not evident in Loma Alta, although this may be simply a function of the facts that (1) relatively small, isolated structures are themselves conspicuously absent in this sector; and (2) some of the circles at Loma Alta are comparatively very small.

Visibility analysis complements the axial analysis of connectivity and access control. In the Agent First Moment analysis, we note that the nuclear core of the site evidences a strong relationship between spatial units. It also appears that within this sector the *guachimontón* complexes and Ballcourt 1 formed an open and integrated configuration of architectural spaces (the stippled pattern in Figures 5.5 and 5.6 indicates higher connectivity values in first point analysis) set apart from the unconstructed environment. Point Second Moment visibility analysis suggests that circulation and transit occurred throughout broader, more open spaces whose access was not strictly controlled (in Figures 5.5 and 5.6, the diagonal line pattern indicates higher connectivity values at the second point). The high first point connectivity values within and between *guachimontones* in the nuclear core suggest that the circles were designed as spaces in which circulation was relatively unrestricted, whereas the high second point connectivity values suggest potential circulatory paths between circles.

On the one hand, this interpretation is supported by the axial line graphs (Figures 5.3 and 5.4), which in many cases suggest convergent, potentially social, nodes within these unrestricted spaces.[8] On the other hand, visibility analysis of the Loma Alta sector indicates less-interrelated spaces. Point Second Moment analysis, however, does suggest a more open space that trends toward greater integration, particularly between Circles A, C, D, and E. As in the nuclear core, this more permeable space coincides with nodes of convergent axial lines. Overall, however, analysis suggests that spatial units in Loma Alta are less connected and display less axial convergence, despite relatively smaller open spaces between units.

In both sectors, the quadripartite residential units adjacent to *guachimontón* complexes reflect a high degree of segregation within the sector-specific spatial syntax (particularly clear via the low connectivity values of these structures in Loma Alta, as represented by the darker gray tones of shading in Figures 5.5 and 5.6). These units are somewhat removed and therefore outside the range of possible transit within the spatial unit as a whole. It may be that the circulation program at the site was designed to restrict access to these areas by those who were not direct users of the internal spaces, especially since no natural or cultural barriers to these spaces are evident. Rather, it is simply the location of these units within the larger spatial order that isolates them. There is a further potential coincidence in the spatial order of the two sectors—open spaces within each sector appear to be zones of convergence between axial lines, and would thus seem to be areas of high integration and syntactic connectivity. In the nuclear core, the Gran Plaza is just such an open space, whereas in Loma Alta a similar open plaza is evident, on a reduced scale, between Circles A, C, and D.[9]

With regard to gamma analysis, we placed zero points according to the convergent nodes of connectivity revealed in the axial line and visibility analyses, as well as natural points of entry as suggested by topography. In the nuclear core, the zero point was located to the southeast of the space between Circle 8 and the quadripartite residential unit directly to its northeast (Figure 5.7).[10] The zero point in Loma Alta corresponds to the natural access point to the sector. As Figure 5.7 demonstrates, in both sectors the first node, corresponding to the first order of depth, is an open space (marked with the number 1 in both sectors). From this point, both axial lines and visibility suggest an immediate connection to units at greater degrees of depth—the Gran Plaza in the nuclear core (node 4 in Figure 5.7a) and Circle A (node 2 in Figure 5.7b). In both cases, these units were those that presented the greatest number of options for transit to other areas and buildings, having the greatest degree of connectivity with spaces at greater depths within the spatial order. In the nuclear core, the gamma graphs reveal a significant degree of connectivity among the built spaces. Circulation from the zero point would flow naturally into the Gran Plaza, which would act as the distributional axis through which other spaces were accessed and connected. This spatial order robustly complements the visibility analysis, and all nodes present significant connectivity, openness, and a relatively lower degree of access control and depth.

The ballcourt complexes in both the nuclear core (nodes 13 and 14 in Figure 5.7a) and Loma Alta (node 15 in Figure 5.7b) do not present significant connectivity, or offer direct access to other structures or open areas, contradicting the visibility analysis. This suggests that these complexes were integrated into an open and accessible configuration. However, their placement within the higher-order spatial configuration—bounded and enclosed by *guachimontones* (and thus segregated from other spaces)—suggests that the ballcourts were at a greater degree of depth within the built environment, and that access to them was more controlled—perhaps as a function of their status as a ritual space. In either case, Ballcourt 2 in the nuclear core (node 14 in Figure 5.7a) is decidedly segregated, and syntax analysis did not reveal evident association with other structures within the architectural program. Finally, gamma graphs suggest that the quadripartite residential units maintain connections with open spaces, but do not present direct connections with *guachimontón* complexes or other monumental construction in either sector.[11]

These graphic results are supported by the quantitative calculations of depth, asymmetry, and integration (Tables 5.1 and 5.2). In the nuclear core, asymmetry values are low, and integration values are high, suggesting a more permeable, accessible built environment. Likewise, lower depth measures (relative to Loma Alta) indicate a more connected spatial order with fewer areas that were inaccessible or lacked connectivity. In Loma Alta, however, quantitative values suggest a relatively less open spatial arrangement, in which depth is greater and access could potentially have been more tightly controlled. Mean values of depth and asymmetry in this sector are almost twice as much as those in the nuclear core, and the integration value is roughly half of that of the site center. Nonetheless, in both sectors the asymmetry value is significantly above, and the integration value below, the control values. Since lower RA and higher i values indicate greater connectivity, these results would suggest that the differences between sectors are more of degree than kind. In other words, the spatial orders in both sectors are not substantially distinct in broad terms, but do present differences relative to each other.

Accounting for Potential Temporal Variability in the Analysis of Space Syntax at Los Guachimontones

A potential drawback of our analyses is the fact that our temporal framework—Late Formative to Early Classic—is admittedly a bit broad. Further, as Beekman's recent analyses (this volume) have shown, there were at least three different construction episodes in the nuclear core of the site during this time frame. Moreover, the majority of construction in the Loma Alta sector did not start until the first century AD, well in to the Tequila III phase, and evidence for activity in Loma Alta prior to Tequila III is minimal (see Beekman, this volume, Figure 2.4d). Thus, not all architectural elements and site sectors coexisted in time, and they may not have been used contemporaneously—at least not until the third or fourth century AD (Tequila IV phase). Since our analyses considered the totality of the built environment in the nuclear core at the site's apogee in Tequila III, we recognized the potential for temporal variability in the architectonic configuration of the site at different points throughout its occupational history to affect the results of our analyses. To assess this possibility, we undertook a second set of analyses, considering only those architectonic elements in the nuclear core that Beekman's research suggests were constructed in the Tequila II phase: Circles 1, 5, 6, 7, and 10, and Ballcourt 1.[12] The results of these secondary analyses are presented below in Figures 5.8–5.10 and Table 5.3.

The results of our secondary analyses reveal a similar pattern of openness and connectivity in the nuclear core of Los Guachimontones during the Tequila II phase. Axial connections between architectonic units are evident, with the exception of Ballcourt 1—although a lack of connectivity and nodal integration between ballcourts and other structures is also evident in later occupational phases, as discussed above. Even Circle 10 (in the southeast corner within Figure 5.8) appears more integrated in the spatial arrangement than it does in later phases (see above and n. 7 supra). Axial lines also reveal an open, connected space in the area where Circles 2 and 3 were later located (to the southwest of Ballcourt 1 in Figure 5.8), which may have functioned as a plaza during the earlier occupational history of the site.

Interestingly, however, visibility analysis of Tequila II–phase constructions does not appear to support this possibility. The area to the southwest of Ballcourt 1 appears somewhat segregated, as indicated by the darker gray shading between *guachimontón* complexes in Figure 5.9. Further, at both Agent First and Point Second Moments, visibility analysis suggests lower connectivity values between architectonic groups (although connectivity within groups does appear high). Yet these seemingly contradictory results may simply be a function of the fact that there were fewer clearly defined architectonic groups at this time, which is unsurprising, given that there were fewer structures in general during the Tequila II phase. In addition, the results of visibility analysis are contradicted not only by the axial line graphs, but also by the node map, gamma graph, and quantitative integration calculations for Tequila II constructions, as discussed below. In this sense, the results of our secondary visibility analysis appear as outliers, in both synchronic and diachronic terms. We suspect that these results stem from the significantly smaller sample population of architecture that dates to this occupational phase.

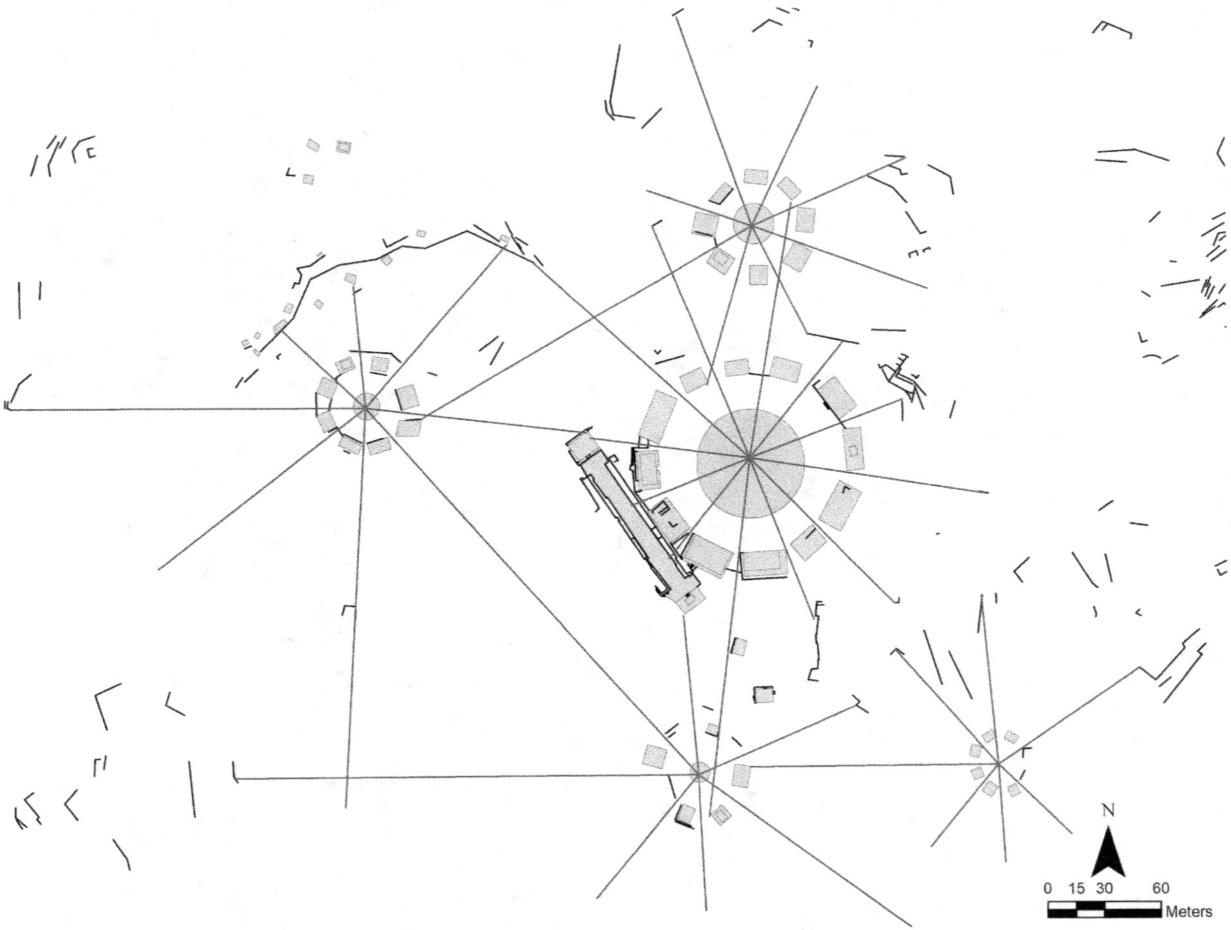

Figure 5.8. Axial line graph for the Nuclear Sector of Los Guachimontones considering only TII phase construction (map modified by David A. Muñiz García with original data courtesy of Verenice Heredia; analysis by the authors).

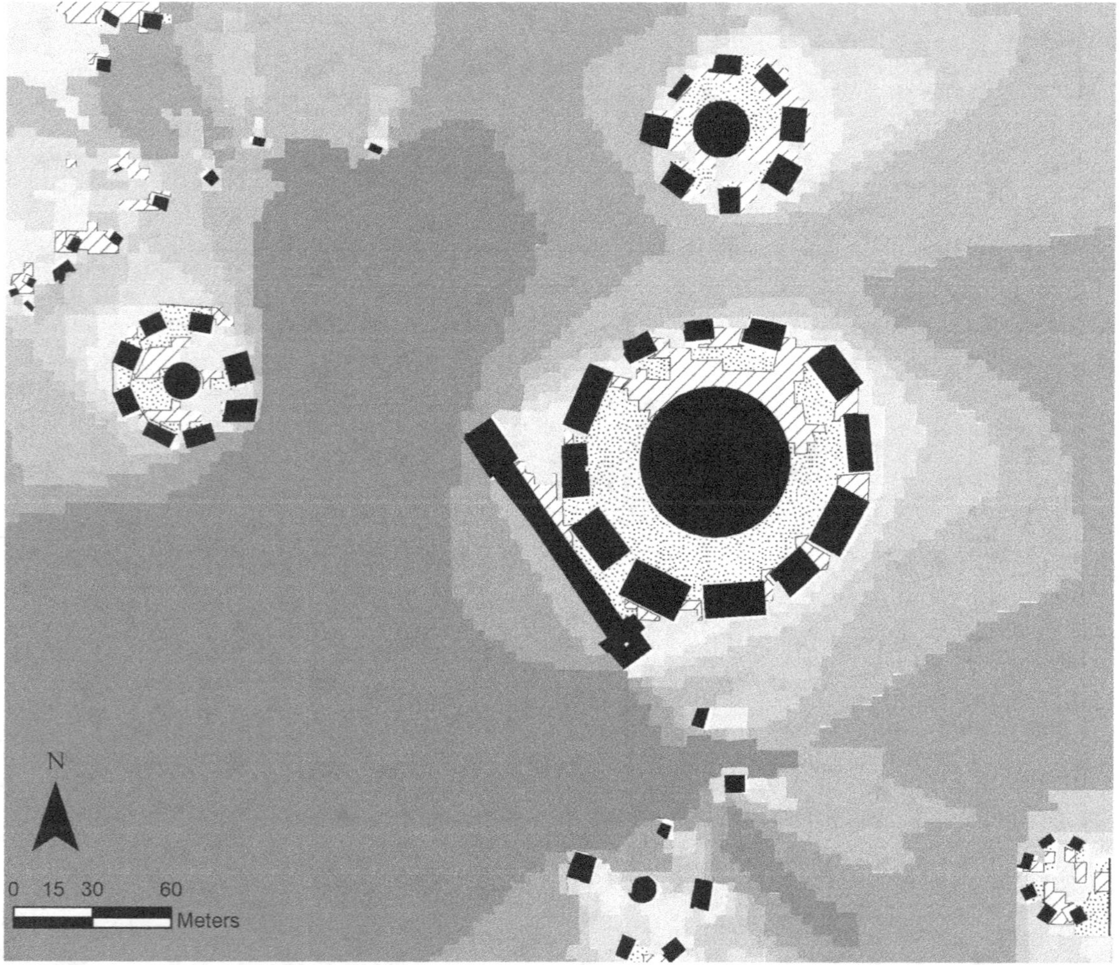

Figure 5.9. Depthmap visibility analysis within the Nuclear Sector considering only TII phase construction. The stippled pattern indicates higher connectivity values and paths of least resistance at the first moment, whereas the diagonal line pattern details circulatory options at the second moment. Darker shades of gray indicate increasing segregation and lower connectivity values (map modified by David A. Muñiz García with original data courtesy of Verenice Heredia; analysis by the authors).

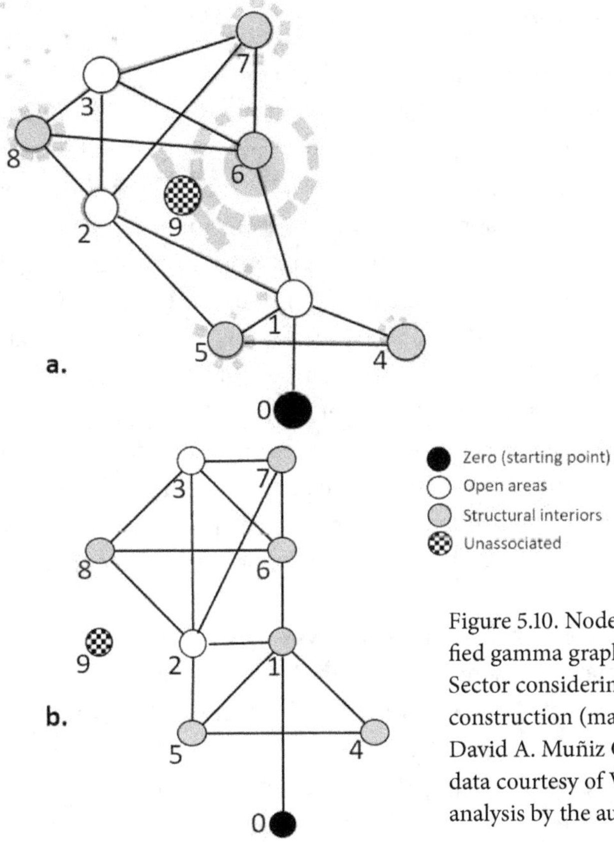

Figure 5.10. Node map and justified gamma graph of the Nuclear Sector considering only TII phase construction (map modified by David A. Muñiz García with original data courtesy of Verenice Heredia; analysis by the authors).

Table 5.3. Integration value calculations for axial nodes in the Nuclear Sector considering only TII-phase construction

Node	TDn	MDn	RA	i	CV
0	28	3.11	0.52	1.89	0.20
1	21	2.33	0.33	3.00	2.28
2	21	2.33	0.33	3.00	1.45
3	24	2.66	0.41	2.40	1.11
4	27	3.00	0.50	2.00	0.53
5	23	2.55	0.38	2.57	0.90
6	22	2.44	0.36	2.76	1.11
7	25	2.77	0.44	2.25	0.70
8	25	2.77	0.44	2.25	0.70
9	90	10.00	2.25	0.44	0.00
Min.	21.00	2.33	0.33	0.44	0.00
Mean	30.60	3.40	0.60	2.25	0.90
Max.	90.00	10.00	2.25	3.00	2.28

Notes: TDn Total depth of a node; MDn Mean depth of a node; RA Relative asymmetry; i Integration value; CV Control Values. Control Values are calculated by assigning each node a total value of 1, thus assuming that the node is equally connected to all other nodes; the CV thus acts as a baseline against which asymmetry and integration values may be evaluated for significance (see Manum et al. 2009).

Gamma analysis tends to support the interpretations gleaned from the analysis of axiality. As in the following Tequila III phase, the node at the first order of depth (node 1 in Figure 5.10) is an open space, in this case corresponding to the area between Circles 7 and 10. From this point, there are immediate connections to spaces and architectonic units at greater degrees of depth, such as Circle 1 (node 6 in Figure 5.10) and, interestingly, the open space to the southwest of Ballcourt 1 (node 2 in Figure 5.10) that we suggest may have functioned as a plaza during the Tequila II occupational phase. Both nodes correspond to units that permit the greatest number of options for circulation and transit to other areas and structures, presenting the greatest degree of connectivity with spaces at greater depths within the spatial order. The gamma graph for the nuclear core during Tequila II thus reveals a high degree of connectivity among an open built environment with a relatively lower degree of access control and depth, complementing the axial analysis.

Our interpretations of the graphic analyses are supported by the quantitative measurements presented in Table 5.3. As in Tequila III, asymmetry values remain relatively low, with relatively high integration values. These values again indicate a more open and accessible spatial order. Relatively lower depth measures also suggest a more connected built environment with comparatively fewer inaccessible or unconnected areas or units, although, as in Tequila III, Ballcourt 1 is an unconnected outlier (see discussion above). Although asymmetry, integration, and TDn values during Tequila II appear higher, lower, and lower, respectively, compared to similar values in Tequila III (see Table 5.1), this is again likely due to a comparatively smaller analytic sample with fewer structures within the built environment. Despite these differences, it is noteworthy that asymmetry and integration values during the Tequila II occupation are still lower and higher, respectively, than analogous measurements in the Loma Alta sector during Tequila III (compare Table 5.2).

The results of the secondary analyses suggest two significant conclusions. First, it would appear that the spatial order of the nuclear core at Los Guachimontones remained relatively constant throughout its occupational history.[13] There appears to have been little diachronic variability in this configuration; our analyses suggest that the sector was always a relatively open and accessible space in which almost all nodes and structures were interconnected. Second, the results of our secondary analyses suggest that temporal variation would not substantially affect the primary analyses, or, therefore, the interpretations derived via these analyses. Of course, our analyses are preliminary, and as we attain a greater understanding of the occupational and construction sequence at Los Guachimontones, it is likely that further analyses of spatial syntax will allow for a fuller appreciation of potential variability in the architectonic configuration of the site, as well as the ramifications of such variation on the interpretation of sociocultural organization derived through the analysis of architectural syntax and discourse.

Architectural Syntax and Social Structure at Los Guachimontones

The results of our analyses are consistent with previous suggestions (for example, Weigand 2008a, 2008b; Weigand and Esparza 2008) that the Loma Alta sector was an elite zone. Syntactic relations within and between architectural spaces in Loma Alta support this interpretation. The built environment of this sector presents spaces to which access is more tightly controlled, in which circulation is relatively more restricted, and which exhibit a relatively higher degree of syntactic variability (that is, less interrelation between spatial units) than in the nuclear core—especially as regards the quadripartite residential units. Throughout the sector, spatial relations—and the social relationships these reflected—are most pronounced in the areas of deepest depth within the spatial order, to which access was most restricted, considering both cultural and natural topographic barriers. In other words, the most connected and permeable spaces appear to be those to which access was most tightly controlled. Of course, a more complete evaluation of other archaeological evidence from the sector is necessary to further test this hypothesis, as is a fuller consideration of diachronic changes in architectural construction sequence and, subsequently, spatial relationships (see, for example, Englehardt and Nagle 2011; Palyvou 1987).

However, the nuclear core of Los Guachimontones traditionally has been considered a more public configuration. Weigand (1993, 1996, 2008a, 2008b) proposed that the architectural types, the sheer number of *guachimontón* complexes, and the monumentality of constructed spaces in the site center are indicative of a space used primarily for public, possibly ritual, functions. The relatively greater degree of connectivity between and access to architectural spaces in this sector supports this interpretation. Our gamma analysis suggests that the spatial units within the nuclear core are significantly integrated, particularly through the access of an open distributional axis,[14] a pattern common at other major Mesoamerican urban centers, and in general (Ardren 2015; Englehardt and Nagle 2011; Magnoni et al. 2014). In addition, such a spatial configuration is conducive to public functions and more open, inclusive social relations, insofar as social groups could freely access and interact within integrated architectural units (compare Hillier and Hanson 1984; Maran 2006).

The apparent segregation of the elite Loma Alta sector (Weigand and Esparza 2008), the quadripartite residential units in both sectors, and the La Joyita residential complex to the northwest of the nuclear core (Figure 5.1) is logical.[15] If we consider the flow of circulation of a large number of inhabitants, it is rational to assume that the areas to which access would be most tightly controlled, and that exhibit less integration with other elements of the built environment, are precisely those private spaces in which people lived. Rather, the more public spaces that exhibit higher interconnectivity and less circulatory restrictions—such as those in the nuclear core—are those that are more conducive to fomenting intergroup social relations and shared experiences.

The patterns of architectural discourse that emerge from our analyses provide a window onto the sociocultural structure of the groups that inhabited the site. Variability in potential patterns of circulation and the control of access to architectonic complexes between discrete sectors—when considered in conjunction with evident homogeneity in architectural forms and a degree of similarity in the spatial order between the nuclear core and Loma Alta—suggest several intriguing possibilities. On the one hand, there is significant variability in the spatial configuration within Loma Alta, as well as between this sector and the nuclear core. Further, access to spaces within Loma Alta—and to the sector itself—appears tightly controlled. These findings would suggest that the groups that inhabited Los Guachimontones were more network-oriented. On the other hand, the absence of significant syntactic variability in the architectural groups of the nuclear core—the largest and most central area of the site as a whole—coupled with greater permeability and a more open spatial layout in this sector, could suggest a more corporate orientation for the society.

In resolving these conflicting interpretations, we recur to the high degree of homogeneity in both architectonic complexes and spatial configuration within and across both sectors. Departing from the supposition that Loma Alta was indeed an elite occupational zone (compare Weigand and Esparza 2008)—an assumption supported by the strict access control to spatial units within the sector—we suggest that on the whole the evidence tends toward a more corporate orientation of the sociocultural structure at Los Guachimontones. As Ardren has convincingly argued in her analysis of architectural patterns in the northern Maya lowlands (2015: 28; compare

Magnoni et al. 2014), the expression of similar organizational or spatial principles in a given site's "elite" monumental architecture[16] signals "an essential conceptual unity between the ruling population and the ruled." Ardren continues: "This is not to say there were not severe status differences in the population, but to highlight that these differences were deliberately muted in the choices evident in elite architecture" (Ardren 2015: 28). A similar pattern may be evident at Los Guachimontones, where homogeneity in architectonic forms and configurations is evident throughout the site, and similar patterns emerge—admittedly, with some significant variability—through analysis of spatial syntax and architectural discourse. The "muting" (or segregation) of social differences in architectural reflections ultimately would suggest flexible intergroup structural relations conducive to more open relationships between distinct social groups.[17]

This suggestion is, of course, tentative, and further research is necessary to explore the possibility, particularly in terms of the overall construction sequence at Los Guachimontones. Our primary analyses present only a synchronic snapshot; a broader diachronic consideration is necessary to more fully appreciate the implications of the conclusions suggested by this preliminary reading of spatial syntax. Further, as mentioned above, analysis of archaeological materials recovered from distinct contexts at the site may clarify our interpretations. Finally, ongoing investigations of burial patterns and practices at the site may shed further light on questions of sociostructural organization (Ian Pawn and Christopher Beekman, personal communications, 2015). As stated at the outset of this chapter, we do not pretend with these analyses to definitively characterize sociocultural structure at Los Guachimontones. Rather, our analyses complement other lines of data (such as ceramics; see Beekman, this volume) in suggesting aspects of social differentiation. Of greater significance, our results demonstrate that there were consistent, deliberate differences between *guachimontones* and other elements of the built environment throughout the site. Accordingly, the analysis of spatial syntax and architectural discourse at Los Guachimontones serves both to establish essential data points and as a point of departure for future investigations that seek to approach these fundamental questions.

Concluding Thoughts

Every human society organizes itself in a specific and particular sociocultural configuration according to its needs and possibilities. This structural arrangement is subsequently negotiated, reified, and reflected in the material culture produced and consumed by the groups that comprise that society, including its architectural spaces and the design of its built environment. In this chapter, we have argued that a reading of architectural discourse permits a fuller understanding of the sociostructural order of the groups that constructed, utilized, and experienced those architectonic complexes. Through various analyses, we have offered a reading of the spatial syntax within and between architectural groups at Los Guachimontones. We interpret syntactical patterns evident in constructed spaces within and between discrete sectors as potentially indicative of multifaceted structural and social relations among the groups that inhabited the site.

Our analyses of spatial syntax in the nuclear core and Loma Alta sector reveal similar internal patterns within the sectors, with some variability between them. Previous readings of spatial and syntactical relationships in the Los Guachimontones architecture (for example, Hollon 2015) suggest a spatial order distinct from that which we have interpreted. Nonetheless, it is our contention that an interpretation of spatial syntax that more fully considers the natural environment—and topography—of the site leads to the creation of hypotheses that are more testable against available archaeological evidence. In our reading, a different logic in the ordering of spatial relationships emerges, revealing both repetitive architectural patterns as well as differences between the sectors. We suggest that the nuclear core presents a more open spatial syntax that was more conducive to flexible circulation than the spatial order evident in Loma Alta. Although the architecture in both sectors is formally similar, methodologies for the analysis of spatial syntax open new possibilities for accessing the deeper sociocultural structures that at once constrained and enabled the underlying meaning of the built environment for those that both constructed and experienced it.

In this sense, we offer the hypothesis that the sociopolitical structure of Los Guachimontones society was oriented more toward the corporate end of the dual-processual continuum, at least at its apogee in the Late Formative and Early Classic periods. The suggestion is, of course, preliminary, and ongoing analyses of material evidence from various contexts at the site will assist in evaluating this proposition. For now, however, we simply wish to highlight how spatial syntax and discourse analyses of the architectural program at Los Guachimontones offer new options for interpreting the sociostructural logic that underpins—and is reflected in—architectural spaces at the site. In the end, the data suggest multiple interpretations, and support no definitive conclusions as yet. Regardless, we would suggest that, employed systematically and in multiple contexts, analyses like the ones from which this chapter emerges, which seek to discern aspects of social organization as reflected in the spatial order of the built environment, offer great potential for furthering archaeological research and contributing to a better theoretical understanding of the relationships between material culture and sociocultural structure in past societies.

Notes

1. "Corporate" and "network" are not static typological categories; rather, it is more productive to conceive of these as points along a continuum that serve as a heuristic device for discussing temporal variability in diverse sociostructural configurations—and the material reifications of these arrangements.

2. http://whc.unesco.org/en/list/1209, accessed March 8, 2016.

3. As Hillier et al. (1987: 366) note, however, such a quantitative picture of a spatial configuration is not a simple qualitative diagram, insofar as it betrays the cultural bias through which ideas are built into things, reflected in "the configurational principles by which a spatial pattern is constituted by its makers into a cultural intelligible."

4. Palyvou (1987: 195) continues, noting that these "residents" and "visitors," respectively, "are probably divided into subgroups according to their position in the socioeconomic structure of the community."

5. Since there is no obvious starting point or "entrance" to the *guachimontón* complexes, axial lines were plotted by extending lines from the center of each complex to the point at which the line is interrupted by a natural or cultural barrier that prevents its continuity (for example, an architectural feature or a steep slope). Although the selection of circle centers as a starting point may seem arbitrary, in fact, having explored and analyzed a variety of options with the software, we determined that a zero point for axial lines in circle centers provides the broadest range of possible connections between architectonic complexes and, consequently, an array of interpretive possibilities. Statistical analyses of potential connections provide a further degree of methodological rigor (compare Hodder and Orton 1976: 6, 12–13).

6. For a detailed explanation of the calculation of these quantitative measures, see Hillier et al. 1987: 364–366 (see also Hillier and Hanson 1984: 92–123, 147–55).

7. This conclusion is supported when considering other zones within the site; secondary analysis revealed low indices of axial line convergence in the northeast and northwest areas of the site, and between those areas and both the nuclear core and Loma Alta.

8. A significant exception is Circle 10 in the nuclear core, which does appear relatively isolated, in terms of both axiality and visibility (Figures 5.3 and 5.5; even more so when considered in conjunction with topographic data). Beekman's analyses (this volume) reveal that Tabachines ware has its greatest concentration in Circle 10, whereas the utilitarian Coarse Colorines occur in its lowest percentage in this same circle. In this sense, our results tend to support his suggestion that this small circle was "built in a marginal location

perhaps more conducive to private ritual." Nonetheless, the node map and gamma graph of the nuclear core (Figure 5.7; see discussion below) do suggest potential connectivity for Circle 10.

9. This area may have served a similar function to that proposed for the Gran Plaza (see Weigand 1993, 1996); ongoing lithic and ceramic analysis may clarify this possibility.

10. This zero point is distinct from that proposed by Hollon (2015), which, perhaps coincidentally, corresponds to the current location of the entrance to the site. Disregarding the modern access road, considering the natural topography, and assuming a minimal degree of direct connection between the nuclear core and the workshops, terraces, and other structures to the south and southeast of the site center (see Figure 5.1), we suggest that this area is a more logical access point to the nuclear core (and beyond). We did run gamma analysis using Hollon's zero point, but the results obtained differed substantially from those suggested by the axial lines and visibility analysis. More important, asymmetry and integration values derived from the analysis using Hollon's zero point did not differ substantially from the control values (see Manum et al. 2009), suggesting that those results lacked statistical significance.

11. This finding is consistent with the visibility analysis, which revealed that the residential units were somewhat segregated in terms of first-moment choices for direct circulation or transit.

12. See Beekman, this volume, Figs. 2.4a–c. Many of these elements were later expanded in the first century AD (Tequila III [TIII]), concurrent with the construction of Circles 2–4 and the major construction episode in Loma Alta. Although the dates for building episodes in Circles 5 and 10 are not entirely clear, we follow Beekman (personal communication, 2018) in assigning their construction to TII. Finally, although C14 dates from Circle 4 indicate activity during TII (see Beekman, this volume, Fig 2.4b), its construction clearly dates to TIII. Since all construction in Loma Alta dates to TIII, there was no need to realize secondary analyses for units in that sector.

13. Although we did not perform a separate analysis for TIV, there were no major construction episodes during this phase. The built environment at Loma Alta presents no significant changes following TIII, and in the nuclear core the only differences between TIII and IV are an increase in activity in Circle 8 and the expansion of Circle 3 at the TIII–IV transition (see Beekman, this volume, Figs. 2.4a–d).

14. Although speculative, the fact that the distributional axis in the nuclear core is an open space (the Gran Plaza)—as opposed to a built structure to which access may have been more easily controlled (Circle A in Loma Alta)—is also potentially indicative of a more public function for the site center, and, conversely, a distinct function of the constructed spaces in Loma Alta.

15. The La Joyita complexes were not considered in our analyses, although extending the axial lines from the nuclear core does not suggest any significant relationships between the structures in this residential zone with those in the site center.

16. In the northern Maya area, this unity in spatial configuration presents itself via the ample presence of interior patios enclosed by architecture (Ardren 2015: 28), whereas at Los Guachimontones the analogous pattern would be the *guachimontón*-ballcourt complexes (and quadripartite residential complexes) present in both sectors.

17. This is doubly true when considering the potential symbolic meanings codified in the built environment at Los Guachimontones—particularly in the *guachimontón* complexes themselves (see Beekman 2003a, 2003b).

References

Al-Sayed, Kinda, and Alasdair Turner

2012 *Agent Analysis in Depthmap 10.14*. Bartlett School of Graduate Studies, UCL, London.

Ardren, Traci

2015 *Social Identities in the Classic Maya Northern Lowlands: Gender, Age, Memory, and Place*. University of Texas Press, Austin.

Bachelard, Gaston

1958 *La Poétique de l'Espace*. Presses Universitaires de France, Paris.

Beekman, Christopher S.

2000 The Correspondence of Regional Patterns and Local Strategies in Formative to Classic Period West Mexico. *Journal of Anthropological Archaeology* 19(4): 385–412.

2003a Agricultural Pole Rituals and Rulership in Late Formative Central Jalisco. *Ancient Mesoamerica* 14(2): 299–318.

2003b Fruitful Symmetry: Corn and Cosmology in the Public Architecture of Late Formative and Early Classic Jalisco. *Mesoamerican Voices* 1: 5–22.

2008 Corporate Power Strategies in the Late Formative to Early Classic Tequila Valleys of the Central Jalisco. *Latin American Antiquity* 19(4): 414–434.

Beekman, Christopher S., and Phil C. Weigand

2008 Conclusiones, Cronología, y un Intento a Síntesis. In *Tradición Teuchitlán*, edited by Phil C. Weigand, Christopher Beekman, and Rodrigo Esparza, pp. 303–337. El Colegio de Michoacán, Zamora, México.

Bermejo, Jesús

2009 Leyendo los Espacios: Una Aproximación Crítica a la Sintaxis Espacial Como Herramienta de Análisis Arqueológico. *Arqueología de la Arquitectura* 6: 47–62.

2015 Aplicaciones de Sintaxis Espacial en Arqueología: Una Revisión de Algunas Tendencias Actuales. *Arqueología de la Arquitectura* 12: e030–e031.

Blanco Morales, Ericka S.

2009 El Juego de Pelota en la Tradición Teuchitlán: hacia una propuesta sobre su función social. Unpublished MA thesis, Centro de Estudios Arqueológicos, El Colegio de Michoacán, La Piedad.

Blanton, Richard E.

1998 Beyond Centralization. In *Archaic States*, edited by Gary M. Feinman and Joyce Marcus, pp. 135–172. School of American Research Press, Santa Fe.

Blanton, Richard E., Gary M. Feinman, Stephen A. Kowalewski, and Peter N. Peregrine

1996 A Dual-Processual Theory for the Evolution of Mesoamerican Civilization. *Current Anthropology* 37(1): 1–14.

Bourdieu, Pierre

1989 Social Space and Symbolic Power. *Sociological Theory* 7(1): 14–25.

Castillo Luján, Feren Alexard

2015 Las Residencias Moches: Un Primer Análisis de la Sintaxis Espacial en las Huacas del Sol y de la Luna, Perú. *Arqueología de la Arquitectura* 12: e036.

El-Bizri, Nader

2011 Being at Home among Things: Heidegger's Reflections on Dwelling. *Environment, Space, Place* 3: 47–71.

2015 On Dwelling: Heideggerian Allusions to Architectural Phenomenology. *Studia UBB Philosophia* 60: 5–30.

Englehardt, Joshua D., and Donna M. Nagle

2011 Variations on a Theme: Dual-Procesual Theory and the Foreign Impact on Mycenaean and Classic Maya Architecture. *American Journal of Archaeology* 115(3): 355–382.

Fernández, Gloria

2015 La Articulación del Espacio Doméstico en las Casas de Patio Central. Un estudio para el Noreste peninsular ibérico entre los siglos IV–II a. C. *Arqueología de la Arquitectura* 12: e033.

Grau, Ignasi

2015 Sintaxis Espacial en el *Oppidum* Ibérico: Reflexiones sobre los Modelos Espaciales y Sociales. *Arqueología de la Arquitectura* 12: e032.

Heredia Espinoza, Verenice Y., Iziar Martínez Rojo, and Juan José Cortés Guzmán

2014 Proyecto Arqueológico Teuchitlán: Mapeo en el Sitio de Los Guachimontones. Report submitted to the Instituto Nacional de Antropología e Historia, Mexico City.

Hillier, Bill, and Julienne Hanson

1984 *The Social Logic of Space*. Cambridge University Press, Cambridge, UK.

Hillier, Bill, Julienne Hanson, and H. Graham

1987 Ideas are in Things: An Application of the Space Syntax Method to Discovering House Genotypes. *Environment and Planning B: Planning and Design* 14: 363–385.

Hodder, Ian

1989 This is Not an Article about Material Culture as Text. *Journal of Anthropological Archaeology* 8(3): 250–269.

Hodder, Ian, and Clive Orton

1976 *Spatial Analysis in Archaeology*. New Studies in Archaeology 1. Cambridge University Press, Cambridge, UK.

Hollon, Kristie

2015 Analysis of Social Space: Analysing Lived Experiences in the Los Guachimontones Central Ceremonial Area. Unpublished MA thesis, Department of Anthropology, University of Colorado–Denver.

Magnoni, Aline, Traci Ardren, Scott Hutson, and Bruce Dahlin

2014 The Production of Space and Identity at Classic Period Chunchucmil, Yucatán, Mexico. In *Making Ancient Cities: Space and Place in Early Urban Societies*, edited by Andrew Creekmore and Kevin D. Fisher, pp. 145–180. Cambridge University Press, Cambridge, UK.

Manum, Bendik

2009 AGRAPH Complementary Software for Axial-Line Analysis. In *Proceedings of the*

7th International Space Syntax Symposium, edited by Daniel Koch, Lars Marcus, and Jesper Steen, Ref. 70. KTH, Stockholm.

Manum, Bendik, Espen Rusten, and Paul Benze

2009 AGRAPH, Software for Drawing and Calculating Space Syntax Graphs. In *Proceedings of the 7th International Space Syntax Symposium*, edited by Daniel Koch, Lars Marcus, and Jesper Steen. KTH, Stockholm. Electronic document, http://spacesyntax.tudelft.nl/media/Long%20papers%20I/agraph.pdf, accessed March 19, 2016.

Maran, Joseph

2006 Architecture, Power, and Social Practice: An Introduction. In *Constructing Power: Architecture, Ideology and Social Practice/Konstruktion der Macht: Architektur, Ideologie und soziales Handeln*, edited by Joseph Maran, Carsten Juwig, Hermann Schwengel, and Ulrich Thaler, pp. 9–14. Geschichte, Forschung, und Wissenschaft 19. LIT Verlag, Hamburg.

Marcus, Joyce

1998 The Peaks and Valleys of Ancient States: An Extension of the Dynamic Model. In *Archaic States*, edited by Gary M. Feinman and Joyce Marcus, pp. 59–94. School of American Research Press, Santa Fe, NM.

Martin, Simon

2001 Court and Realm: Architectural Signatures in the Classic Maya Southern Lowlands. In *Royal Courts of the Ancient Maya*. Vol. 1, *Theory, Comparison, and Synthesis*, edited by Takeshi Inomata and Stephen D. Houston, pp. 168–194. Westview Press, Boulder, CO.

Mathews, Jennifer P., and James F. Garber

2004 Models of Cosmic Order: Physical Expression of Sacred Space among the Ancient Maya. *Ancient Mesoamerica* 15(1): 49–54.

Mountjoy, Joseph B., and Phil C. Weigand

1975 The Prehispanic Settlement Zone at Teuchitlán, Jalisco. *Actas del XLI Congreso Internacional de Americanistas*, Vol. 1, pp. 353–363. Mexico City.

Norberg-Schulz, Christian

1980 *Genius Loci: Towards a Phenomenology of Architecture*. Rizzoli, New York.

Ohnersorgen, Michel, and Mark Varien

2008 Arquitectura Formal y Organización de Asentamientos en el Antiguo Occidente de México. In Tradición Teuchitlán, edited by Phil C. Weigand, Rodrigo Esparza, and Christopher S. Beekman, pp. 191–234. El Colegio de Michoacán, Zamora.

Palyvou, Clairy

1987 Circulatory Patterns in Minoan Architecture. In *The Function of the Minoan Palaces: Proceedings of the Fourth International Symposium at the Swedish Institute in Athens, 10–16 June, 1984*, edited by Robin Hägg and Nanno Marinatos, pp. 195–203. Skrifter Utgivna av Svenska Institutet i Athen 4°, XXXV. Svenska Institutet i Athen, Stockholm.

Rapoport, Amos

1988 Levels of Meaning in the Built Environment. In *Cross-Cultural Perspectives in Nonverbal Communication*, edited by Fernando Poyatos, pp. 317–336. C. J. Hogrefe, Toronto.

1990 *The Meaning of the Built Environment: A Nonverbal Communication Approach.* University of Arizona Press, Tucson.

Small, David B.

2009 The Dual-Processual Model in Ancient Greece: Applying a Post-Neoevolutionary Model to a Data-Rich Environment. *Journal of Anthropological Archaeology* 28(2): 205–221.

Smith, Sean M.

2008 Comparación entre Círculos Monumentales y no Monumentales. Aproximación a una Tipología. In *Tradición Teuchitlán*, edited by Phil C. Weigand, Christopher S. Beekman, and Rodrigo Esparza, pp. 123–142. El Colegio de Michoacán, Zamora.

2009 El Caso Teuchitlán–Loma Alta: Un Ejemplo de Patrón de Asentamiento Mesoamericano. Unpublished MA thesis, Centro de Estudios Arqueológicos, El Colegio de Michoacán, La Piedad.

Tilley, Christopher

1991 *Material Culture and Text: The Art of Ambiguity*. Routledge, London.

Weigand, Phil C.

1979 The Formative-Classic and Classic-Postclassic Transitions in the Teuchitlán-Etzatlán Zone of Jalisco. In *Los Procesos de Cambio en Mesoamérica y Áreas Circunvecinas, XV Mesa Redonda, Vol.* 1, pp. 413–423. Sociedad Mexicana de Antropología, Mexico City.

1993 *Evolución de una Civilización Prehispánica: Arqueología de Jalisco, Nayarit, y Zacatecas.* El Colegio de Michoacán, Zamora.

1996 La Evolución y Ocaso de un Núcleo de Civilización: La Tradición Teuchitlán y la Arqueología de Jalisco. In *Las Cuencas del Occidente de México: época Prehispánica*, edited by Eduardo Williams and Phil C. Weigand, pp. 185–246. El Colegio de Michoacán, Zamora.

2008a The Teuchitlán Tradition and the Excavations at the Guachimontones de Teuchitlán, Jalisco. In *Urbanism in Mesoamerica,* edited by Guadalupe Mastache Alva, Robert Cobean, Ángel García Cook, and Kenneth G. Hirth, pp. 557–592. Pennsylvania State University Press, State Park.

2008b La Tradición Teuchitlán del Occidente de México: Excavaciones en los Guachimontones de Teuchitlán, Jalisco. In *Tradición Teuchitlán,* edited by Phil C. Weigand, Christopher S. Beekman, and Rodrigo Esparza, pp. 29–62. El Colegio de Michoacán, Zamora.

Weigand, Phil C., and Christopher S. Beekman

1998 The Teuchitlán Tradition: Rise of a Statelike Society. In *Ancient West Mexico: Art and Archaeology of the Unknown Past,* edited by Richard F. Townsend, pp. 35–52. Art Institute of Chicago, Chicago.

Weigand, Phil C., and Rodrigo Esparza

2008 Informe de Excavaciones 2003–2006 en el Complejo Arqueológico Guachimontones "La Tradición Teuchitlán del Occidente de México." Manuscript on file, El Colegio de Michoacán, La Piedad.

6

Constructing the Pre-Hispanic Landscape in the Santiago Bayacora Basin, Durango

DAVID ARTURO MUÑIZ GARCÍA AND KIMBERLY SUMANO ORTEGA

This study presents a critical review of the settlement patterns recognized to date in relation to the occupation of pre-Hispanic groups in the central-west region of the modern state of Durango in western Mexico. It also proposes visualizing settlement patterns in the region through the perspective of landscape archaeology (Criado Boado 1993; Ingold 1993; Knapp and Ashmore 1999; Parcero 2002; Tilley 1994), in which distribution over a given landscape may be viewed as part of a society's power strategies (Foucault 1979, 2003). To that end, we analyze a series of settlements that pertain to the Chalchihuites culture that existed between AD 550 and 1250 in the Santiago Bayacora River Basin (Figure 6.1).

First, we consider the internal dynamics at the El Encinal site, where what was once considered a system of neighboring sites is now being interpreted as one large site with differentiated sectors and, possibly, specialized functions. Second, we examine the possibility that different sites in this region may have taken on distinct roles within a dynamic of interaction. In both cases, we argue that in societies that are characterized by scarcity of usable materials and manpower—such as the Chalchihuites culture—the spatial distribution of sites and associated sectors could potentially function as a means of displaying power (Criado Boado 1991; Parcero 2002).

Our study utilized GIS tools, the characteristic approach of landscape archaeology, and Foucault's conception of the exercise of power (1970, 1979, 2003). Together, their application led us to develop the argument that, for the Chalchihuiteños of Durango, power was distributed spatially and exercised through the appropriation of place.

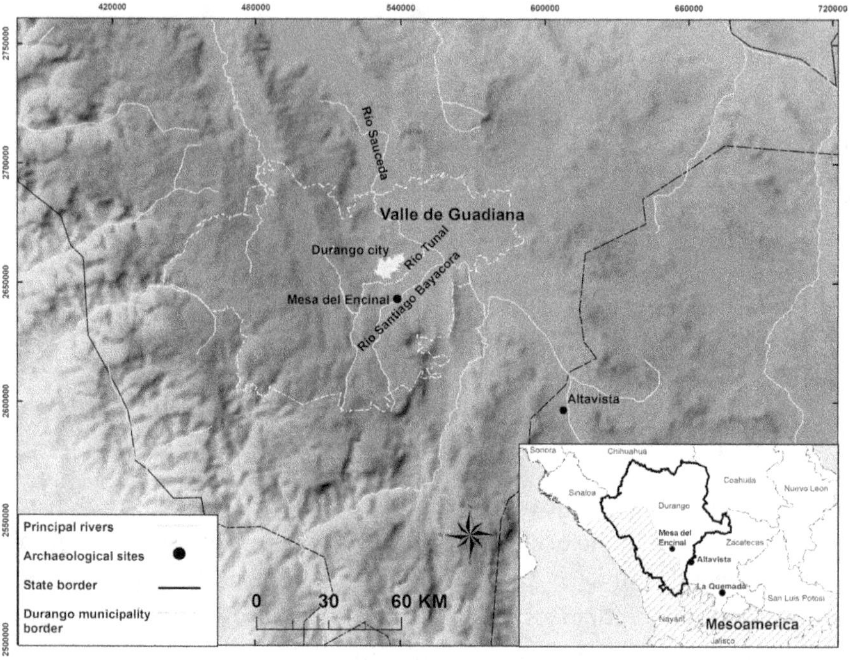

Figure 6.1. Map of the Santiago Bayacora Basin (SBR) study area within northern Mesoamerica (map by David A. Muñiz García).

Applying the Concept of Landscape in the Santiago Bayacora Basin

One of the central challenges of archaeology is to understand and explain why sites were established in certain places and not others and, concomitantly, why humans chose to live in one place instead of another. The choice of settlement location may be attributed to a combination of factors—including symbolic perceptions (Rapoport 1988, 1990; Tilley 1994: 5)—the assessment of which leads groups to make rational decisions. While the natural environment, technological possibilities, availability of raw materials, and the significance of place are all indispensable elements in the selection of settlement sites, practicality must also be considered in the decision-making process (Thomas 2001: 17).

The idea of landscape provides an important conceptual means of ordering the elements mentioned above because it underscores their interrelations (Tilley 1994). The interaction established between humans and nature is of primary importance, dialectical, and in constant change. The daily practice of the interrelation of communities and nature transforms physical spaces, allowing for their appropriation: "humans transform physical spaces into significant places" (Anschuetz et al. 2001: 159). For landscape archaeology, nature, like space, is not a constraining frame (Knapp and Ashmore 1999: 2), rather it is a system of signification through which social life is reproduced and transformed by exploring structurally organized processes (Knapp and Ashmore 1999 Rapoport 1969, 1988, 1990).

Landscape is thus a social construction. One of its characteristics is a specific historical context in which not one, but multiple, landscapes exist, landscapes that overlap and mutually transform one another (Anschuetz et al. 2001; Knapp and Ashmore 1999; Llobera et al. 2011; Thomas 2001; Tilley 1994). The fundamental premise of this approach is the idea that human groups construct their reality and that this construction is reflected in their material culture, through a system of knowledge-power given by their historicity (Foucault 1970: 12).

To clarify the basic postulates of landscape archaeology and how it differs from other key paradigms in the discipline, we must first understand that landscape *is not* a synonym for natural environment. Rather, we envision landscapes as synthetic, cultural products generated through people's beliefs, values, and everyday activities. It is through cultural practice that communities transform physical spaces into significant places, which become the stage for a given community's activities. However, landscapes are not only constructions made by human populations—they also constitute the very milieu in which humans survive and sustain themselves (Anschuetz et al. 2001).

Thus, we adopt the definition of landscape as "the material product and cultural creation that results from the everyday social action of human communities governed by a certain system of knowledge-power [that] is as dynamic and enduring as [that system]" (Parcero 2002: 16, translation ours). In our research, we seek to understand how the everyday actions of the ancient peoples of the Santiago Bayacora Basin allowed them to appropriate the natural environment and transform it into the significant place of a settlement. We assume that this relation left behind material traces that can be identified by certain constant characteristics, including: the location of settlement in horizontal and vertical stratigraphic relations to topoforms, the dimensions of structures, the spatial distribution of architectural elements, the intervisibility of those elements inside the settlement, and, finally, ceramic materials.

Archaeology in the Santiago Bayacora River Basin

Compared to more well-known areas of Mesoamerica (for example, the Maya area, Oaxaca), relatively little archaeological work has been done in the vast reaches of northern Mexico (Braniff 2009; Hers 2004; Reyes 2004). This fact in part accounts for the paucity of archaeological understanding of such topics as settlement patterns in regions like the Guadiana Valley in central-west Durango, especially its Chalchihuiteño occupation from AD 550 to 1250. Since being defined by Charles Kelley (Kelley and Abbott 1971), the Guadiana branch of Chalchihuiteño culture has been conceptualized as a development that maintained close links to southern Mesoamerica. However, analyses of the elements cited to support such affirmations—including ceramic iconography and the nature of monumental architecture—reinforce arguments that devalue the agency of these peoples and perpetuate a vision of these societies as passive recipients of a nebulous outside influence. Sadly, this vision has guided studies of Chalchihuiteño society all too often. This paper sets out to challenge this perspective.

The spatial delimitation traced herein is based on the hydrographic basin of the Santiago Bayacora River (hereafter referred to as the "SBR Basin") in the central-western area of the Guadiana Valley, which includes the capital of the state of Durango (Figure 6.1). The temporal spectrum considered for Chalchihuiteño occupation proposed by Kelly (Kelley and Abbott 1971: 3–4) spanned the years AD 550–1350 (see reviews by Hers 2004: 530–540, and Punzo 2008a: 10). More recent analyses employing absolute dating methods based on thermoluminescence (Punzo 2008b, 2016) suggest an occupational sequence spanning AD 550–1250. We have chosen to use this latter chronology.

The data we analyze were gathered during several years of fieldwork and laboratory studies by the team of archaeologists in the Archaeology of Central-West Durango Research Project (PIACOD for its initials in Spanish), under the auspices of the Centro INAH Durango and directed by José Luis Punzo (see Punzo 2004, 2005, 2006, 2007, 2008a, 2009, 2010). The fieldwork methodologies employed varied slightly from season to season and place to place, but in general involved archaeological surveys based on random stratified sampling (Cerrato 2011) and collecting materials in relation to stratigraphy. Transects were performed following topoforms, after previous study with GIS tools to assess their potential for containing deposits. An additional strategy was to use features of the terrain as possible indicators of sites. Upon finding vestiges, the approach changed to a more intensive form of recognition and the recording of sites. This methodological approach led us to increase the number of sites recorded in the SBR Basin from just 1 in 2004, to 35 fully identified and registered archaeological sites in 2010. The research team also identified almost 100 more sites in the wider Guadiana Valley.

Of the sites identified by PIACOD, 54 percent show Chalchihuiteño occupation, 23 percent date to the following Tepehuan occupation (AD 1250/1350–1563), 4 percent are modern, and 19 percent have not yet been identified. Recognizing diagnostic sherds was an important aspect of surface surveys that allowed us to make an initial temporal classification of sites and discern possible functions, although later it became necessary to refine that chronology and achieve a better understanding of the construction systems at the settlements. Therefore, a second phase of PIACOD was undertaken, which included a series of excavations in distinct sites in the valley, including the Cerro del Gato and Plan de Ayala sites, both in the SBR Basin.

Vertical and Horizontal Stratigraphy of the Sites

The location of settlements on the landscape reflects past social realities because "architecture emerges when space begins to be formed and structured by physical elements: fundamentally, horizontal and vertical ones" (Mañana et al. 2002: 17, translation ours). Various seasons of fieldwork revealed distinct and differential occupations of space during the Chalchihuiteño period. To understand these distinctions, we decided to divide settlements according to vertical and horizontal stratigraphy with respect to the valley's topoforms. The maximum density values of monumental architecture coincided with higher topographic areas located toward the interior of the topoforms (entrances and midsections of the canyons); this finding contrasted with flat zones or those with rugged terrain (inside the valley, or at the bottom of canyons), which had lower settlement densities.

Vertical Stratigraphy

The first category of our analysis was site location in relation to vertical position, a characterization given by the topoform and elevation of settlements. The division proposed is: (1) mountaintop (high reaches of the foothills of the sierra); and, (2) hillsides, understood as the slopes that run down from the high points of the sierra toward the valley or to the bottom of the canyon, including subdivisions of low and high mesas, as well as "hat-type"

Table 6.1. SBR site names, the corresponding abbreviations used to refer to them throughout this chapter, possible cultural affiliations, and vertical and horizontal positions

Name	Abbreviation	Cultural Affiliation	Horizontal Position	Vertical Position
Cerro del Gato	CEG	Chalchiuiteña	Entrance	Mountaintops
Cerro del Chiquihuitillo	CCH	Chalchiuiteña	Entrance	Hat-type hills
Cerro de la Cal	CAL	Chalchiuiteña	Entrance	High mesas
Cordón del Huarache1 1	CDH1	Chalchiuiteña	Entrance	Hillside
Cordón del Huarache 2	CDH2	Chalchiuiteña	Entrance	Hillside
Cordón de la Presa	COP	Tepahuano	Bottom	Mountaintops
Mesa del Encinal 1	MEN1	Chalchiuiteña	Entrance	High mesas
Mesa del Encinal 2	MEN2	Chalchiuiteña	Entrance	Hillside
Maravillas de Arriba 2	MA2	Tepahuano	Bottom	Hillside
Maravillas de Arriba 3	MA3	Tepahuano	Bottom	Hillside
Maravillas de Arriba 4	MA4	Tepahuano	Bottom	Hillside
Maravillas de Arriba 5	MA5	Tepahuano	Bottom	Hillside
Maravillas de Arriba 6	MA6	Tepahuano	Bottom	Hillside
Maravillas de Abajo	MDA	Tepahuano	Channels	Hillside
Maravillas	MA	Chalchiuiteña	Channels	Hillside
Mesa del Alguacil 1	MEA1	Chalchiuiteña	Entrance	High mesas
Mesa del Alguacil 2	MEA2	Chalchiuiteña	Entrance	Hillside
Plan de Ayala	PAY	Chalchiuiteña	Entrance	Low mesas
Pilar de Zaragoza	PIL	Chalchiuiteña	Entrance	Low mesas
Pilar de Zaragoza 3	PIL3	Chalchiuiteña	Entrance	Hillside

Source: Data from PIACOD, compiled by authors.

hills (*tipo sombreretillo*). The results of the distribution of sites in the SBR Basin by vertical stratigraphy indicate that hillside sites accounted for 50 percent of the total; those on high mesas, 22 percent; on mountaintops, 7 percent; hat-type, 7 percent; and low mesas, 14 percent. The fact that half of the sites are of the hillside type, and that this coincides with the distribution of the Tepehuan sites, may reflect their productive quality or potential, as it was feasible to build contention terraces and generate spaces for agricultural activity in those areas. The high mesa and hat-type hill seem to be the most common types of Chalchihuiteño settlements, and they include various architectural complexes, especially in the early periods (Ayala–Las Joyas phase, AD 550–950). In contrast, the low mesas show evidence of Chalchihuiteño occupation predominantly in later periods (Tunal-Calera phases, AD 950–1250).

Since access to water and the potential for creating terraces on the high mesas and hat-type hills would have involved considerable difficulty, subsistence there would have been more challenging. Thus, we speculated that the explanation of why 29 percent of sites are found in those zones could involve factors of an ideological nature. We know that occupying high zones was common in early periods, especially during processes of consolidating sociopolitical positioning. This reasoning allowed us to perceive that the most uniform high mesa, and the one with the best visual position in the SBR Basin, is La Mesa del Encinal, which is precisely the location of two of the largest, most complex sites registered: MEN1 and CEG (see Table 6.1 for an explanation of site abbreviations).

Several low mesas dot the Santiago Bayacora Basin, but only two show signs of pre-Hispanic populations: Pilar de Zaragoza (PIL) and Plan de Ayala (PAY). The range of altitudes at which settlements were found is broad: from 1,890 (PIL) to 2,070 meters at Cordón de la Presa (COP). And this range spans varying environmental and edaphological conditions, since in an area almost 8 kilometers long, elevation varies by 180 meters, and at least three distinct ecological niches with their respective ecotonal zones can be identified. It is potentially significant that the sites dating to the Chalchihuiteño occupation have a slightly broader altitude spectrum than the non-Chalchihuiteño sites, which were located between 1,870 and 1,920 meters above sea level. Finally, it is noteworthy that 90 percent of the Chalchihuiteño sites are located toward the northern margin of the basin (Figure 6.2).

Horizontal Stratigraphy

The second category of analysis was horizontal stratigraphy, which refers to the distribution of settlements in relation to the wide basin that forms at the mouth of the valley. The interior of this geographic form contains distinct environmental conditions that could be key factors in selecting areas for settlement. This diversity led us to classify horizontal spaces as: bottom, channel, and entrance. Most of the sites were near the entrance (68 percent), almost all of them Chalchihuiteño. The few sites along the basin channels (14 percent) are mostly Tepehuan, while those in the canyon bottom (18 percent) are either Tepehuan or modern (Figure 6.3). Once the sites were separated temporally and located in terms of their horizontal and vertical position, we considered two other indicators: the dimensions of settlements, and their architecture. This involved quantitative analyses (relative size in m^2 and number of structures/complexes), which were later

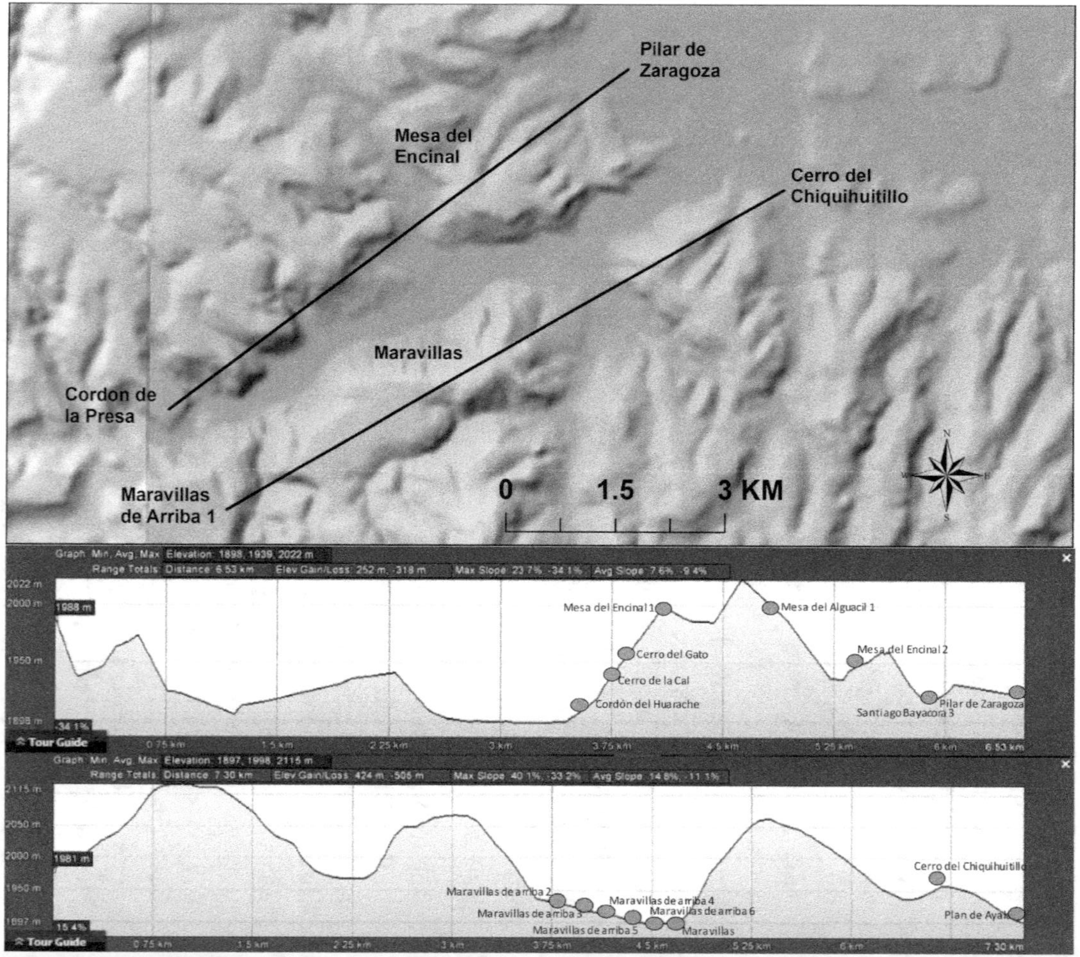

Figure 6.2. Vertical stratigraphy in the Santiago Bayacora Basin. Altitudinal profiles (*top*: north; *bottom*: south) are shown with the location of some of the archaeological sites mentioned in the text (satellite image from Google Earth; map by David A. Muñiz García with PIACOD information).

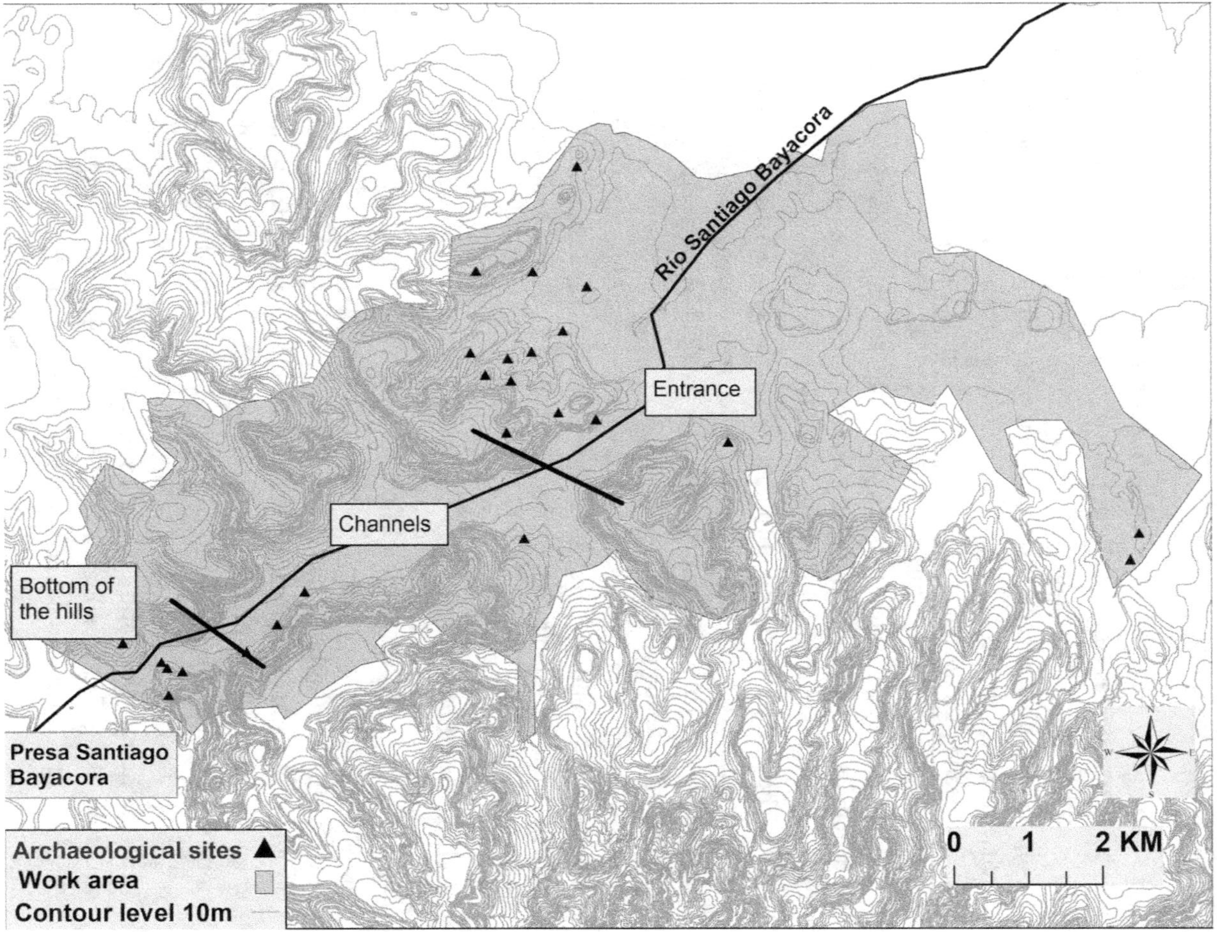

Figure 6.3. Horizontal stratigraphy in the Santiago Bayacora Basin (map by David A. Muñiz García with PIACOD information).

compared to the results of our qualitative studies (that is, hierarchy of sites, possible configuration of architectural complexes, and spatial distribution).

Other Indicators

Dimensions

Site dimensions are transcendental indicators, as long as they are associated with other variables (such as the density of materials and the variability of structures), since they speak to the relative importance of certain places compared to other sites in the Guadiana Valley. Traditionally, the largest archaeological deposits in terms of dimensions have been identified as the capital heads of a social system. Figure 6.4 summarizes our data on the dimensions of the Chalchihuiteño sites.

Of the five largest sites, two are very similar: Pilar de Zaragoza (PIL) and Plan de Ayala (PAY). Both sit upon low mesas at the entrance to the valley and have evidence of intense occupation in the late Tunal-Calera phase. The Cerro del Chiquihuitillo site (CECH) is on a hat-type hill, also near the entrance, but it was occupied earlier, in the Ayala–Las Joyas phase (AD 550–950). These three sites share one particularly important feature: toward the interior of their topoforms, there are habitational and civic-religious zones. The Mesa del Encinal 1 site is on a high mesa where we have only found elite civic-religious and habitational spaces, while Cordón del Huarache is a habitational site on a hillside. All other sites are also habitational.

Architectural Elements

Other aspects that are essential for understanding the sociopolitical dynamics of the SBR Basin are the types and distribution of constructions. The classification of buildings in the Guadiana Valley that our team proposed (Muñiz and Murguía 2010) revealed that the construction techniques employed there always utilized materials available in the surrounding area; none were brought in from outside. Various aspects of the architecture in the valley have been proposed for analysis, including the materials used, structure design, the nature of the complexes, and a hypothetical reconstruction using GIS (see Figure 6.5). We found that rooms are the most common structures. People may have performed domestic functions there, used them for storage or even for ritual purposes. There are single, double, triple, and quadruple rooms, but the latter are the least frequent, and were

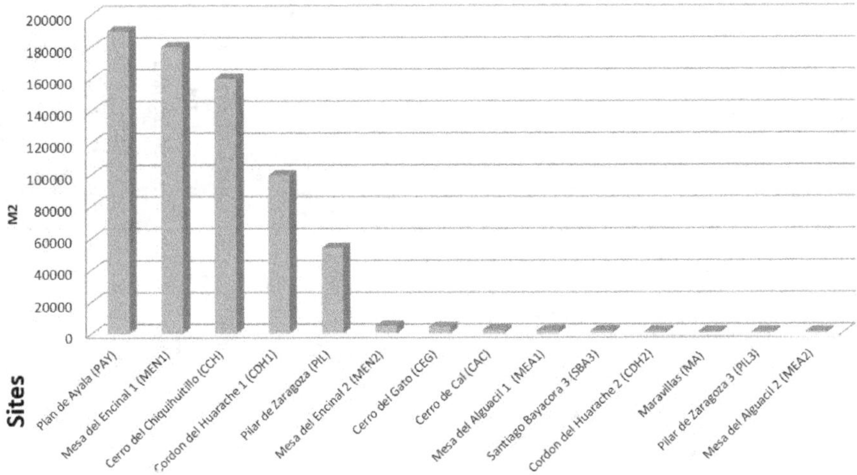

Figure 6.4. Dimension hierarchy in the Santiago Bayacora Basin, showing the total dimensions of sites in the SBR (map by David A. Muñiz García with PIACOD information).

found only at Mesa del Encinal 1 and La Ferrería (Punzo 2007, 2011). Triple rooms are uncommon but were uncovered at La Ferrería and Plan de Ayala. Double rooms are a distinctive feature of Chalchihuiteño occupation in Durango (Tsukada 2006: 48–49), especially in the Ayala–Las Joyas phase (Punzo 2004, 2008b), while singles are the minimal construction unit, and are present at all sites with Chalchihuiteño architecture.

The orientation of passages and hallways clearly reflects how movement occurred within a complex. We understand passages as spaces of greater length than width that would have distributed access and flow satisfactorily from one structure to another. Hallways may have similar dimensions, but they communicate distinct distribution spaces inside a specific structure. In the SBR Basin, we identified the presence of hallways at the following sites, at least: Cerro del Gato (CEG), Cordón del Huarache 1 (HUA1), Cerro del Chiquihuitillo (CCH), and Plan de Ayala (PAY).

The largest concentrations of architectural elements were found at MEN1, CCH, and PAY, but upon adding the variables of patios and plazas, we found that sites like MEN, CCH, and HUA exhibit a distinct pattern, perhaps reflecting the functions of each one. For example, PAY and CCH are more-or-less constant in their variables, so this may indicate a plurifunctional space, while other sites may have been devoted to distinct, more specific functions.

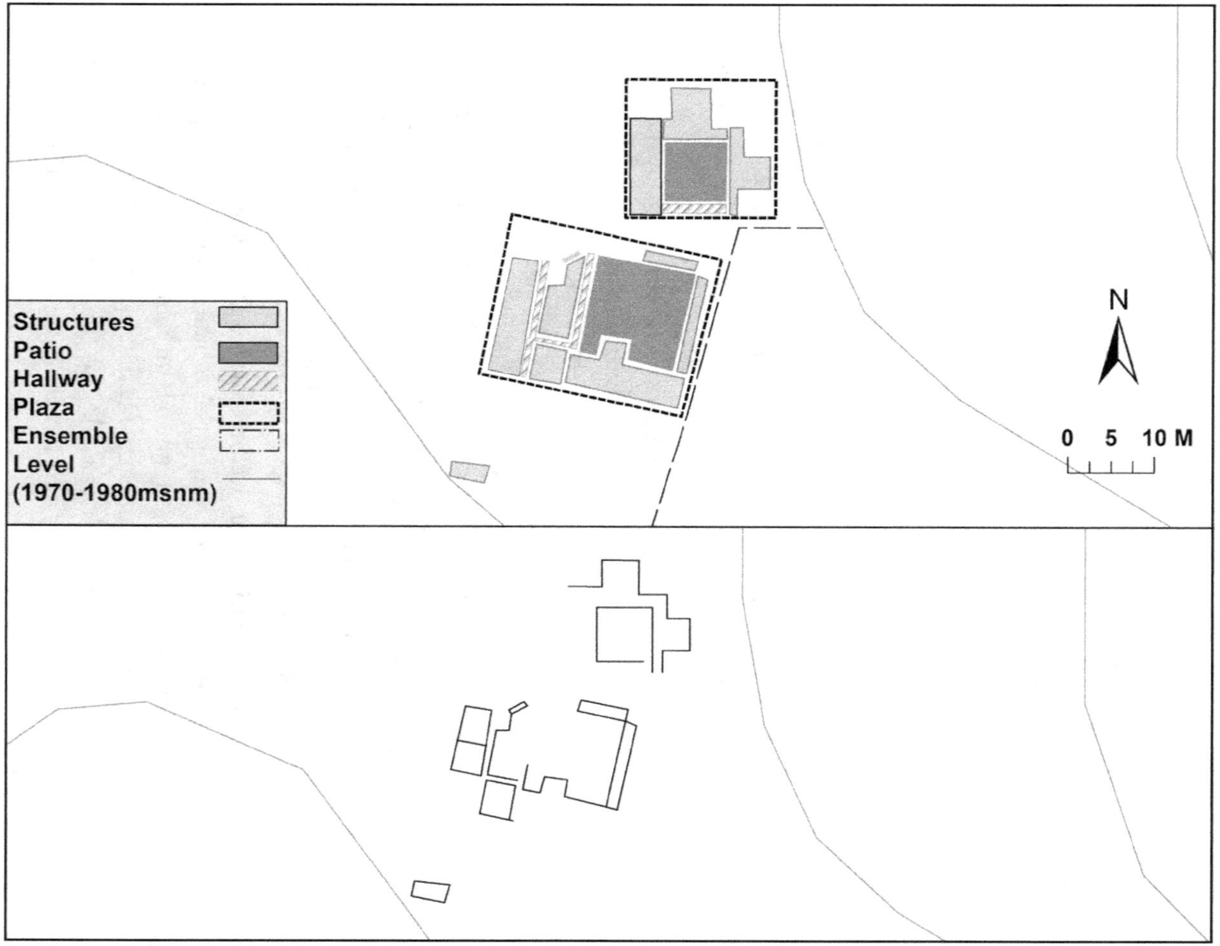

Figure 6.5. Comparative maps of the Cerro del Gato site: *bottom*, original map (Muñiz and Murguía 2010); *top*, redrawn to detail proposals for structure clusters (map by David A. Muñiz García with PIACOD information).

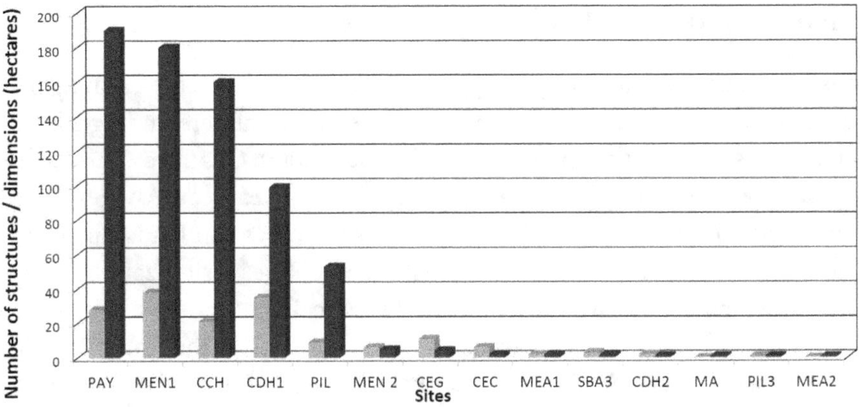

Figure 6.6. Comparison of sites in the Santiago Bayacora Basin by structure and dimension (map by David A. Muñiz García with PIACOD information).

If plazas are defined, in part, as spaces for public functions (Arancón 1992), then the absence of these features at sites like CAL, SBA3, and CDH2—given the proximity of such sites to MEN, which has more plazas than any other site in the SBR Basin—may indicate relations of dependence or predominance between sites. It is interesting to note that CCH and PAY share a constant, similar behavior in terms of their proportions of structures, patios, plazas, and complexes. On this basis, we classified those sites as "self-sufficient" or, at least, not dependent in material or ideological terms.

Finally, we compared the two categories of dimensions (in m^2) *versus* number of structures (as a quantitative parameter of architectural activity). Results showed differential behaviors. First, PAY is the most extensive site in the SBR Basin, but does not have the greatest number of structures, circumstances that present an interesting anomaly. The two sites with more structures are MEN1 and CDH1. While CCH and PIL show similar behavior in terms of proportions, they have low structure-density in proportion to their size. The other sites—the smallest ones in terms of spatial distribution—have behaviors that are proportionally similar to MEN1 and CDH1, marked by a balance between dimensions and the number of structures found (Figure 6.6).

Visibility and Visibilization of Sites

Visibility and visibilization are operative categories that serve as analytical tools for recording archaeological phenomena and understanding the social processes of the peoples studied (Bermejo 2009; Chatford Clark 2007; Criado Boado 2003; González 2001; Parcero and Fábrega Álvarez 2007; Smith 2007). They are constituent elements of analysis in landscape in relation to lived spaces. Criado Boado (1999: 9) defines visibilization "as the way in which an archaeological element is perceived, while visibility is the panorama [visible] from it" (translation ours). He adds that intervisibility is "the visual relation between [these] elements." This difference is fundamental, since the presence, degree, or absence of visibility and visibilization may impact the roles that actors are able to play in developing social practices. Ideally, "being seen" entails a disadvantageous position in relation to the "one who sees" (observer) (Criado Boado 1991; Foucault 1979; Parcero 2002). Thus, the act of making something visible allows for a window into the society that created it, especially, that society's conception of space.

To perform intervisibility tests, we used ArcMap 9.3 software on a digital elevation model with digital shadowing to construct the architectural maps modified in the Autodesk-Autocad platform following the proposal mentioned above (Figure 6.4). We then ran the viewshed application for each Chalchihuiteño site in the SBR Basin. After resolving the problems entailed in the visibility test (see González 2001), this procedure generated the resulting maps. For all cases, we interpreted the data shown on the maps (Figure 6.7).

Defining the parameter "how far I can see" is insufficient to understand the value of a given settlement's position in the landscape, for this spatial location has a complement, namely, determining "from where I can be seen." Furthermore, the interrelation among settlements involves two additional parameters: "who can see me," and "how much can they see." For this reason, this kind of analysis must be complemented by visibilization. With this in mind, we proceeded to perform the exercise in ArcMap again. Once we generated the entire set of maps, we conducted a reverse analysis to identify the points (sites) from which it was possible to see the object site analyzed. For each settlement, we determined the sites that could observe it. For example, to ascertain the visibilization of MEN1, we examined the viewshed maps of all other sites and made a list of the ones from which it was possible to see MEN1. To compile these data, we generated an

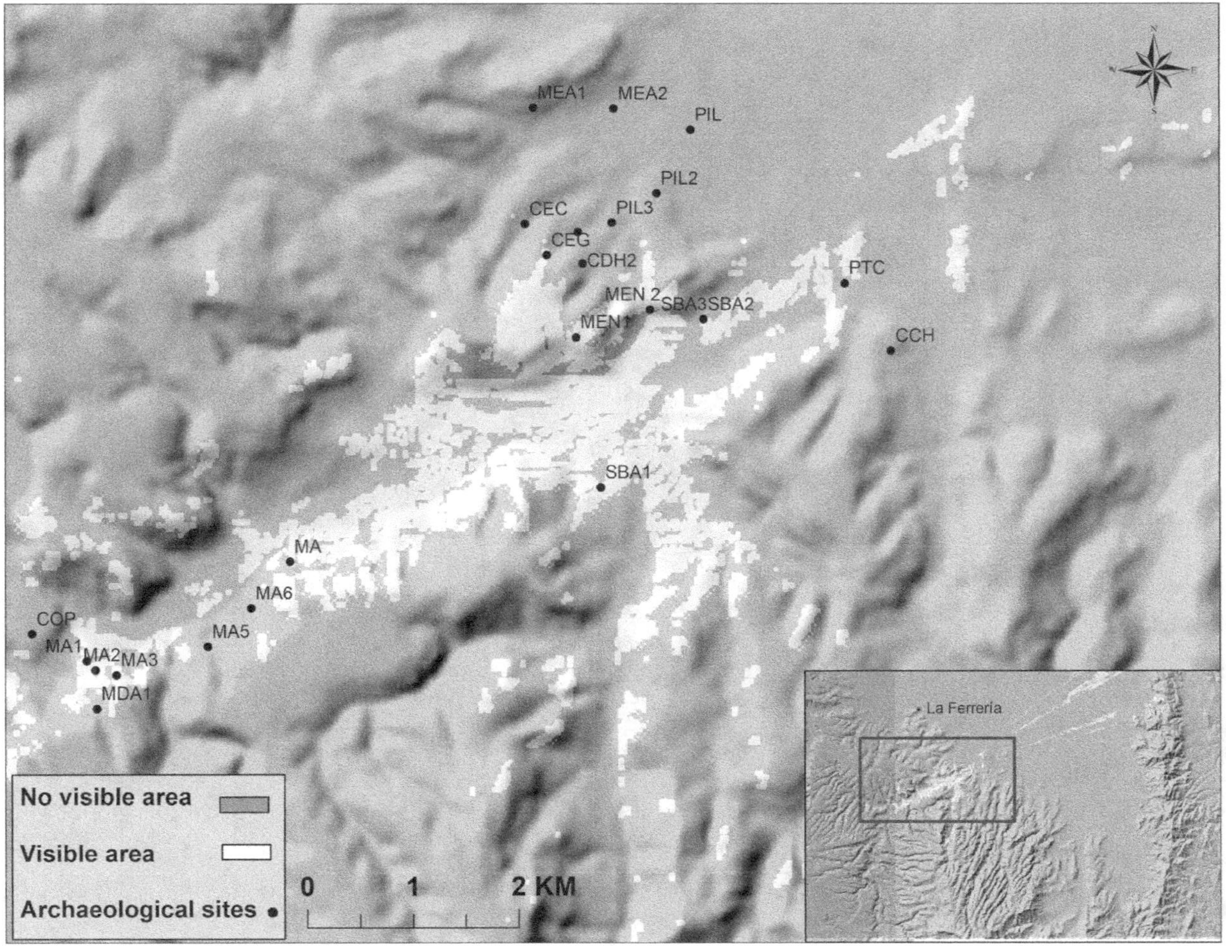

Figure 6.7. Map showing details of the visibility application for the SBR (map by David A. Muñiz García with PIACOD information).

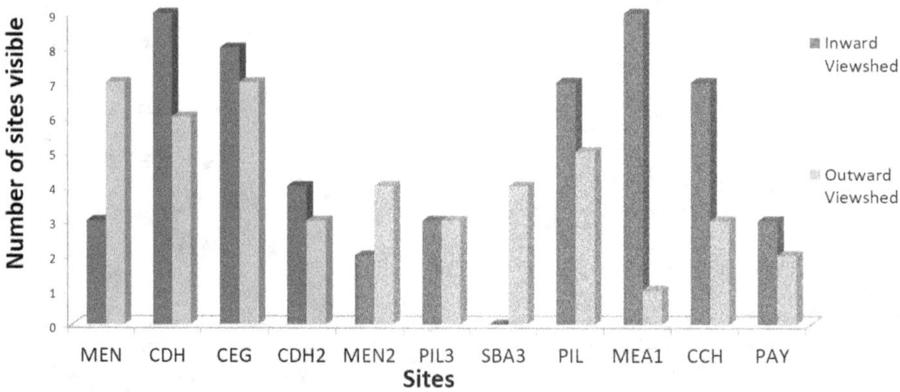

Figure 6.8. Levels of intersite visibility in the SBR Basin, showing the number of visible sites (0–9) and indicating the number of sites visible in the inward and outward viewsheds.(map by David A. Muñiz García with PIACOD information).

attributes table in Excel, and then graphed the results to construct more general interpretations (Figure 6.8).

These tests presented a revelation that led us far from our original hypothesis of a dispersed settlement pattern made up of multiple small sites. These results reflect, in general, a broad balance between visibility and visibilization at most sites, except MEA1 and CCH, which can see more sites than those from which it can be seen. This suggested a separation between two groups of sites: one whose constituent sites can be constantly intervisibilized among themselves, found primarily around Mesa del Encinal and its foothills; and a second that lies in the extreme northwest and southeast of the entrance to the canyon. The latter have a broad visual capacity, and CCH shows a certain independence of structures. Finally, two sites appear to be visually "isolated" from the rest. Based on these findings, we propose three large groups:

1. MEN1-MEN2-SBA3-CAL-CEG-CDH1-CDH2-PIL3-PIL2-PIL: concentrated around Mesa del Encinal with a broad intervisual relation.
2. CCH-MEA1: in the extreme northwest and southeast with a great capacity for vigilance (MEA1), as well as vigilance and independence (CCH).
3. PAY and SAL: distant from the SBR Basin with little visual contact between them, but visual contact with sites on the Calabazas River,

and some degree of independence in terms of constructions and location.

Analysis of Materials

The final aspect of our work consisted of analyzing diagnostic ceramic materials, since sherds provided a means of locating sites temporally. We infer the presence of decorated materials as an indicator of cultural affiliation. This analysis was based on data generated by PIACOD (especially the data reported in Punzo 2010). The number of sherds recovered left only four sites susceptible to this analysis: CEG, MEN1, CCH, and PAY. Although the number of sherds recovered at these sites varies greatly, results generally coincide with the grouping scheme proposed above.

Both PAY and CEG were excavated, but the number of decorated sherds found varied greatly (100 at PAY, but only 2 at CEG), perhaps reflecting differences in the functions performed and/or a temporal spectrum. The tendency in these materials suggests that CEG may have been a sector of a larger site that encompassed seven of the sites recorded, grouped around Mesa del Encinal and its foothills. This sector could correspond to a habitational unit for the elite that exercised surveillance over five other sites—or sectors of the same site—where the civic-religious, public, and storage functions reserved for the MEN1 site—or sector—were concentrated. Those are the kinds of activities that may leave more abundant material remains (decorated sherds) and, perhaps, produce quantities and varieties similar to those at PAY. However, the data obtained also open the possibility that the temporal spectrum of the Mesa del Encinal sites, including CEG, could fall in an early moment of the Ayala phase, reaching its apogee in that period, before continuing into the Las Joyas phase (AD 850–950); that is when activity declined and perhaps ceased between the Tunal and Calera phases (circa AD 950–1250).

Discussion

The results of this and other research show that PIACOD succeeded in generating an amount of concrete information broad enough to apply distinct research strategies in an attempt to resolve several questions (see, for example, Gómez 2013; Muñiz and Murguía 2010; Rangel 2014; Sandoval 2011; Vidal 2011). Our research was undertaken to understand why the Chalchihuiteño sites are distributed in a certain way. Since our initial hypothesis was that intervisibility would prove to be an essential factor in site distribution in the SBR Basin, we assumed that the rest of the material evidence would revolve around this prime indicator.

However, the results of our analyses focused more on describing the settlement pattern in the SBR Basin than the visual capacity of the sites located there. This unexpected result is not a problem in itself, but it does reinforce the idea that the archaeology of northern Mexico is distinct from that of other areas of Mesoamerica, and therefore requires distinct research strategies. New data derived from our investigations allow us to search for interpretative models that are applicable to the specific characteristics of this study region, thereby achieving a more nuanced understanding of the ancient inhabitants of Durango and their cultural processes. The interpretations presented below include proposals based on this new approach.

Settlement Pattern

The vertical stratigraphy indicates that the sites where recurring habitations occurred were on hillsides, a trait shared by the Chalchihuiteño, Tepehuan, and modern groups. This is understandable if we consider that processes of erosion and deposition meant that constructing contention terraces on the smooth slopes facilitated the accumulation of valuable sediments suitable for agricultural activities. The other topoforms described (high mesas, low mesas, mountaintops, and hat-type hills) were used almost exclusively by Chalchihuiteño groups.[1] This pattern may be a function of the general tendency of Chalchihuiteño groups to appropriate nature and transform it into an extension of their constructions (or vice versa). Structures in the different topoforms may reflect distinct architectural discourses (see Sumano and Englehardt, this volume)—for example, administrative, religious, public, or habitational functions, among others. Further, each distinct topoform may have been linked to each site's function and the people who lived there.

If vertical stratigraphy reveals the differential use of the terrain in relation to form and elevation, horizontal stratification helps us understand the sense of proximity and distance that inhabitants of the SBR Basin might have had. Generally speaking, that which is closest to us is most familiar, better-known, and more clearly identified than things that are farther away. This is a basic principle of regional filiation. In modern terms, we would say that we feel more closely identified with those in our own neighborhood than with residents of other, more distant parts of a town or city. Applying this perspective to the spatial distribution of sites in the Guadiana Valley, which was long thought to be a complex, patchwork distribution, we were able to classify the settlements in two ample groups, distributed in relation to the parameters of horizontal stratigraphy, as described above. The first group includes the Tepehuan and modern settlements, which are distributed throughout the basin and in the bottom of the canyon, with a clearly discernible preference for the latter space. The second group includes the Chalchihuiteño sites and a few others whose cultural affiliation is not yet clear. These are grouped around the canyon entrance.

The analysis of construction techniques provided the key for accessing the next level of analysis: landscape. Earthen architecture predominated in the valley, although some important structural elements were made of stone. Without doubt, the landscape was marked by earth-covered walls. Whether simple houses, or large complexes built around patios and plazas, structures were distinguished by earth-based coverings, probably in different colors.

Turning to visibility, we found that two sites—CCH and MEA1—would have visually dominated all others, but they are separated by a considerable distance and occupy topoforms of difficult access. Despite the difference in size and quality of the constructions at MEN1, this site does not appear to have exerted any significant dominion over the other sites. Rather, its control seems to have been more localized (CEG compared to CDH1, CDH2; MEN1 compared to MEN2 and SBA3; see Figure 6.7). Nonetheless, the accumulation of data led us to consider that these relations were not isolated but formed part of a complex strategy of visual domination shared by many sites that now appear to have been sectors of a single settlement (rather than independent sites, as was long believed to be the case). In this scenario, distributional relations would reflect the internal dynamics of one large site at Mesa del Encinal, with CCH and MEA participating as external agents, while PAY—and SAL—would seem to have remained outside of this dynamic among sites located at the canyon entrance. PAY would have been visible from afar and would have been able to see a few sites from its location, but it seems likely that it was wrapped up in its own particular dynamic.

After interconnecting these elements of analysis, the scheme that emerges suggests that MEN1, MEN2, SBA3, CEG, CDH1, CDH2, CEC, and PIL3 actually form a single site. In this view, all the sites immediately around Mesa del Encinal can be grouped into one large site, which we propose naming, simply, El Encinal. The differences among the functions of the constructions at the nearby sites—or, better stated, sectors or complexes—become understandable if they are considered parts of the same settlement. The data on dimensions and material densities approximate what we would expect to find at a rector—or hegemonic—site, and its characteristics make it comparable to the large, first-tier sites of La Ferrería and Navacoyan.[2]

A landscape is formed gradually through the organic interrelation of the constructed and natural environments (Rapoport 1969). In contrast to southern Mesoamerica, where geography provided the base and framework for constructions that appear to have imitated nature, and where private (visually restricted) space was an artificial construct, the limited manpower in the SBR Basin meant that geography was not simply a framework. Rather, the physical environment, in reality, formed an organic and integral part of the constructed environment, such that the choice of the location for an individual structure fostered the concealment or exhibition (display) of certain parts of the construction.

Organic integration occurs on distinct scales, at both the macrolevel of visible architecture, and the level of the not visible. Examples of such invisible integration are the cavities carved into bedrock for burials, where the relation and use of the contained space is dynamic and participative, as in the two interments at PAY and those at El Nayar (Punzo and Luna Ramírez 2007; Punzo and Murguía 2008). We also come to understand that "nonvisible" space plays an important role in a symbolic sense—in addition to the burials we have evidence from the entrance thresholds that are not necessarily exhibited, since they were interred in the systemic context (as at El Nayar). That which "is not seen" is just as important—or perhaps more so—than that which is readily apparent. And this takes us back to the macrolevel of architecture and landscape. The coming-and-going between strategies of concealment and exhibition, organically integrated into the geography, reveals the importance of structures and the spaces generated to demonstrate, or legitimize, the social position of part of the society in the "knowledge-power" dynamic (Criado Boado 1991; Foucault 1970, 1979) of the Chalchihuiteño groups in the SBR Basin.

We can see this in the SBR Basin if we recall that the group of sites comprised of MEN1-MEN2-SBA3-CEG-CDH1-CDH2-CEC-PIL3-PIL2—which we now consider to be part of the single site of El Encinal, rather than as a series of distinct sites—may respond to an internal logic of inhibition/exhibition that would signal the importance, function, or position of each group of inhabitants inside one sole site. In this interpretation, CDH1, CDH2, SBA3, PIL3, PIL2, CEC, and MEN2 would have been habitational units constantly "exhibited" but with a limited capacity to "see" other areas (compared to MEN1 and CEG). The primordial function of these spaces may have been agricultural production, because they are located in areas closer to the sources of water and on gentle slopes or, at least, ones that favored the deposition of sediments (for example, CDH1, CDH2, PIL2, and PIL 3 are all very close to modern zones where extensive cultivation takes place). CEG, in contrast, would have been a habitational unit with a clearly marked strategy of internal concealment/exhibition; although the habitational sites would have been able to see some of CEG's walls, they would not have been able to observe activities of everyday life there because its spatial positioning and organization produced concealment in the form of private, segregated spaces.

Yet, despite the walls and roofs on the units at, for example, CDH1, people at CEG would have been able to observe activities in the patios—the focal space in all domestic constructions in Mesoamerica (Arancón 1992; Manzanilla 1986)—as well as production areas, which were spaces open to constant exhibition. This "exposure" would have made it possible to control—and tax—the activities conducted in those lower spaces. The position of dominance that the inhabitants of CEG enjoyed would not have been very different from that of those in the Casa de los Dirigentes at the La Ferrería site (Figure 6.9), and suggests a possible strategy that might have been employed frequently in the exercise of power in the region. This pattern may contrast with other examples in northern Mexico, such as La Quemada or Altavista in Zacatecas. Nonetheless, in those sites natural topoforms such as a mountain slope were incorporated into the constructed landscape (Figure 6.10).

MEN1, however, has characteristics more indicative of exhibition than of visual dominion over other sites, but this case is similar to that of CEG, in that "what is seen" is highly focalized. The architectural features of MEN1 resemble those of the zone that encompasses the Pyramid, Ball Court, and Big House at La Ferrería. Therefore, MEN1 could have been the sector where the elite resided and civic-religious functions were performed.

If our assumption of El Encinal as a single site is correct, then its spatial distribution can only be classified as dispersed. It would be logical to think that the complexes at CDH1, CDH2, SBA3, PIL3, PIL2, and MEN2 were residential in nature, with secondary agricultural functions, and that the "empty" spaces between them were cultivated fields, perhaps with still undiscovered complexes, as suggested by the constant presence of materials. The condition of these "constantly exhibited" sites with modest architecture, in addition to the indicators of horizontal and vertical stratigraphy,

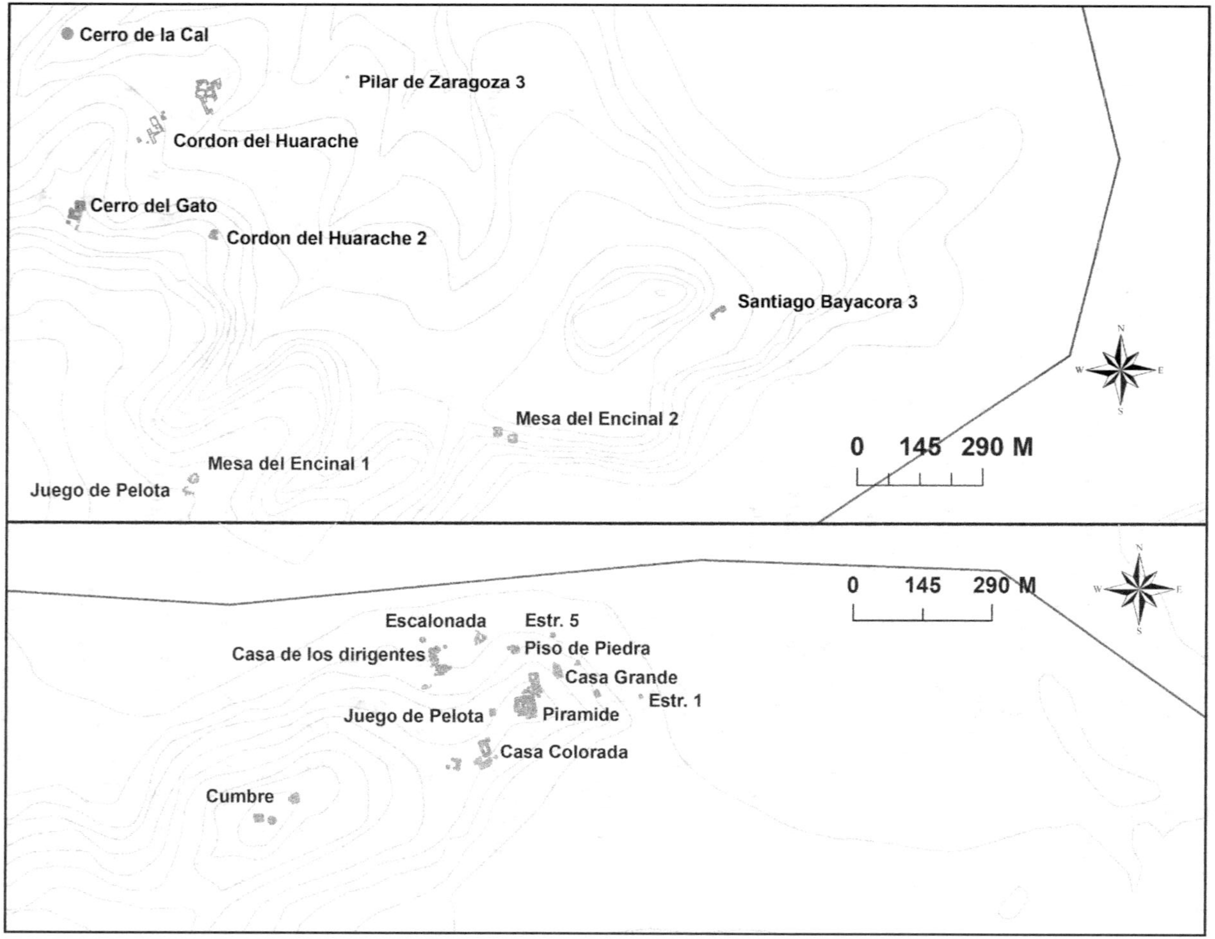

Figure 6.9. Comparative maps of El Encinal (*top*) and La Ferrería (*bottom*), detailing the distribution of internal architectonic elements (map by David A. Muñiz García with PIACOD information).

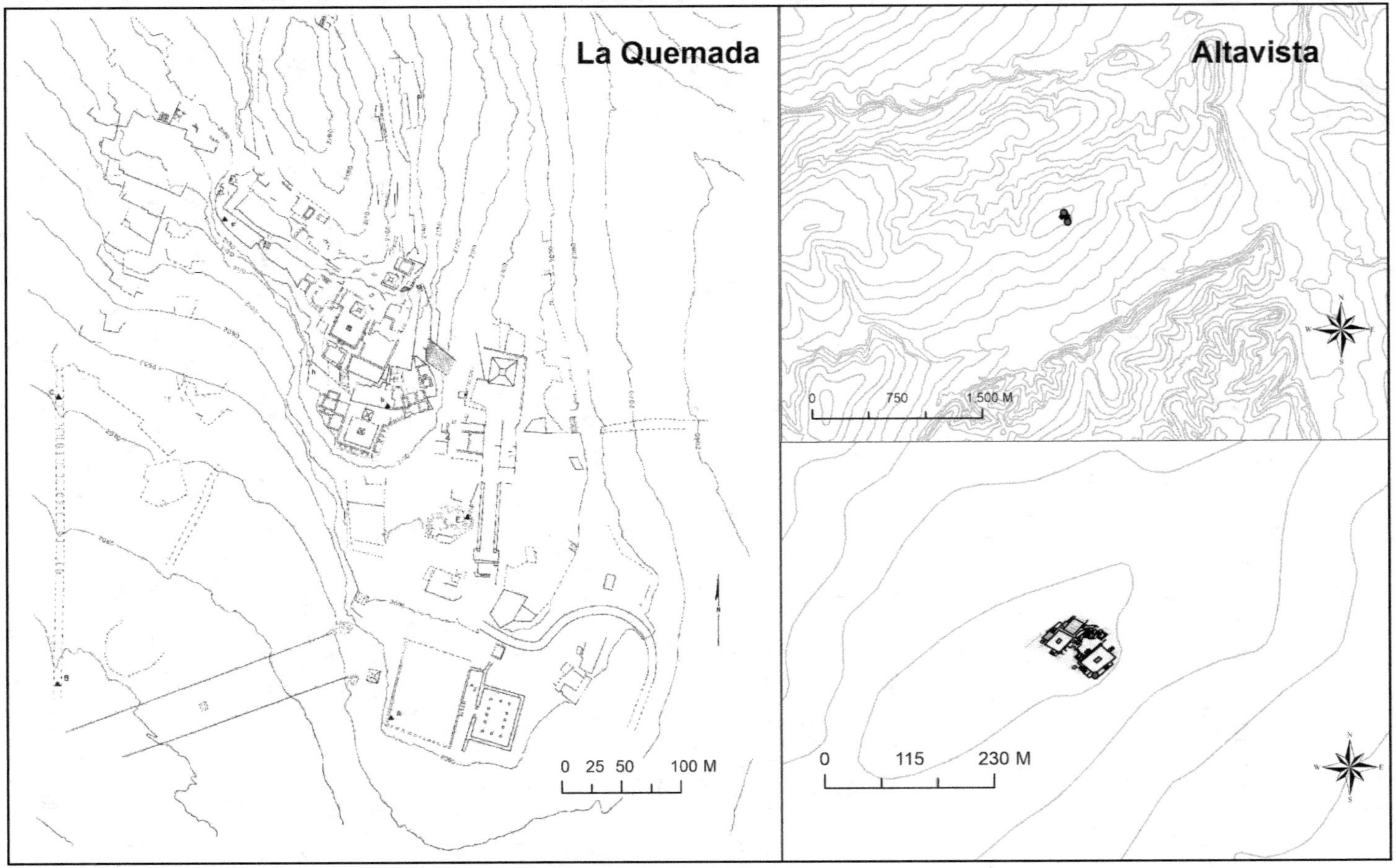

Figure 6.10. Comparative map of other sites in the northern area: *left*, La Quemada (Nelson et al. 1992: 300); *right*, Altavista (map by David A. Muñiz García with information from Nelson et al. 1992: 300; and Medina and García 2010).

dimensions, and associated materials employed in this study, all seem to indicate a domestic character associated with agricultural functions (Figure 6.11).

By contrast, the sites with different degrees of concealment, permanent visual dominion over the aforementioned group, and qualitatively more structured architecture, are CAL and CEG. We consider these sites to be elite residences, in part because distribution there is much more concentrated. In this scheme, the sites set aside from this dynamic are MEA—with "sentry" functions—and CCH, which displays a certain degree of spatial independence and is located on the opposite side of the canyon. However, both of these sites are connected visually and form an imaginary line that would have delimited the space of El Encinal. The only sites that would be completely excluded are PAY and SAL.

Appropriation of the Landscape

The Chalchihuiteño groups that inhabited the Santiago Bayacora Basin appropriated the landscape in distinct ways, some reflecting strategies that are remarkably similar to those utilized in other areas of Mesoamerica. One consists of the visual restriction of certain spaces, generally those devoted to private rituals or the activities of elites (as may have been the case at Los Guachimontones, Jalisco; see Sumano and Englehardt, this volume). But one important difference is the number of resources available in the SBR Basin to generate these effects of concealment compared to sites in other areas, such as Xochicalco (Hirth 1984), Monte Alban (Joyce 2009), or El Tajín (Ladrón de Guevara 2005), among others. The Chalchihuiteño groups generated these strategies through a complex architecture with sets of walls and elevations that were in harmony with the natural surroundings. To achieve such harmony, especially given the scarce manpower available, they imitated their natural environment to produce landscapes with the characteristics similar to those at sites further south; that is, with visual restriction—visibilization—and the capacity of observation, or visibility. In this way, the groups that inhabited the SBR Basin molded a discourse of power based on how and where constructions were made.

The sites at the entrance to the Santiago Bayacora Canyon appear to have been established at a relatively early moment of occupation in the valley's history (that is, the Ayala and Las Joyas phase, AD 550–950). This precociousness allowed them broad visual control over the microspace appropriated (the canyon and its slopes) and generated a cultural landscape that reflected the relation of knowledge-power that those groups sought to

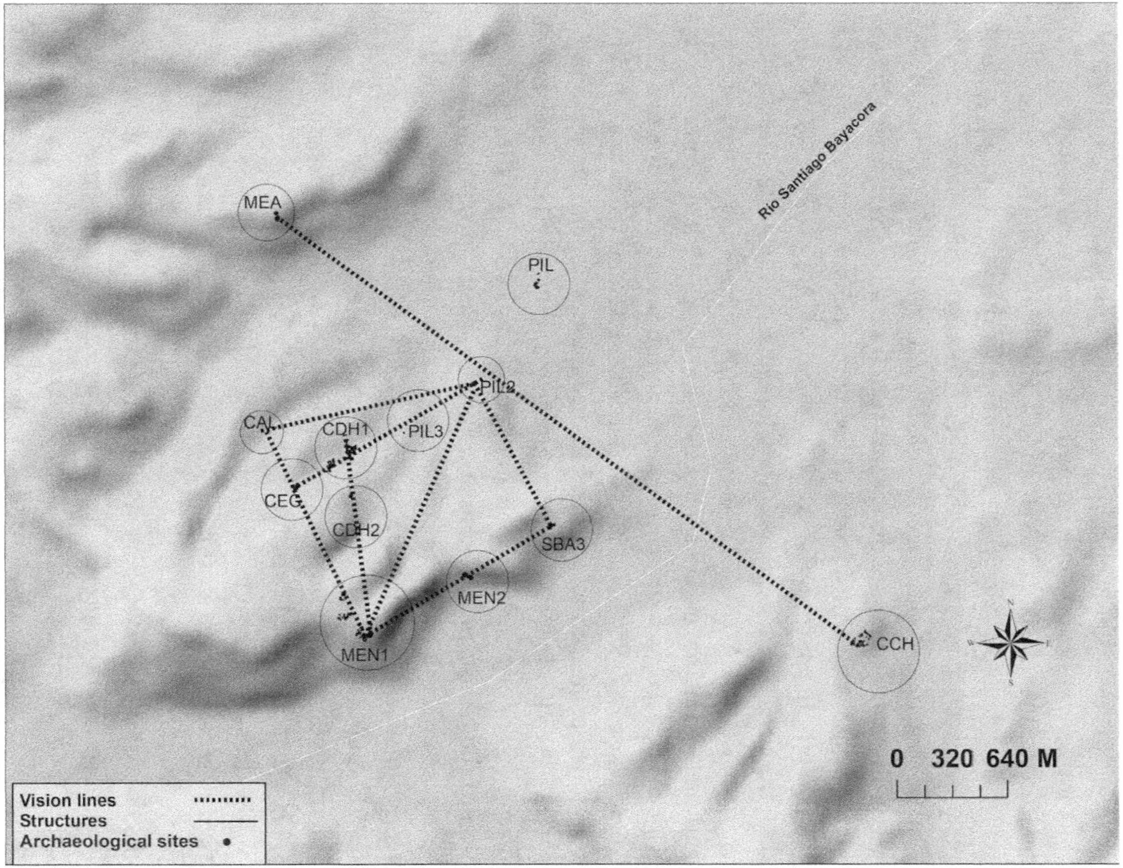

Figure 6.11. Proposal for the establishment of the El Encinal site. The longest dotted line represents the visual line that coincides with the entrance to the canyon and connects the sites of Mesa del Alguacil (MEA) with Cerro del Chiquihuitillo (CCH), both of which are excluded from the possible dynamic at El Encinal, but may mark its boundaries. The other dotted lines mark the visual connections between the possible sectors of El Encinal that were originally recorded as distinct sites. This illustrates their coupling with the position on the landscape, topography, the possible maximum extension of sectors, and a distribution similar to that of sites like La Ferrería (map by David A. Muñiz García with PIACOD information).

establish. In order to manifest a nonverbal discourse of power in physical form, the ancient inhabitants of the SBR Basin opted to modify the privileged geographic position of the Cerro del Chiquihuitillo and the southwestern area of the Mesa del Encinal to obtain visually restricted spaces.[3] As the visibilization tests applied in this study show, the visual capacity that an observer would have had from almost any point in the valley to identify the activities performed inside the two sites was practically nil. By producing visually restricted zones with terraces and walls—including those of residential units—the inhabitants of MEN 1, CEG, and CCH strengthened an image of power in the imagination of other inhabitants of the SBR Basin.

Another aspect of potential significance is the greater density of decorated ceramic material at MEN1 and CCH. Although PAY contains material that is diagnostic for the early phases, it does not coincide with the position in the landscape of the other sites of the SBR Basin; nor is the density of materials similar, since the ceramic sequence evident at PAY is generally representative of diagnostic types from all of the distinct Chalchihuiteño phases identified in controlled contexts, indicating the site's temporal depth. In contrast, the sites (or sectors) of El Encinal have materials only from the first two phases. Therefore, it seems that the sites at El Encinal—CCH, MEA, and PAY—emerged during the Ayala phase (AD 550), and that El Encinal entered into crisis, for as-yet-undetermined reasons, which produced its (partial or total) abandonment toward the end of the Tunal phase (AD 950); this allowed sites like PAY and CCH to ascend to a position of prime importance within the regional settlement hierarchy.[4]

The PAY site does not manifest the first and most visible aspect of the discourse of power of the Chalchihuiteño groups based on the appropriation of the landscape and consisting of a privileged position with regard to aspects of exhibition and concealment. This leads us to surmise that at its first moment of occupation, PAY was a secondary site (perhaps equal to PIL), but that, at some point in the Tunal-Calera phase (AD 950–1250), gained predominance in the Guadiana Valley as one of two sites—with CCH—that continued to function in the SBR Basin.

The density of decorated ceramic material is one indicator of this situation. Based on the excavations at PAY, we know that occupation there was continuous from founding to abandonment. This continuity is reflected in the floors with foundations and in the absence of any evidence of rupture in the construction stages. Clearly, PAY was present throughout the Chalchihuiteño occupation of the valley. When it gained prominence within the dynamic of intersite relations, it had to appropriate the landscape (which was being abandoned by the El Encinal sites or sectors) in a distinct manner in order to justify its new position. Because PAY did not enjoy the ad hoc conditions of sites like MEN1-CEG and CCH to generate a discourse of power through its own geography (high mesas and hat-type hills at the entrance to the canyon, compared to a low mesa with less visibility and greater visibilization), it had to *create* them. That is, its inhabitants had to modify the space it occupied so that it would generate a discourse similar to that of MEN1-CEG and CCH. To achieve this, its population turned to the second level of analysis that we have proposed: spatial distribution.

The more widespread use of architectural complexes to "close" interior spaces and restrict visibilization from other settlements in the valley may be related to the higher number of architectural complexes with interior patios and the scarcity of plazas. During the Tunal and Calera phases, PAY must have been the most important site in the SBR Basin and so had to show itself as such, despite the limitations of its position in the landscape. The means of doing so involved not only intensifying its architectural activity (while MEN1-CEG and CCH occupied better positions on the landscape, their architectural adaptation was more modest, since it was less necessary). It also involved increasing its architectural complexity to reflect its position in the power system. Thus, it carried forward the general strategy of appropriating space and generating a landscape that mirrored a position in the system of knowledge-power, while simultaneously adapting the tactics for producing that landscape by conditioning its existing geographic reality. This situation contrasts with the first period of occupation in the SBR Basin, in which the choice of geographic position was a form of power in and of itself.

Concluding Thoughts

Our final reflections revolve around settlement patterns. MEA was a satellite site that functioned in different time periods, perhaps performing activities of vigilance and control. MEN is a single site with distinct sectors that was occupied principally in the earlier Ayala–Las Joyas phase with dimensions similar to those of La Ferrería and Navacoyán. CCH is a "nucleated" site that functioned in the Ayala–Las Joyas phase and, to a lesser degree, during the Tunal-Calera phase. It may have been a satellite of MEN or competed with it. PIL is a small site occupied during the Tunal-Calera phase, while PAY is a "nucleated" site that gained hegemony toward the Tunal Calera phase.

Based on the results obtained, we propose that in the first moment of Chalchihuiteño occupation in the Guadiana Valley, the strategy for appropriating space was designed to substitute the use of topoforms for the scarce manpower available as the means of exhibition and concealment. That is, inhabitants employed visibility as a strategy of power. Reflecting this strategy, settlements were located at the entrance to the canyons that connected the valley with the sierra. Those sites, including El Encinal, La Ferrería, and El Nayar, would have shared similar conditions and, possibly, functions in civic and administrative terms, controlled food production and trade routes, and regulated daily activities. All three would have had relations with Navacoyán, which appears to have been a center of a religious nature, although with structures and functions similar to those at the other sites.

At some point toward the end of the Las Joyas phase (AD 550–950), this system suffered a crisis, and there is evidence of a demographic movement. In the case of the SBR Basin, most of the population appears to have moved toward PAY, with a smaller group going to PIL. CCH and MEN continued to be inhabited, but at a much more modest level. The strategy of appropriating space also changed, as preference was given to the low mesas over the valley floor, but at sites close to the canyons. This continued during the Tunal and Calera phases. Finally, between AD 1250 and 1350, the Chalchihuiteño populations abandoned their settlements in the valley and moved into the sierra (Punzo 2006).

What we originally thought were various dispersed sites appear, in reality, to have been parts of a single site. This suggests, once again, that the logic that dominates our interpretations of Mesoamerican settlements needs to be rethought in the contexts of northern Mexico. While a similar discourse may hold for the structural part of this thought, environmental conditions, human resources (manpower), and adaptive capacity all led to the generation of appropriation-of-space strategies that were distinct from those characterizing settlements in southern Mesoamerica. The Chalchihuiteños may have shared a system of knowledge-power with the rest of Mesoamerica, but the northern landscape is distinct, so the way in which human populations appropriate it must have also differed. What is important now is to continue working intensely to advance our appreciation and understanding of variability in ancient Mesoamerican lifeways.

Acknowledgments

We thank all members of the INAH Durango and PIACOD who generated the information discussed in this chapter through participation in the 2004–2010 field seasons. We also thank the project director José Luis Punzo. Special thanks to Joshua Englehardt and Agapi Filini for comments on this text. All errors of omission or fact are the sole responsibility of the authors.

Notes

1. The most important exception is the COP site, the mountaintop location of which was used as a sentry outpost during the Cristero War of the early twentieth century.

2. Two sites of considerable size in the Guadiana Valley. La Ferrería is the only site open to the public in the state of Durango.

3. More than "visually restricted," we might call the strategy employed at CCH and the MEN1 sector "selectively restricted visibilization," or perhaps "partial exhibition." These terms seek to conceptualize how the constructions "let themselves be seen," but only those parts that were convenient for strengthening the idea of their presence, while concealing the qualitatively more important parts so that others could not "observe" what occurred there, although they could have "seen" it.

4. Although at these sites activities may have also decreased at this time.

References

Anschuetz, Kurt F., Richard H. Wilshusen, and Cherie L. Schieck

2001 An Archaeology of Landscapes: Perspectives and Directions. *Journal of Archaeological Research* 9(2): 152–197.

Arancón García, Ricardo

1992 La Plaza, Generadora del Espacio Urbano Mesoamericano. *Cuadernos de Arquitectura Mesoamericana* 16: 19–40.

Bermejo, Jesús

2009 Leyendo los Espacios: Una Aproximación Crítica a la Sintaxis Espacial Como Herramienta de Análisis Arqueológico. *Arqueología de la Arquitectura* 6: 47–62.

Cerrato Casado, Eduardo

2011 La Prospección Arqueológica Superficial: Un Método no Destructivo para una Ciencia que sí lo es. *Arte, Arqueología e Historia* 18: 151–160.

Chatford Clark, David L.

2007 Viewing the Liturgy: A Space Syntax Study of Changing Visibility and Accessibility in the Development of the Byzantine Church in Jordan. *World Archaeology* 39(1): 84–104.

Criado Boado, Felipe

1991 Construcción Social del Espacio y Reconstrucción Arqueológica del Paisaje. *Boletín de Antropología Americana* 23: 5–25.

1993 Límites y Posibilidades de la Arqueología del Paisaje. *Spal* 2: 9–55.

1999 Del Terreno al Espacio: Planteamientos y Perspectivas para la Arqueología del Paisaje. *CAPA, Grupo de Investigación en Arqueología del Paisaje* 6: 1–77

2003 Arquitectura como Materialización de un Concepto. La espacialidad megalítica. In *Arqueología de la Arquitectura*, edited by Felipe Criado Boado, pp. 103–111. Instituto de Estudios Gallegos Padre Sarmiento, Santiago de Compostela, Spain.

Braniff, Beatriz

2009 La Historia Prehispánica de Sonora. *Arqueología Mexicana* 97: 32–38.

Foucault, Michel

1970 *La Arqueología del Saber*. Siglo XXI, Buenos Aires.

1979 *El Ojo del Poder*. Ediciones de la Piqueta, Madrid.

2003 *Vigilar y Castigar: Nacimiento de la Prisión*. Siglo XXI, Barcelona.

Gómez Ambriz, Emmanuel Alejandro

2013 La Iconografía Cerámica Chalchihuiteña: Análisis Iconográfico de las Imágenes Centrales en Espiral. Unpublished undergraduate thesis, Escuela Nacional de Antropología e Historia, Mexico City.

González Acuña, Daniel

2001 Análisis de Visibilidad y Patrones de Asentamiento Protohistóricos. Los Yacimientos del Bronce Final y Periodo Orientalizante en el Sureste. *Archeologia e Calcolatori* 12: 123–142.

Hers, Marie-Areti

2004 Arqueología de Durango, Destellos en el Olvido. In *Introducción a la Arqueología del Occidente de México*, coordinated by Beatriz Braniff, pp. 525–545. Universidad de Colima, Consejo Nacional para la Cultura y las Artes, Instituto Nacional de Antropología e Historia, Colima and Mexico City.

Hirth, Kenneth G.

1984 Xochicalco: Urban Growth and State Formation in Central Mexico. *Science* 225(4662): 579–586.

Ingold, Tim

1993 The Temporality of the Landscape. *World Archaeology* 25(2): 152–174.

Joyce, Arthur A.

2009 *Mixtecs, Zapotecs, and Chatinos: Ancient Peoples of Southern Mexico*. Wiley-Blackwell, Malden, MA.

Kelley, J. Charles, and Ellen Abbott

1971 *An Introduction to the Ceramics of the Chalchihuites Culture of Zacatecas and Durango, México: Part I, The Decorated Wares*. Mesoamerican Studies 5. University Museum, Southern Illinois University, Carbondale.

Knapp, Bernard, and Wendy Ashmore

1999 Archaeological Landscapes: Constructed, Conceptualized, Ideational. In *Archaeologies of Landscape: Contemporary Perspectives*, edited by Wendy Ashmore and Bernard Knapp, pp. 1–30. Wiley-Blackwell, Malden, MA.

Ladrón de Guevara, Sara D.

2005 *Imagen y Pensamiento en El Tajín*. Universidad Veracruzana, Jalapa.

Llobera, Marcos, Pastor Fábrega Álvarez, and César Parcero Oubiña
2011 Order in Movement: A GIS Approach to Accessibility. *Journal of Archaeological Science* 38: 843–851

Manzanilla, Linda
1986 Introducción. In *Unidades habitacionales mesoamericanas y sus áreas de actividad*, edited by Linda Manzanilla, pp. 9–20. Universidad Nacional Autónoma de México serie antropológica 76. Mexico City.

Mañana Borrazás, Patricia, Rebeca Blanco Rotea, and Ayán Vila Wurxo
2002 Arqueotectura 1: Bases teórico metodológicas para una arqueología de la arquitectura. *Tapa* 25: 1–93.

Medina González, Humberto, and Baudelina García Uranga
2010 *A 100 años de su descubrimiento: Alta Vista.* Instituto Nacional de Antropología e Historia, Gobierno del Estado de Zacatecas, Quadrante Plástico, Mexico City.

Muñiz García, David, and Ana Iris Murguía
2010 La Arquitectura Chalchihuiteña del Valle de Guadiana. In *Historia de Durango Tomo I: Época Antigua*, coordinated by José Luis Punzo Díaz and Marie-Areti Hers, pp. 210–231. Instituto de Investigaciones Históricas, Universidad Juárez del Estado de Durango, Durango.

Nelson, Ben A., Andrew Darling, and David Kice
1992 Mortuary Practices and the Social Order at La Quemada, Zacatecas, Mexico. *Latin American Antiquity* 3(4): 298–315.

Parcero Oubiña, César
2002 *La Construcción del Paisaje Social en la Edad de Hierro del Noroeste Ibérico.* Unpublished doctoral thesis, Departamento de Historia I de la Universidade de Santiago de Compostela, Spain.

Parcero Oubiña, Cesar, and Patsor Fábrega Álvarez
2007 Diseño Metodológico para el Análisis Locacional de Asentamientos a través de un SIG de Base "Raster." In *La Aplicación de los SIG en la Arqueología del Paisaje*, edited by Ignasi Grau Mira, pp. 69–91. Universidad de Alicante, Alicante, Spain.

Punzo Díaz, José Luis
2004 *Informes Técnicos de las Investigaciones Arqueológicas del Área Centro-Oeste de Durango.* Report submitted to the Instituto Nacional de Antropología e Historia, Mexico City. Manuscript on file, Centro INAH Durango.
2005 *Informes Técnicos de las Investigaciones Arqueológicas del Área Centro-Oeste de Durango.* Report submitted to the Instituto Nacional de Antropología e Historia, Mexico City. Manuscript on file, Centro INAH Durango.
2006 *Informes Técnicos de las Investigaciones Arqueológicas del Área Centro-Oeste de Durango.* Report submitted to the Instituto Nacional de Antropología e Historia, Mexico City. Manuscript on file, Centro INAH Durango.
2007 *Informes Técnicos de las Investigaciones Arqueológicas del Área Centro-Oeste de Durango.* Report submitted to the Instituto Nacional de Antropología e Historia, Mexico City. Manuscript on file, Centro INAH Durango.
2008a *Informes Técnicos de las Investigaciones Arqueológicas del Área Centro-Oeste de Durango.* Report submitted to the Instituto Nacional de Antropología e Historia, Mexico City. Manuscript on file, Centro INAH Durango.

2008b Contributions to Mesoamerican Settlement Chronology in the Sierra Madre in Durango, México. Paper presented at the 73rd Annual Meeting of the Society for American Archeology, Vancouver, BC.

2009 *Informes Técnicos de las Investigaciones Arqueológicas del Área Centro-Oeste de Durango*. Report submitted to the Instituto Nacional de Antropología e Historia, Mexico City. Manuscript on file, Centro INAH Durango.

2010 *Informes Técnicos de las Investigaciones Arqueológicas del Área Centro-Oeste de Durango*. Report submitted to the Instituto Nacional de Antropología e Historia, Mexico City. Manuscript on file, Centro INAH Durango.

2011 Construyendo, morando y pensando: los habitantes de las casas en acantilado de la Sierra Madre Occidental. In *Durango: Identidad, Paisaje y Patrimonio*, edited by Stanislaw Iwaniszewski and Silvina Vigliani, pp. 341–354. Instituto Nacional de Antropología e Historia, Escuela Nacional de Antropología e Historia, y Dirección de Estudios Históricos, Mexico City.

2016 Revisando la Cronología en la Frontera Norte de Mesoamérica, Estado de Durango, México. *Arqueología Iberoamericana* 29: 38–43

Punzo Díaz, José Luis, and Ana Iris Murguía

2008 *Informe Técnico del Estado de Conservación de los Bienes Culturales Inmuebles de la Zona Arqueológica de La Ferrería*. Manuscript on file, Centro INAH Durango.

Punzo Díaz, José Luis, and Ángel Luna Ramírez

2007 The Chalchihuites Chronology Revisited. Paper presented at the 72nd Annual Meeting of the Society for American Archaeology, Austin, TX.

Rangel Estrada, Diego Antonio

2014 Identificación de Zoomorfos en los Materiales Arqueológicos de la Cultura Chalchihuites, Rama Guadiana. Unpublished undergraduate thesis, Escuela Nacional de Antropología e Historia, Mexico City.

Rapoport, Amos

1969 Some Aspects of the Organization of Urban Space. *Response to Environment* 18: 122–139.

1988 Levels of Meaning in the Built Environment. In Cross-Cultural Perspectives in Nonverbal Communication, edited by Fernando Poyatos, pp. 317–336. C. J. Hogrefe, Toronto

1990 The Meaning of the Built Environment: A Nonverbal Communication Approach. University of Arizona Press, Tucson.

Reyes Valdez, Antonio

2004 Pimas, Pápagos y Tepehuanes: Relación Lengua-Cultura entre los Pueblos Tepimanos del Noroeste de México y el Suroeste de los Estados Unidos. Unpublished MA thesis, Facultad de Filosofía y Letras, Universidad Nacional Autónoma de México, Mexico City.

Sandoval Mora, Cindy Cristina

2011 La Aplicación de la Petrografía en la Caracterización y Proveniencia de las Cerámicas Chalchihuiteñas de las Ramas Guadiana y Súchil del Sitio Arqueológico La Ferrería en Durango, Dgo. Unpublished undergraduate thesis, Universidad Autónoma de Zacatecas, Zacatecas.

Smith, Michael E.
2007 Form and Meaning in the Earliest Cities: A New Approach to Ancient Urban Planning. *Journal of Planning History* 6(3): 3–47.

Thomas, Julian
2001 Archaeologies of Place and Landscape. In *Archaeological Theory Today*, edited by Ian Hodder, pp. 165–186. Polity Press, Cambridge, UK.

Tilley, Christopher
1994 *A Phenomenology of Landscape: Places, Paths, and Monuments*. Berg, Oxford.

Tsukada, Yoshiyuki
2006 Grandes Asentamientos Chalchihuiteños de la Sierra Madre Durangueña: Estudio Comparativo entre Cañón de Molino y Hervideros. In *La Sierra Tepehuana: Asentamientos y Movimientos de Población*, coordinated by Chantal Cramaussel and Sara Ortelli, pp. 45–56. El Colegio de Michoacán, Instituto de Investigaciones Históricas, Universidad Juárez del Estado de Durango, Zamora.

Vidal Aldana, Cinthya Isabel
2011 *El Intercambio en el Noroccidente Prehispánico. La Relación entre la Rama Guadiana de la Tradición Arqueológica Chalchihuites y la Tradición Aztatlán, entre el 600–1300 d.c.* Unpublished undergraduate thesis, Escuela Nacional de Antropología e Historia, Mexico City.

PART 3

DIVERSITY

Refined Theoretical Perspectives

7

What the Teuchitlán Tradition Is, and What the Teuchitlán Tradition Is Not

VERENICE Y. HEREDIA ESPINOZA

The origin and development of social complexity is a theme that continues to be at the center of archaeological discussion. In far west Mexico, the lack of theory to explain the rise of social complexity has led scholars to misuse models or apply them as simple adornments to a vastly descriptive, culture-historical narrative. This is clearly seen in the characterization of the famous Teuchitlán Tradition (Weigand 1993) as a segmentary state. The Teuchitlán Tradition (circa 350 BC–AD 450/500), also referred to as the Teuchitlán culture (see Beekman, this volume) is an early expression of social complexity that was distinct from, in terms of material culture, yet at the same "level" of development as other contemporaneous Mesoamerican cultures. Given its supposedly "unique" features (for example, circular architecture), but also considering its shared "Mesoamerican" characteristics (for example, ballcourts and iconography), the Teuchitlán Tradition has eluded adequate definition and explanation in anything but questionable neoevolutionary terms and via the misuse of anthropological models, the main topic of this chapter.

Scholars working the area described it as a statelike society, a protostate, a quasi-urbanized civilization (Weigand 2008; Weigand and Beekman 1998), and a chiefdom (López Mestas 2011), in order to place it in an evolutionary linear scheme and on a par with the rest of Formative Mesoamerica. In addition, it has been repeatedly categorized as a segmentary state (Weigand 2009, 2010; Weigand and Beekman 1998), such as those found in Central Africa. Although the use of anthropological models to generate a hypothesis to be tested archaeologically is common practice in our discipline, their application is not always as adequate.

The segmentary state model to refer to the Teuchitlán Tradition (TT) has permeated the published literature of this region in West Mexico for over two decades (Esparza 2015; Weigand 2010; Weigand and Beekman 1998) with little modification. Our understanding of this important society and its origin, development, and eventual demise continues to be a topic of great interest. In light of new groundbreaking and sophisticated archaeological theory on alternative pathways to complexity (Blanton and Fargher 2008: Fargher et al. 2011; Fargher and Heredia 2016), our understanding of the nature of the TT is greatly improving (Beekman 2008; Heredia 2017). In this chapter, I examine the primary literature on the segmentary state (Southall 1988, 2004) to understand the real implications of the model; in this vein, rather than treating this political formation as a "type," I focus on how political actors and their factions create these political structures. Once I describe the strategies used by political actors, I then extract the archaeological correlates to evaluate the extant evidence regarding the Teuchitlán Tradition as it refers to the model. I conclude that this culture does not conform to the strategies and processes of factional politics typical of segmentary states, and that a major problem in the application of this model—particularly as applied to the Tequila valleys of central Jalisco—are its treatment as a romanticized "type" of political structure that fits neither chiefdom nor state. Finally, I offer some thoughts on our understanding and provide future avenues for research on the origins, nature, and workings of the TT.

Segmentary States

Segmentary state is an intermediate category that does not fit the typical definition of either a chiefdom or a state. In this sense, the use of the term "state" as segmentary is an oxymoron (Marcus and Feinman 1998: 7), because it is not a true state. Southall (2004) coined the term to define the political structure of the Alur of Central Africa, whose political strategies used kinship and ritual to integrate local descent groups into communities and, in turn, communities into regional polities that recognized a single suzerain as the symbolic head of the "nation." But the basic concept has wider application for anthropologists interested in state formation processes. Importantly, a segmentary state is a loosely organized polity, often multiethnic and volatile. In the anthropological literature, Geertz (1980) referred to them as theater states and Tambiah (1977) called them galactic polities (see also Demarest 1992).

Segmentary states have particular characteristics that set them apart from other political structures. Primarily, they include a central territory surrounded by semiautonomous and peripheral areas, each headed by a leader. The central territory has a king whose political power does not extend beyond the central territory that he controls directly, although he has ritual preeminence over a large territory. Thus, the definition by Southall (1988: 52) of a segmentary state is "one where the spheres of ritual suzerainty and political sovereignty do not coincide"; this translates into the very nature of this political structure where the only "unifying" element is a common ritual and the acknowledgment of a ritual hegemon. Secondary leaders or kings have control over their own territories politically and economically, they can extract tribute from their immediate territories and are unrestrained from waging war with other polities. The central king can call upon other leaders in times of conflict, but these leaders are semiautonomous economically and politically from the center. The relationship between the king at the central territory and those leaders in surrounding areas is based on patron-client relationships, where gifts in exotics or luxury goods create bonds and debt among them. In essence, the segmentary state only holds its unity through ritual, while the rest is grounded on fragile and feeble ties that are ever-changing, depending on personal alliances, interests, and situations; at their core, they are based on network or exclusionary strategies (Blanton et al. 1996). In this chapter, I argue that these segments act as factions, which are the unifying feature of segmentary states and the basic unit of political integration.

Factional Political Strategies

Factions are "structurally and functionally similar groups which, by virtue of their similarity, compete for resources and positions of power or prestige" (Brumfiel 1994: 4). Factions therefore include a mix of people, and do not represent a particular group (such as craft specialists, priests, charismatic leaders, age sets, high-status individuals, and so on). They are highly consistent with recent agency perspectives because the goals of political actors involve gaining and maintaining control of a faction, and then subduing or dominating other factions to build pyramid-shaped political hierarchies (Brumfiel 1994: 3). Factions are therefore disruptive and temporary, but they can also unite at times when cooperation is most advantageous (Byland and Pohl 1994: 117).

Factions have weak horizontal, economic, and political ties. Political decisions are not centralized, nor is the economy monopolized by a single group. Thus, secondary leaders are unrestricted, in their own territories, with respect to political decisions involving migration, warfare, and alliance building (Southall 1988). The territory of the state is thus constantly changing as these processes occur. Economically, each faction is independent from each other, and it can be said that the economy of a unit does not go beyond the faction to which it belongs. Economic ties to other factions are therefore weak, in general, since each unit duplicates the activities of the other (low integration). Moreover, these structures are fragile because they are based on personal authority and influence (contra infrastructural power, where the state is capable of enforcing its policies throughout its territories, affecting all levels of society [Mann 1984: 185]).

Political actors or leaders use a range of strategies to gain control of a faction and to subordinate other factions. These strategies include kinship (Southall 1988; 2004), ritual (Tambiah 1977), theatrics (Geertz 1980), economic links (Stein 1980), marriage alliances, descent rhetoric, gifting, feasting, and warfare (Chance 2000; Fox et al. 1996; Fargher et al. 2011), among others. In order to understand factional political strategies and thus "segmentation" of the state, I focus on the "integrative" strategies used by political actors to build and maintain factions, as well as to dominate or subdue other factions.

Network Strategies as Integrative Strategies of Segmentary States

Fargher et al. (2011) and Fargher and Blanton (2012), have elaborated on the deployment of network strategies to build states that exhibit segmentary or factional structures. Network strategies are those in which "preeminence is an outcome of the development and maintenance of individual centered exchange relations established primarily outside one's local group" (Blanton et al. 1996: 4). Individuals in this political economic pattern seek to establish a powerbase, particularly by increasing the number of followers, whose loyalty has to be reciprocated through gifting. Thus, power is open to negotiation and must be certified through reciprocal gift exchange and marriage alliances, and their ability to create obligated followers (clients) based on unbalanced exchange relationships (for example, patron-client relationships). One way is through the distribution of prestige and luxury goods among other kinds of resources they may control. Their goals can only be achieved given the degree to which they have cultivated the loyalty of clients and the degree to which potential dissidents believe the ruler has the power to punish them successfully if they rebel or do not comply. Individuals in this form of political economy may also make use of descent rhetoric that legitimizes their domination or the right to rule. Hence, rulers or individual statuses are constantly reproduced and legitimated in addition to not having to respond to the demands of other members of the political community (Fargher and Blanton 2012).

Given the nature of the factional political economy and the inability of the king or leader to effectively extract tribute within and beyond their immediate territories, kings resort to making personal alliances and establishing patron-client relationships. These strategies are employed both within and between factions. Thus, the wealth and the economic support that they acquire are completely dependent on their abilities to create and maintain these relationships. Leaders in segmentary states adopt a range of strategies to expand a power base albeit temporary and weak. Classic Maya, Postclassic Valley of Puebla, and Postclassic Mixteca Alta in Oaxaca offer representative examples of network strategies in segmentary structures (see Fargher et al. 2011). I briefly discuss two examples below.

Classic Period Maya

Classic period Maya society has long been a quintessential example of network strategies (Blanton et al. 1996) that exhibits a segmentary organizational pattern (Fox 1989; Fox et al. 1996; but compare Chase and Chase 1996). The strategies used by Maya kings during the Classic period involved patrimonial rhetoric, as it was crucial for them to legitimate and maintain their positions of power (Fox et al. 1996; Schele and Miller 1986). Maya noble houses were, in essence, the focus of political power and authority, and their political economy relied heavily on factionalism, a prestige goods system, marriage alliances, warfare, and ritual authority of the kings. The iconographic and epigraphic records heavily emphasize these themes associated with exclusionary strategies. Warfare was fought between elites and rulers among the Classic Maya (Fox et al. 1996: 799; Martin and Grube 2008: 16; Webster 1998: 337). Hence the elite used conflict to compete for and gain status and not for the defense of entire communities. Thus, many battles occurred within polities or between neighbors attempting to dominate each other (especially during the Terminal Classic). There is also extensive evidence of violence involved in competition for internal control of the faction, manifested in coups within royal houses (Sharer with Traxler 2006: 299–301) and the arrival of interlopers or foreigners who took control of these houses. Key to understanding the nature of warfare is the occurrence of defensive sites. The Maya protected core areas or epicenters with defensive walls (Webster 1998: 324–325), but entire settlements were not defended, suggesting that rulers were interested in protecting their personal property and their factions, and not local populations.

Maya kings are also known to have controlled the distribution of water by building reservoirs underneath palaces and temples (Scarborough et al. 2012), which allowed for symbolic control of water and therefore contributed to the legitimation of rulers. Water control can also be linked to agricultural land, which, by extension, was also controlled by the nobility. However, again, these systems serviced only the areas immediately surrounding elite precincts and were not a public good that serviced entire regions. Thus, peasants directly linked to rulers may have experienced more agricultural security than independent peasants, given the highly unstable rain patterns in the Maya lowlands. So they would have benefited from their loyalty to these rulers. Therefore, these Maya rulers were fundamentally able to control resources (water and land) and use them to attract followers (both noble and commoner) and supporters (patron-client relationships), thereby creating factions.

The materialization of these strategies is evident in carved stelae and hieroglyphic writing focusing on the lineages and personal achievements of rulers, especially warfare and conquest (Martin and Grube 2008). It is also evident in the presence of elaborate burial monuments filled with prestige goods, paintings on walls, and pottery showing exclusive feasting and hierarchical relationships between rulers and ruled. In addition, symbolic representations of power and authority took the form of elaborate residential structures (palaces) and jaguar thrones. Power was more directly expressed via monopolies on ritual spaces and ritual interaction with supernatural forces, as well as through the construction of elaborate reservoirs in and around central plazas that could be used to irrigate land directly under the ruler's control and geographically adjacent to central monumental architecture (Scarborough 1998). These are all manifestations of the strategies in segmentary structures.

Siqui in the Mixteca Alta

My colleagues and I (Fargher et al. 2011) have discussed the Postclassic Mixteca Alta petty states or *ñuu* in terms of a segmentary nature. The *ñuu* was the basic political unit at the center of which was a hereditary ruler (*yya*) who possessed extensive patrimonial lands, attached laborers, and palaces, who claimed rights over the subdivisions of the state called *siqui* or *siña* (Pastor 1987: 35, 52, 54; Spores 1984: 65; Terraciano 2001: 105, 347–348). The *yya* appears to have been all-powerful; their power and authority derived from claims to supernatural ancestors, the control of luxury goods trade, and their ability to build and maintain factions (Dahlgren 1990: 144; Pastor 1987: 53; Spores 1984: 65, 75). Mixtec creation stories state that the original *yya* were born from natural objects (trees, rocks, rivers, and the sky) inhabited by supernatural spirits, and that deified cult heroes provided them with assistance in founding their *ñuu* (Jansen 1990: 102–103; Pohl 2003a: 64; 2003b: 206; Pohl and Byland 1990: 115–116). Factions were constructed and maintained using gift exchanges (especially involving land, dependent populations, and codex-style polychrome serving vessels), reciprocal feasting, and patrimonial rhetoric (network strategies in Blanton et al. 1996; Dahlgren 1990: 281–284; Spores 1967: 16–18, 1974: 301, 302–305, 2007: 102). Documents show that the *yya* claimed thousands of laborers as personal property as well as the best irrigated bottomlands (Spores and Balkansky 2013: 109–111). Alliance building was a prominent aspect of the political process among the Mixtecs. Powerful alliances involved the marriage of two *yya* from different *ñuu*, which created a sort of super-*ñuu* called a *yuhuitayu* (Spores 1974: 302–306; Terraciano 2001: 347–348). Yet, each *yya* retained absolute control over his/her respective *ñuu* (Spores 1974: 303–304; Terraciano 2001: 104). Materialization of the factional strategies used by the *yya* include the construction of enormous palaces, such as that built in Yucundaa or Teposcolula, the capital of one of the most powerful *yuhuitayu* at the time of Spanish contact (Spores and Robles 2007). The presence of codex-style polychromes in palaces and restricted spaces meant that these items were intended to be used by the elite, their close kin, and their followers. Themes on codices are focused on the personal lives of rulers, their descent, conquest, marriage alliances, and feasting.

Archaeological Correlates of Segmentary States

Theoretically, the strategies of segmentary states, that is, factional politics, should be materialized in the archaeological record and should correlate with the evidence of the integrative strategies mentioned above. Leaders and kings who adopt network strategies make use of different "funds of power" which include warfare, competitive feasting, writing and imagery focusing on individual acts (usually involving martial themes) and genealogy, monopolies over prestige goods, descent reckoning and ancestor worship, and control of knowledge (Blanton 1998: tables 5.1, 5.2). These strategies are associated with specific behaviors that have material correspondences in the archaeological record (Table 7.1).

Table 7.1. Archaeological correlates of factional strategies

Prestige goods	Monopoly in luxury goods (e.g., exotic and fancy items), materialized in a limited distribution; they appear only in specific contexts, such as high-status residences and elite tombs
Warfare	Fortified elite precincts or site epicenters; warrior imagery
Kinship and descent reckoning	Ancestor worship; genealogical reckoning recorded in codices, stone sculpture, ceramics, or other forms of written communication
Competitive feasting	Highly decorated serving wares associated with high-status residences and tombs (funerary feasts)
Architecture	Monumental "private" architecture: Focus on closed and inaccessible plazas, ritual spaces, and luxurious residences (e.g., palaces). Limited evidence of communal spaces to hold large rituals inclusive of entire communities. "Public" spaces inscribed on private spaces (e.g., "public spaces" embedded in palaces; repetition in ritual elements but high degree of idiosyncratic variation in spatial arrangements and especially façades of monumental architecture at each level in the hierarchy and at each center.

Source: Based on Blanton (1998); Blanton et al. (1996); Fargher and Blanton (2012).

Significance of *Guachimontón* Architecture

The Teuchitlán Tradition is the material expression of one of Mesoamerica's early complex societies. Its atypical circular architecture is a composite of three features: (1) a central circular structure also called an "altar"; (2) a circular, sometimes sunken, patio surrounding the "altar"; and (3) a series of rectangular platforms, usually eight, surrounding the patio. These complexes are known as *guachimontones* and have received a great amount of attention (Weigand 1993; Weigand 2008). Other architectural elements found consistently associated with this circular architecture, such as ballcourts and high-status residences, have also been the focus of study (Blanco 2009; Herrejón 2008; Smith 2009; Weigand and Garcia de Weigand 2005). A significant amount of ink has been spilled as to the function of the circles (Beekman 2003a, 2003b; 2008; Jennings 2008; Johns 2014; Tyndall 2008). Many times, explanations rely heavily on the ceramic models found in shaft tombs that vividly depict the activities that were carried out in these spaces (Townsend 1998: figs. 2, 27). The ceramic models and excavations (see below) indicate that community-wide ritual ceremonies were performed in the *guachimontones*, where large numbers of people gathered around the central altar in the patio and in the platforms that surround the patio and central altar. Ceramic models are very informative in this respect since they illustrate that, at the central altar, a pole was erected and used in a way analogous to the modern *volador* ceremony. Weigand (1996: 98) has suggested that this ritual ceremony is associated with an early form of the wind god *Ehecatl*. Excavations at various central altars at the Los Guachimontones site (Figure 7.1) partially support the presence of a "pole print" (Beekman 2003a; Esparza and Weigand 2008). Beekman (2003a, 2003b) proposes an alternative and complementary hypothesis whereby the ceremonies carried out at the *guachimontones* represent harvest ceremonies. Both interpretations fit well in that the wind god brings rain and thus is intimately associated with fertility. It is clear that the ceremonies carried out at these circular complexes, at least the largest ones at the largest sites, had some sort of ritualistic nature associated with fertility, specifically with the corn harvest. An iconographic study of fine ceramics shows that pan-Meosoamerican symbols and themes associated with water, sacrifice, ritual, fertility, and cosmovision are frequent (Heredia and Englehardt 2015). All of these themes are more in tune with more corporate-oriented strategies (Blanton 1998; Blanton et al. 1996).

Repetition of these circular complexes throughout the Tequila valleys, as well as the frequent occurrence of accompanying architectural elements such as ballcourts, has been interpreted as evidence for a repetition of the same rituals at all levels of the civic-ceremonial hierarchy (Heredia 2011). *Guachimontones* are ubiquitous in the Tequila valleys, but they have also been found beyond the limits of this region in Zacatecas, Colima, Michoacán, and Querétaro (Cabrero 1989; Cabrero and López Cruz 2002; Olay and Sánchez 2015; Weigand 2009: 60–61). Due to this widespread distribution in circular architecture, scholars have proposed the segmentary state model (Weigand and Beekman 1998; Weigand 2009).

Previous Interpretations of the Teuchitlán Tradition as a Segmentary State

As a way to explain the presence of *guachimontón* architecture beyond the core, Weigand (2009) has suggested that the Teuchitlán Tradition was analogous to a segmentary state, which contained a core and a periphery. The core was centered in the valleys around the Tequila Volcano and extended for about 2,500 square kilometers at the apex of the tradition (circa AD 200), while the periphery extended into an undefined territory, but still

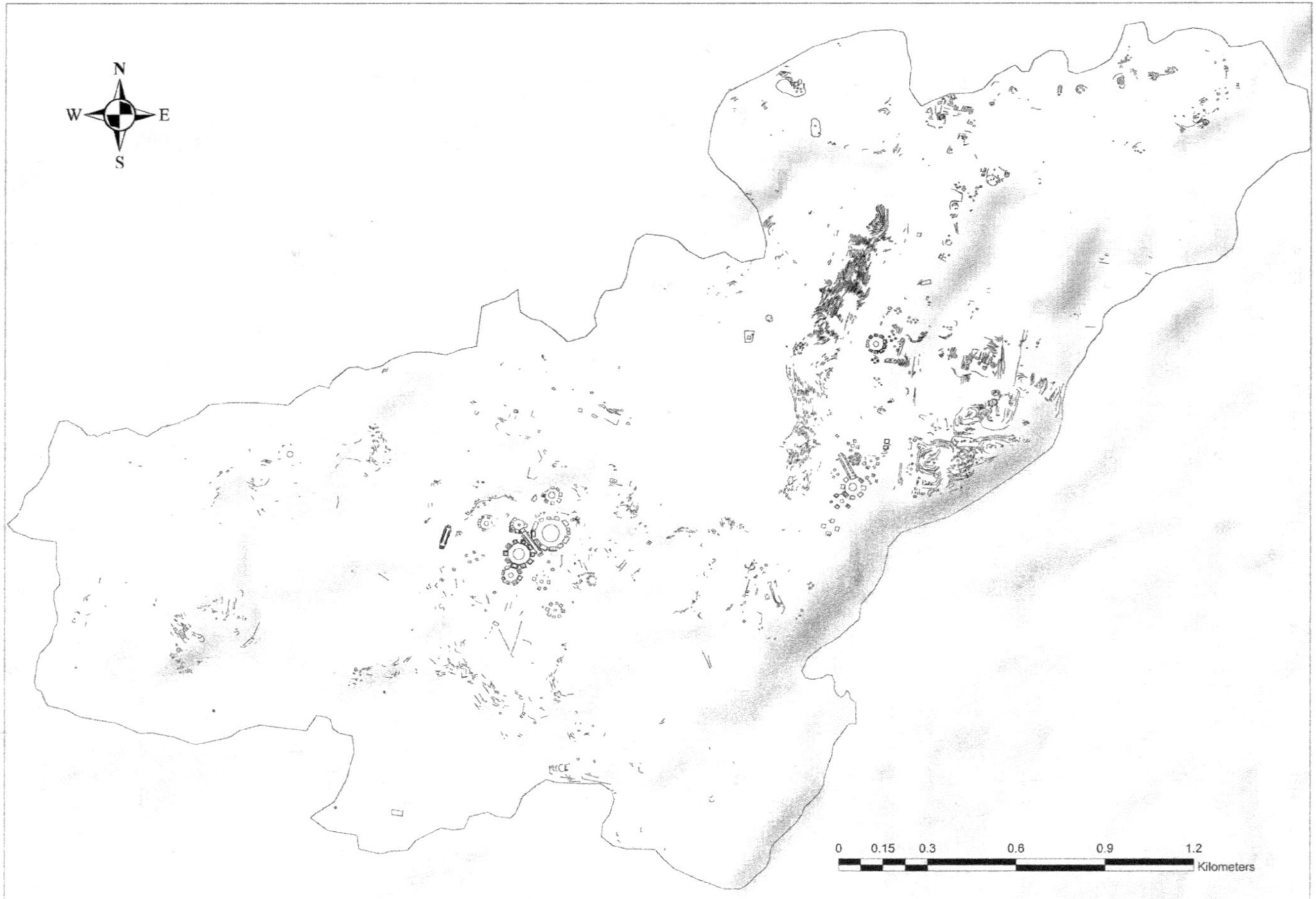

Figure 7.1. The site of Los Guachimontones (map by author).

including all those areas with circular architecture (Figure 7.2). Yet, in his writings, Weigand (2009; see also Beekman 1996b) proposes that, at the core, the Teuchitlán Tradition was probably unified as a single entity, while at the periphery authority was in the ideological realm, contradicting and juxtaposing two different types of political structures (unitary vs. segmentary). For example:

> The system we have postulated can be characterized as a segmentary state, in which the prestige of its ceremonialism, rather that outright force, was utilized to cement the social system and for its expansion into the neighboring region. (Weigand 2008: 583)

He continues:

> For the areas outside the nuclear zone it is appropriate to use Southall´s model (1988) of "segmentary states" as an explicative element: using the ceremonial power of the core area, secondary elites were established in areas with highly prized resources (Colima, Sinaloa, Nayarit) or in charge of trade routes into El Bajio and Zacatecas. It was the ritual hegemonic presence more than something based on force and weapons; symbiosis and collaboration more than control. In other words, it was a "unitary state." (Weigand 2009: 61, my translation)

The above definitions and archaeological correlates of the segmentary state in Jalisco are not accurate, because what are being referred to are shared corporate cognitive codes and ritual practices. In addition, it is treated as a type of "state" or political structure that has fixed features. Further, by definition, segmentary states are built around factions contending for power as I have established above, but Weigand makes no mention of factions; therefore, his application of the model is restricted to the presence of *guachimontones* within and beyond the core as a reflection of these shared rituals rather than describing a segmentary state. In this chapter, I wish to move away from typologies and instead focus on factions and their strategies. It is my intention to begin a discussion on the connection between the models (in this case, the segmentary state) used to describe and interpret the Teuchitlán Tradition and the extant published data.

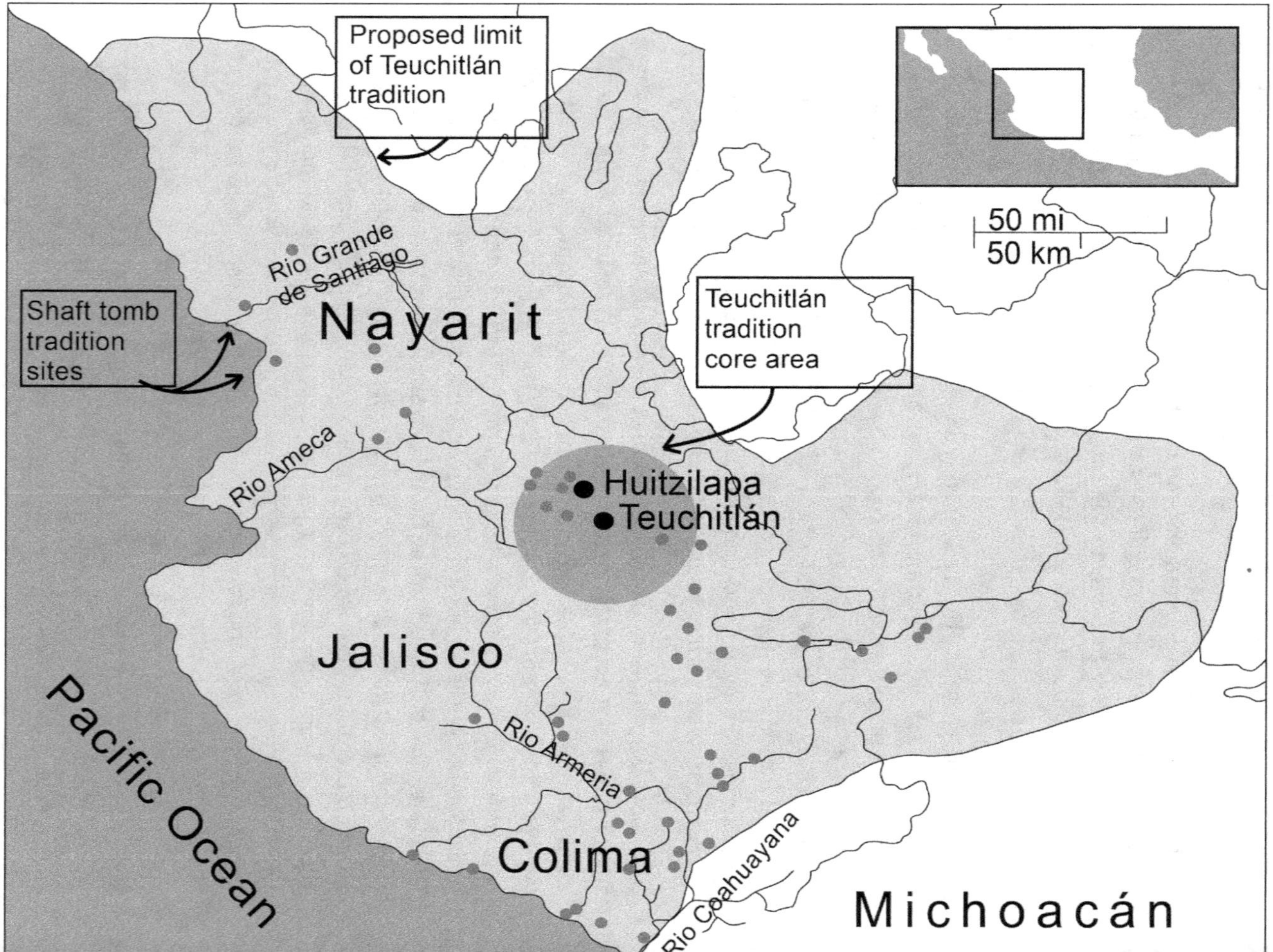

Figure 7.2. The proposed core and periphery of the Teuchitlán tradition (map by Martha A.

What Evidence Is There That the Teuchitlán Tradition Was a Segmentary State? An Evaluation of Current Data

Unlike previous assertions as to the nature of the Teuchitlán Tradition, here I bring together several lines of published data and current fieldwork to evaluate the segmentary state model in the Tequila region. The data come from different scales of analysis including iconographic studies (ceramics and figurines), excavations (architectural features and associated artifacts), and regional surveys (settlement patterns and architecture). Poorly provenienced artifacts, such as the highly coveted figures associated with shaft tombs, also provide a valuable source of information. I assemble these lines of evidence to put the segmentary state model to the test with the goal of improving our understanding of the underlying processes.

Architecture

If factional strategies were prevalent in the Tequila region, they should be materialized in "private" or inaccessible architectural types where faction heads would have held events for a restricted number of people. Thus far, no one has identified palaces or anything analogous to elite administrative offices. Yet there is little doubt that residential architecture varies in size and arrangements, which is taken as a reflection of social stratification. Square or rectangular groups of four structures or platforms around enclosed patios characterize high-status residential architecture (López Mestas 2011). Segregation of high-status families from the more communal, civic places is evident in their location adjacent to or separated from *guachimontones* (Figures 7.3–7.6). Although one such complex was excavated at Los Guachimontones (Herrejón 2008) and one in Huitzilapa (Ramos and López Mestas 1996), there is little to suggest that these were palaces geared toward administrative purposes; instead, they were the residences of important, wealthy, or influential families. This architecture is commonly associated with elite residences due to their size and the objects found with them, such as large anthropomorphic figures and fancy lithic and ceramic artifacts. However, I posit that they are not necessarily those who manage or control society's institutions (elites in the sense of Chase and Chase 1992: 3). In essence, these are high-status residences, some of which contain rich burials, such as the case for Huitzilapa (see below). Further, these residences are located near civic-ceremonial buildings. There is a clear separation between an individual or group residence and civic-ceremonial spaces meant

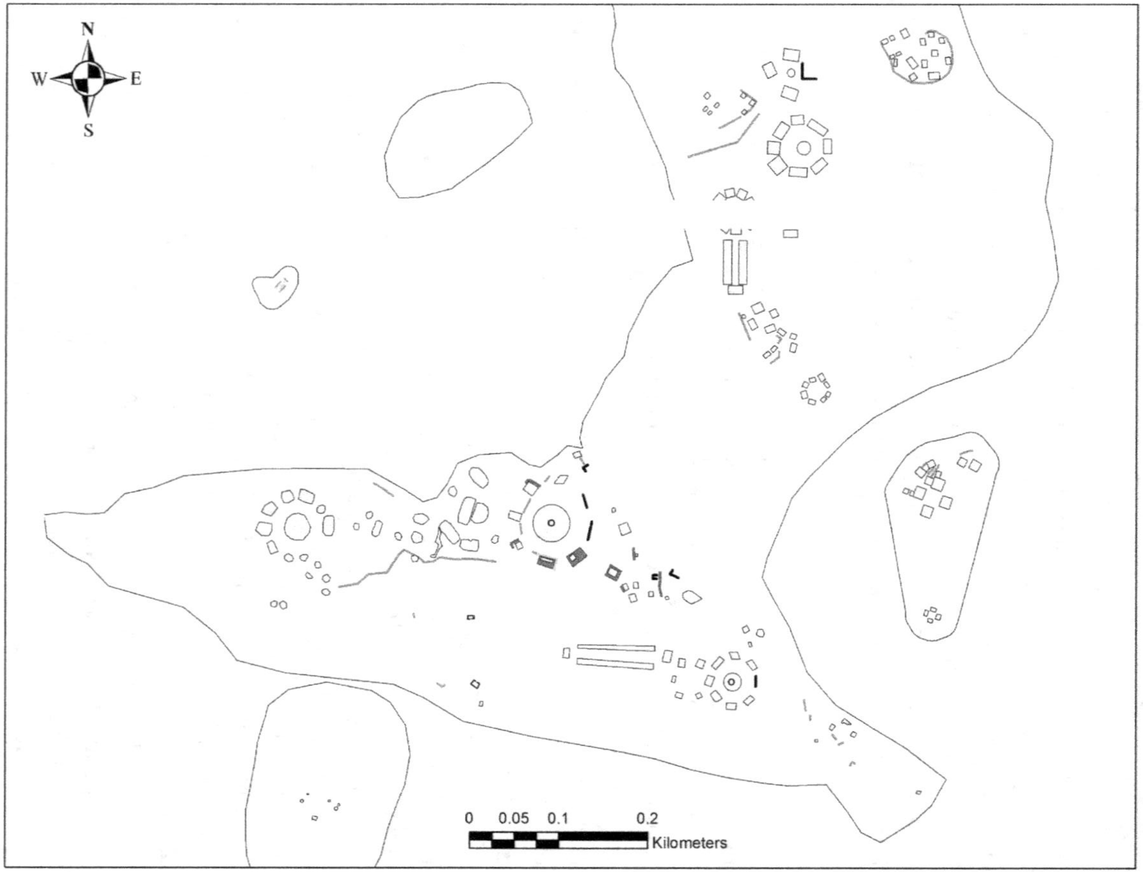

Figure 7.3. Guachimontón and associated four-structure groups at Santa Quiteria (map by author).

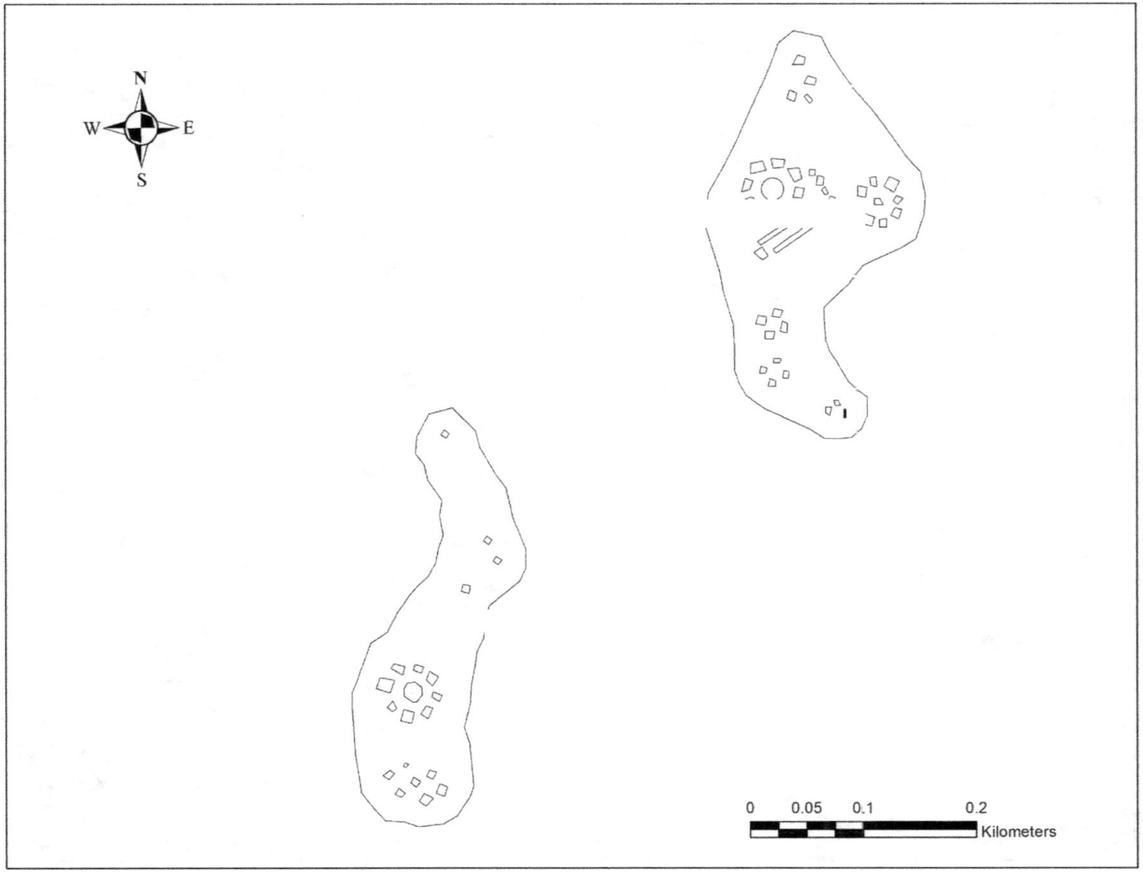

Figure 7.4. Guachimontón and associated four-structure groups at AMA-CUM-06 and AMA-CUM-12 (maps by author).

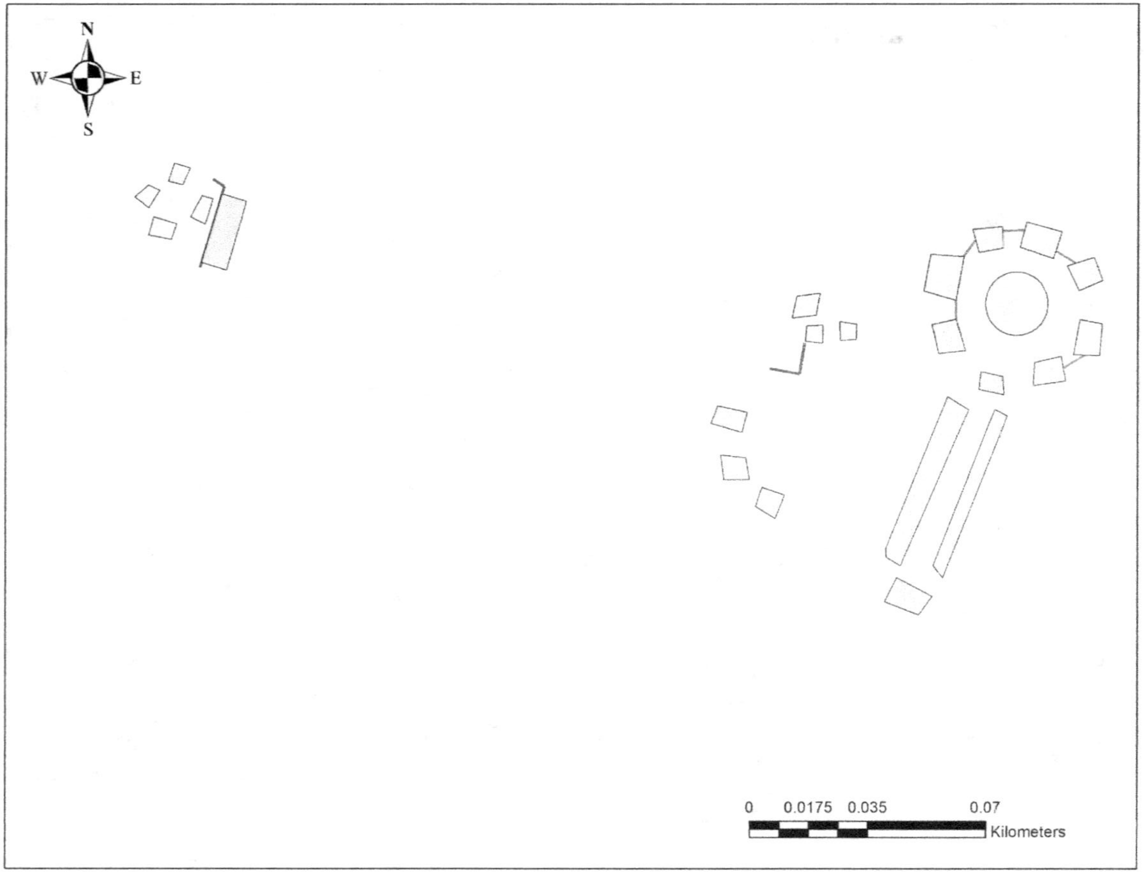

Figure 7.5. Guachimontón and associated four-structure groups at AMA-RIT-35 (map by author).

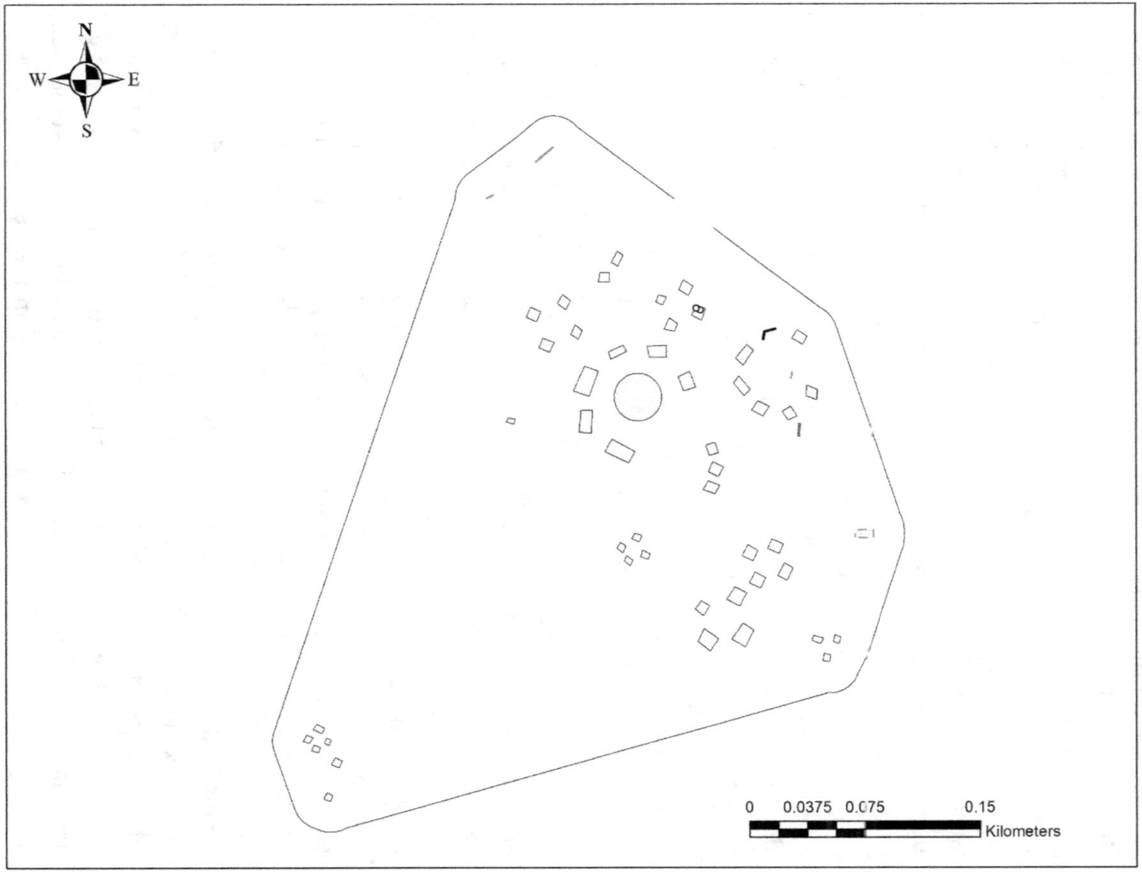

Figure 7.6. Guachimontón and associated four-structure groups MAG-HUI-02 (map by author).

for community-wide events, further demonstrating the division between influential individuals from the public sphere in society.

The extant evidence suggests that *guachimontones*, with their public character, were not used by the elite or high-status groups (factions) as arenas for competitive display and conspicuous consumption. The platforms around the *guachimontones* were not permanent residences, but rather, they were occupied when important rituals were enacted in celebration of a harvest or rain (Beekman 2003a, 2003b). Excavations at Los Guachimontones (Esparza and Weigand 2008: 34–35) and Llano Grande, and at Navajas (Beekman 2008) show that different crews were responsible for the construction of the distinct platforms in a single *guachimontón*, further supporting more collective forms of governance rather than one focused on a single individual (for example, a king or queen), as is the expectation in a segmentary, factional structure. Thus, it appears that different corporate groups occupied the platforms around the central circular altar or temple (Beekman 2012) during important ceremonies. The pattern of the *guachimontón* indicates a system of shared power rather than the concentration of power in the hands of one group or individual (Beekman 2008).

The ceramic models, found in shaft tombs, indicate that the platforms supported structures where people rested, ate, and observed the ceremonies played out in the patios and central altars (Townsend 1998: figs. 2 and 27). Thus, the *guachimontones* and the activities carried out on them are counter to factional strategies expected in segmentary states. However, new research by Sumano (2016) and Sumano and Englehardt (this volume), based on a spatial analysis at Los Guachimontones, suggests that different sectors of the site varied in accessibility, and hence some *guachimontones* may have been limited to serving specific neighborhoods or groups, while others are relatively open and suggest more inclusive, site-wide, community-wide, or region-wide activities. This line of research needs further assessment with excavated data.

Finally, if the Teuchitlán Tradition was a segmentary state, the architecture in secondary centers should express deviations from the conventionalities of a *guachimontón*. Sites in the different levels of the administrative hierarchy of a segmentary state should not conform to a standard architectural plan, as each faction, diverging from the major centers, would want to stand out from the rest. A regional survey in the Tequila region suggests that the circles repeat themselves at secondary and even third-level sites (Heredia 2017), showing a high degree of uniformity. The overall pattern then is a repetition in public architecture associated with collective or corporate rituals, at least as it refers to the *guachimontón* groups. Further, the plans of primary centers vary in complexity but the elements such as *guachimontones*, four-mound residential groups, and ballcourts are present throughout. These findings run counter to the expected pattern of a segmentary state, where faction leaders would strive to personalize the architecture. While the same architectural complexes are expected, they should have decorations or patterns that make them unique; ritual and political activities would be duplicated—but the architecture would be idiosyncratic—and public spaces would be embedded in private spaces and vice versa. In the Maya lowlands, burial pyramids are very common, but each reflects the identity of the individual buried within it. A good example is Temple I at Tikal which has a massive relief of the ruler to which it is dedicated sitting atop the temple constructed at the apex of the pyramidal platform (Sharer with Traxler 2006: 303). This does not seem to be the case in the Tequila region.

Warfare

Factions make use of warfare as a strategy to gain control over other factions, to maintain the status quo, and to obtain economic gains. In their surveys, Weigand (1996) and Heredia (2017) did not identify any defensive sites or sites enclosed by high defensive walls that may suggest warfare conflict within the Tequila region. However, potential evidence for external warfare is suggested by the establishment of sites (some with defensive walls) located in strategic locations at the natural entrances to the Tequila valleys (Beekman 1996b). Navajas in the southeast, Tepopote in the east, and Llano Grande in the west suggest that at some point during the Early Classic (AD 200–450/500), the valleys were guarded and/or access to and movement throughout the valleys was controlled by these sites (Beekman 1996b; Weigand and Beekman 1998: 44). If we look at the evidence closely, the presence of these sites shows a concern for community protection rather than the protection of a few individuals or families, as is evident in some Classic period Maya sites, where walls, moats, or barricades were built for the protection of select groups or individuals. What these sites suggest is a collective pattern in which the sites represent a public good for the protection of the entire region or entire communities from external foes and not as a sign of interfaction conflict.

Conversely, the establishment of these sites in key locations and in mountains may have facilitated the control of long-distance trade (Beekman 1996b: 144; Weigand 2008: 583). This hypothesis has yet to be tested archaeologically especially for the periods under concern in this chapter. From the published literature, long-distance exchange ,in terms of obsidian moving out of the Tequila region, is better documented for the El Grillo phase (circa AD 450/500–900) and the Atemajac I and II phases (AD 900–1600) (Esparza and Tenorio 2004; Millhauser 2002; Tenorio et al. 2015; Trombold et al. 1993). The presence of these sites in strategic locations, on mountains at the natural corridors that connect the valleys to other regions, indicates a concern with exterior foes and threats, in addition to a more cooperative rather than conflictive social milieu. This interpretation conforms to corporate political economic strategies rather than fragmentation, and therefore does support a factional political landscape.

The famed large warrior, prisoner, and ballgame player figures associated with shaft tombs in West Mexico provide an additional line of evidence to discuss warfare. Large hollow figures found in shaft tombs are a valuable information source (Beekman and Pickering 2016) on the importance of warfare and conflict in the region. Some shaft tombs include warrior figurines and depictions on murals of battle scenes and captives (Beekman 2016; Olay and Sánchez 2015). There are interesting interpretations on the significance and meaning of these figures. These artifacts and representations certainly indicate that warfare may have constituted an important activity for some individuals and could have been a source of prestige. Examples of warriors with animal masks are also present and these have correlates in many other places in Mesoamerica (Beekman 2011). Beekman (1998) argues that elite identity is clearly manifested in offerings found in the shaft tombs, which include figures of warriors, battle scenes, trophy heads, and the taking of prisoners, which point to a bellicose environment among multiple lineages *up to* the Late Formative (300 BC–AD 200). Furst (1966: 367) presented a different interpretation of these figures and suggested that these supposed warriors are representations of shamanic guardians of the earth, but his interpretation has been largely abandoned in light of contrary evidence. Osteological evidence for possible conflict or warfare was identified in the Huitzilapa shaft tomb (Ramos and López Mestas 1996), where one of the individuals showed a deformity in the right elbow that the authors attribute to the use of the *atlatl* (López Mestas 2011: 437).

Figurines and human remains thus suggest the presence of warfare in western Mesoamerica, but the combination of the available evidence suggests that such warfare was oriented toward the exterior and not within the region. Further, it is also possible that the evidence is more indicative of symbolic warfare rather than actual armed conflicts. I suggest an additional, complementary, interpretation: that the figures depicting warriors represent individuals who gained status by defending the valleys from outside attacks. This reading is in tune with the lack of defensive structures within the valleys, which would be indicative of internal competition among factions. Given the extant evidence, warfare (if present) was interregional rather than intraregional.

Monopoly on Prestige Goods

Evidence of a prestige goods system in Late Formative–Early Classic period western Mesoamerica comes primarily from the shaft tombs, especially those that contain rich offerings such as those found in Huitzilapa (Ramos and López Mestas 1996), El Arenal (Corona Nuñez 1955) and San Sebastián (Long 1966) in the Magdalena Lake Basin. The Huitzilapa tomb, dated to about AD 1, is one of the better dated and investigated of all western Mesoamerican shaft tombs. Huitzilapa is located northwest of the Tequila Volcano, and it is considered a third-level site during the Late Formative to Classic periods (Ohnersorgen and Varien 1996: table 1). Although its architecture and size are modest compared to other sites in the Tequila region, Huitzilapa and several surrounding sites form a cluster of settlement clearly independent from others in the region (Heredia 2017).

At the Huitzilapa site, the shaft tomb was found beneath one of four platforms of a high-status residential complex, adjacent to *guachimontón* architecture (López Mestas and Ramos 2006: fig. 2) and not in the center or in the outlying platforms of a *guachimontón*. This four-platform, or cruciform, group likely housed an important family; it contained the remains of at least six individuals buried in two chambers, who were related by kinship ties based on epigenetic evidence and a shared congenital deformity (López Mestas and Ramos 2006: 62). Since not all individuals interred in the tomb were accompanied by similar amounts of burial furniture, the objects found indicate that certain individuals had wider connections or were able to obtain exotic goods from distant places. Conch shell from the Pacific Coast, jadeite from Guatemala (López Mestas 2007), and thousands of other locally produced accoutrements were part of the assemblage at Huitzilapa. Clearly, the tomb and the goods within it indicate status and wealth differentiation.

The evidence from the tombs does suggest access to luxury goods by certain groups. However, shaft tombs are not overtly displayed as one would expect in a segmentary state. The fact that some individuals had access to luxury or rare items and were interred with them is not evidence that they were used in public displays as a strategy to gain followers and/or strengthen their factions. It does not mean that these individuals possessed sociopolitical or ritual power because these goods were hidden from public view. Therefore, they are not public displays of wealth meant to be manipulated for power. Ceramic types and funerary furniture have been cited as restricted to the elite. Oconahua Red-on-Cream, of the Tabachines Ware, is consistently reported to have been a high-status and special-purpose ceramic type (Aronson 1993; Beekman 1996a: 479). However, it is not restricted to elite or high-status contexts, as it has been found in low-status residences at Los Guachimontones (Weigand 2004). On the other hand, Tabachines ceramics do not seem to have been used as overt status markers; in other words, the evidence does not show a use for status enhancement or aggrandizement. Tabachines wares tend to be decorated with geometric forms, many of which have been identified as pan-Mesoamerican symbols related to fertility, water, and earth (Heredia and Englehardt 2015). All of these themes run contrary to what one would expect in a factional landscape and are more attuned to the corporate strategies described by Blanton and colleagues (1996).

Kinship and Descent Reckoning

Thus far, no inscriptions that name rulers or portrayals of specific individuals are reported for the Tequila region. The large hollow figures found in association with shaft tombs (Beekman and Pickering 2016) may have abstractly depicted specific individuals (Beekman 2016: 97), possibly heads of corporate groups or even an ancestor. Unfortunately, most of these large figures are without secure archaeological contexts, and they can only be assigned, spatially, to general regions based on their decoration or attire, which mark important differences in style across West Mexico (Townsend 2000). These individuals may represent important people. But the question lingers: did they have personal power? Individuals buried in shaft tombs were without question important and had the ability to acquire goods from distant regions, but they do not seem to represent deified ancestors. Burials underneath houses can indicate ancestor veneration, which can then be used to argue for descent reckoning, where households traced their descent back to specific individuals. Reuse of the tombs by the same kin group can point to claimed ownership of a particular space. From the extant data, there is no secure evidence to suggest that groups in the Tequila region engaged in descent or kinship reckoning (Ramos and López Mestas 1996).

Competitive Feasting

Competitive feasting is a well-known political strategy to seek prestige and wealth, and to attract followers (Blitz 1993; Bray 2003; Dietler and Hayden 2001; LeCount 2001; Levine 2011). Butterwick's (2006) analysis of ceramic models indicates that food consumption was an integral part of the Teuchitlán Tradition. But was feasting of a competitive or an integrative nature? Although Butterwick (2006: 94) argues that feasting was competitive among elites, the data are not in her favor. On the contrary, the evidence suggests that *guachimontones* were arenas for communal feasting. Moreover, comparison of the assemblages between circles and a high-status residence does not show major differences in ceramic types. Highly decorated or high-status ceramic vessels appear in similar (low) proportions in these contexts (Beekman, this volume).

Two plain and simple decorated ceramic types constitute the major proportion in the excavated assemblage, which is highly suggestive of integrative or collective feasting geared at promoting community integration rather than individual aggrandizement (Tyndall 2008). Systematic, controlled excavations in *guachimontón* complex platforms at Navajas, a site in the southeastern corner of the Tequila valleys, recovered evidence that food production was among the activities carried out in these spaces (Johns 2014). Recent analysis of ceramic collection recovered at several *guachimontón* circles at the Los Guachimontones site does not show large proportions of decorated ceramics that would suggest competitive feasting. Thus, it seems likely that *guachimontón* complexes were public spaces and not residences of specific individuals or corporate groups (Beekman 2008). In sum, the archaeological evidence for competitive feasting in the region is scant and thus such activities probably were not a main strategy used by political leaders to gain a prestige base. However, feasting as it is shown in models played a critical part in the corporate strategies geared toward community integration.

Discussion

In the previous pages I have scrutinized the available data and tested against the expectations for factional political strategies associated with segmentary states, a label that has persisted in the literature despite the lack of data to support it. The results indicate that the network or exclusionary strategies used by factions or segments highly contradict the hypothesis that the people of the Teuchitlán Tradition were organized around a segmentary structure. The data point to a much more complex political landscape, one that can greatly benefit from current developments in archaeological theory on collective governments and corporate strategies known as alternative pathways to complexity. Nevertheless, the data available provide information on multiple scales to start a discussion on the nature of this early expression of social complexity.

In the original model, segmentary states are ruled by a king, who controls a small, central territory. Previously, the central territory was assumed to be the area around Los Guachimontones, as this is the largest site in the entire region. No king or analogous figure has been identified, here or at other regional sites. In addition, the lack of obvious palaces and sumptuous architecture suggests that kings as we know them from the typical segmentary states are not present in the region. Thus, faction heads appear nonexistent or, at the least, they are not conspicuous, and a king whose seat of power would have been Los Guachimontones in Teuchitlán is not evident. Therefore, a major component of the model seems to be missing. In addition, the multiple platforms around the altar were constructed, occupied, and used by different groups (Beekman 2008). This is an important point to highlight as these individuals or groups appear to have mobilized people to build a platform in a single *guachimontón*, hence the conclusion about power sharing (see Beekman 2008). Who these groups are, and whether they represent groups in a single site or whether they are corporate groups representing region-wide political units, is a subject that requires future attention.

At the regional level, the presence of sites on natural defensive locations points to the application of collective policies for the benefit of all people within the valleys. However, this does not translate into the presence of a single centralizing state or political structure; settlement patterns are much more complex. Previously, scholars "assumed" the valleys around Tequila Volcano within the valleys acted as the "core" of this segmentary state, with a hegemonic capital centered at Los Guachimontones. The regional survey by Heredia (2017) north of the Tequila Volcano identified a pattern indicative of a fragmented landscape (politically and economically), yet with a shared corporate code as represented by *guachimontón* architecture. The regional surveys in the north (Heredia 2017) and west (Heredia et al. 2018) show a multiplicity of clusters that are interpreted as semi-independent political units that were not part of a unified polity. The survey south of the Tequila Volcano (Weigand 1996) indicates a settlement pattern that may differ from the rest of the region. However, I contend that despite the differences in survey methods between Weigand and subsequent surveys, the presence of clusters is evident. Weigand (1985: 79) mentioned six areas of major settlement that may be interpreted as distinct clusters within the region, comparable to those that Beekman and I have identified in our systematic surveys. In addition, the differences in size of *guachimontones* and site size do not seem to correspond to a well-developed administrative hierarchy. Spatial analyses (Trujillo 2019) indicate that, in general, major sites and densely occupied areas were positioned in strategic locations in relation to water, fertile land, and obsidian mines. This line of evidence supports the presence of small political units that controlled a small surrounding territory based on the presence of these natural resources.

In previous presentations on this topic, I have been asked the question: if the TT is not a segmentary state, then what is it? This question is welcome, but it also belies a continued evolutionary/neoevolutionary mindset that lingers among many colleagues. Asking "what is it?" invites us to couch societies in terms of static categories, a practice that many scholars seek to avoid. Rather than focusing on a set of characteristics, scholars are increasingly focusing on the human actions (be they economic, political, or ideological) that produced them. In this sense, the TT, with its repetitive *guachimontón* architecture, the hidden nature of elites and "elite" or ostentatious architecture and artifacts, the apparent lack of conflict, and the overall corporate ideology, contradicts the segmentary state model.

More than three decades have passed since Weigand coined the term Teuchitlán Tradition to describe the cultural manifestations left behind by the peoples in the Tequila valleys. In the last decade, our understanding of the internal structure and functioning of this society has greatly increased, thanks to the regional surveys, excavations, and artifact analysis of multiple scales and sites. Now, with more than 900 square kilometers of systematic and full-coverage surveys (Beekman and Heredia 2017; Heredia 2017; Heredia et al. 2018), the settlement patterns discerned show a series of site clusters that are separated from the others by clear empty spaces. Each cluster holds some sort of unity through the presence of ballcourts. The distribution of ballcourts, their location within sites, and the artifacts associated with them indicate that they had an important role in promoting community solidarity at the local and regional levels (Beekman and Heredia 2017). Thus, the hypothesized densely packed region, with a centralized single structure analogous to a state, is not holding up to the empirical evidence.

Before the completion of the major regional surveys in the area, I carried out a preliminary analysis using the distribution of artifacts in a small 30 square kilometer area. The repetitive nature of the architecture, as well as the recurring types of ceramics and lithic artifacts, led me to suggest, preliminarily, a lack of a division of labor and therefore the absence of a market system (Heredia 2011). These initial observations are now contradicted by new data. For example, petrographic analysis of decorated types of ceramics from multiple sites in the northern Tequila region tested the hypothesis that highly decorated ceramic types were controlled in their production and distribution. The sample included those types that could potentially be used for political purposes—to legitimize a select number of people in positions of power and/or authority. The results of the petrographic analysis indicate that ceramics were produced in different locations throughout the Tequila region. In addition, it suggests that the same paste was used to produce different ceramic types, fine and coarse irrespectively. Thus, there was a multiplicity of potential workshops in the region that supplied vessels to the population. The conclusion drawn is that there must have been some sort of exchange mechanism between different communities in the area.

Given these results, along with the settlement pattern data, the existence of an exchange mechanism, and perhaps a market system, is now highly probable for the Tequila region. Potential pottery-producing communities are distributed in such a way that people benefited from multiple suppliers, and this trade mechanism would have been an important institution that integrated these communities at a regional level. Further, the multiple production locales of these decorated wares were not controlled in their distribution and consumption. My conclusion is that even though some of the ceramic types are associated with shaft tombs, these objects were not politically charged and instead were available to a wider population and in other than funerary or high-status contexts. An iconographic analysis of decorated vessels shows that the symbolism can be related to a pan-Mesoamerican ideology focused on themes of cosmology, fertility, ritual legitimation, and sociopolitical power shared by many groups (Heredia and Englehardt 2015: 24).

An additional line of evidence for the presence of an exchange mechanism that connected communities in the Tequila region and beyond comes from the distribution of obsidian from different sources. X-ray fluorescence analysis of 259 obsidian objects from sites in the northern Tequila region indicates that people in different settlements had access to objects from multiple obsidian sources (Carrión 2019), some of which fall outside the Tequila region. This line of evidence suggests the existence of an exchange mechanism that connected people widely. Further research on this topic will certainly advance our understanding of economic circuits.

In sum, people were linked though economic networks in addition to the known shared ideology materialized in the *guachimontón* architecture. The existence of a market system needs further testing, but multiple lines of independent research are suggesting that an important mechanism by which various communities were linked was the exchange of goods and information, as evidenced in ceramics, iconography, lithics, and architecture. Thus, rather than factional competition, a strong community ethos is clear in the archaeological record.

Conclusions and Future Directions

Weigand and Beekman (1998) proposed the segmentary state model to describe the distribution of *guachimontón* architecture beyond the limits of the Tequila region. The model was not only underexploited and never fully explained or unpacked, but it was used rather freely to describe a material cultural phenomenon, an unclear one at that, and one that many neoevolutionists ignored, precisely because it did not conform to static typological categories. Nevertheless, the model became somewhat doctrinal, and continues to be regurgitated in publications as if it was an established truth (Esparza 2015; Olay and Sánchez 2015; see Cardona 2016 for a critique of the misuse of terminology in the Tequila region). Thus, despite recent research that has developed new theoretical models that account for the multiple ways in which societies develop (Blanton et al. 1996; Blanton and Fargher 2008; Fargher et al. 2011), the dogmatic segmentary-state position has never been questioned, understood, or tested fully. In some measure, this is due to the poor use of theory to explain social process in this region (but see Beekman 2016 and Heredia 2016, 2017 for exceptions).

In this chapter, I have argued that segmentary states should display the strategies proposed for the various cross-cultural examples in which these social formations have been identified. In such states, the focus is on individuals and/or factions that pursue network or exclusionary strategies in order to gain a political base. I propose that we abandon the use of segmentary state to refer to the Teuchitlán Tradition, as well as the use of neoevolutionary typologies to characterize any society in this region, because such models and typologies do not explain the sociocultural process behind a given archaeological configuration. Rather, they only give us stale descriptions on the distribution of architecture and artifacts that limits our understanding of the past and how societies functioned, in both western Mesoamerica and beyond. Thus, I suggest that we step back and reexamine the evidence to identify agent-based processes that can explain the nature of the archaeological evidence, rather than repeating dubious, *a priori* conclusions.

Future studies of political structure in the Tequila region should focus on generating additional systematic regional data with comparable field methods, thus making possible comparisons within and beyond the area. And since poor chronological control hinders a better understanding of the region as a whole (but see Beekman, this volume), stratigraphic excavations and absolute dates are crucial to getting a better grip on the regional trajectory and growth of political structures. An emphasis on institutions such as markets, religion, government, for example, will be a much fruitful line of future research. Beekman (2008, 2016) and Heredia (2017) are adopting alternative models to investigate the dynamic nature of the Teuchitlán culture as well as subsequent cultural developments in the region (Heredia 2016). The combination of regional and site-specific data (alternating between multiple scales of analysis) will greatly enhance our comprehension of the region in the years to come as new theory is providing valuable perspectives from which to investigate the alternative trajectories of Tequileños during pre-Hispanic times.

References

Aronson, Meredith

1993 Technological Change: West Mexican Mortuary Ceramics. Unpublished PhD dissertation, Department of Materials Science and Engineering. University of Arizona, Tucson.

Beekman, Christopher S.

1996a The Long-Term Evolution of a Political Boundary: Archaeological Research in Jalisco, México. Unpublished PhD dissertation, Department of Anthropology, Vanderbilt University, Nashville, TN.

1996b Political Boundaries and Political Structure: The Limits of the Teuchitlán Tradition. *Ancient Mesoamerica* 7(1): 135–147.

1998 Lineage Tombs and Lineage Relations. Paper presented at the American Anthropological Association Meetings, Philadelphia, PA.

2003a Agriculture, Pole Rituals and Rulership in Late Formative Central Jalisco. *Ancient Mesoamerica* 14(2): 299–318.

2003b Fruitful Symmetry: Corn and Cosmology in the Public Architecture of Late Formative and Early Classic Jalisco. *Mesoamerican Voices* 1: 5–22.

2008 Corporate Power Strategies in the Late Formative to Early Classic Tequila Valleys of Central Jalisco. *Latin American Antiquity* 19(4): 414–434.

2011 Mesoamerican Symbols of Authority in the Shaft Tomb Figures of West Mexico. Paper presented at the 76th Annual Meeting of the Society for American Archaeology, Sacramento, CA.

2012 Current Views on Power, Economics, and Subsistence in Ancient Western Mexico. In *The Oxford Handbook of Mesoamerican Archaeology*, edited by Deborah L. Nichols and Christopher A. Pool, pp. 495–512. Oxford University Press, Oxford.

2016 Built Space as Political Fields: Community vs. Lineage Strategies in the Tequila Valleys. In *Alternative Pathways to Complexity: A Collection of Essays on Architecture, Economics, Power, and Cross-Cultural Analysis*, edited by Lane F. Fargher and Verenice Y. Heredia Espinoza, pp. 59–78. University Press of Colorado, Boulder.

Beekman, Christopher S. and Verenice Y. Heredia Espinoza

2017 Los Juegos de Pelota de Jalisco: ¿Competencia o Integración? *Arqueología Mexicana* XXV(146): 64–69.

Beekman, Christopher S., and Robert B. Pickering (editors)

2016 *Shaft Tombs and Figures in West Mexican Society: A Reassessment.* Gilcrease Ancient Americas Series, Gilcrease Museum, Tulsa, OK.

Blanco Morales, Ericka S.

2009 *El Juego de Pelota en la Tradición Teuchitlán: hacia una propuesta sobre su función social.* Unpublished MA thesis, El Colegio de Michoacán, La Piedad.

Blanton, Richard E.

1998 Beyond Centralization. In *Archaic States*, edited by Gary M. Feinman and Joyce Marcus, pp. 135–172. School of American Research Press, Santa Fe, NM.

Blanton, Richard E., and Lane F. Fargher

2008 *Collective Action in the Formation of Pre-Modern States.* Fundamental Issues in Archaeology. Springer, New York.

Blanton, Richard E., Gary M. Feinman, Stephen A. Kowalewski, and Peter N. Peregrine

1996 A Dual-Processual Theory for the Evolution of Mesoamerican Civilization. *Current Anthropology* 37(1): 1–14.

Blitz, John H.

1993 Big Pots for Big Shots: Feasting and Storage in a Mississippian Community. *American Antiquity* 58(1): 80–96.

Bray, Tamara L. (editor)

2003 *The Archaeology and Politics of Food and Feasting in Early States and Empires.* Kluwer Academic/Plenum, New York.

Brumfiel, Elizabeth M.

1994 Factional Competition and Political Development in the New World: An Introduction. In *Factional Competition and Political Development in the New World*, edited by Elizabeth M. Brumfiel and John W. Fox, pp. 3–13. Cambridge University Press, Cambridge, UK.

Butterwick, Kristi
2006 Comida para los Muertos: El Arte de los Banquetes en el Occidente. In *Perspectivas del Antiguo Occidente de México*, edited by Richard F. Townsend, pp. 93–109. Art Institute of Chicago, Secretaría De Cultura del Gobierno de Jalisco, and Guachimontones y Naturaleza, A.C., Chicago and Guadalajara.

Byland, Bruce E., and John M.D. Pohl
1994 Political factions in the transition from Classic to Postclassic in the Mixteca Alta. In *Factional Competition and Political Development in the New World*, edited by Elizabeth M. Brumfiel and John W. Fox, pp. 117–126. Cambridge University Press, Cambridge, UK.

Cabrero, María Teresa
1989 *Civilización en el Norte de México: Arqueología de la Cañada del Río Bolaños (Zacatecas y Jalisco)*. Universidad Nacional Autónoma de México, Mexico City.

Cabrero, María Teresa, and Carlos López Cruz
2002 *Civilización en el Norte de México* II. Universidad Nacional Autónoma de México, Mexico City.

Cardona Machado, Hector J.
2016 Las Narrativas de la Autoridad: Activación del Pasado en el Sitio Arqueológico Los Guachimontones, Teuchitlán, Jalisco. Unpublished MA thesis, Centro de Estudios Arqueológicos, El Colegio de Michoacán, La Piedad, Michoacán.

Carrión Méndez, Hugo
2019 ¿Cambios o continuidades? El Paisaje político a partir del paisaje lítico en la región del norte del Volcán de Tequila durante las fases Tequila II a IV (350 a.C.–500 d.C.) y El Grillo (500–900 d.C.). Unpublished MA thesis, Centro de Estudios Arqueológicos, El Colegio de Michoacán, La Piedad, Michoacán.

Chance, John K.
2000 The Noble House in Colonial Puebla, Mexico: Descent, Inheritance, and the Nahua Tradition. *American Anthropologist* 102(3): 485–502.

Chase, Arlen F., and Diane Z. Chase
1992 Mesoamerican Elites: Assumptions, Definitions, and Models. In *Mesoamerican Elites: An Archaeological Assessment*, edited by Diane Z. Chase and Arlen F. Chase, pp. 3–17. University of Oklahoma Press, Norman.
1996 More than Kin and King. *Current Anthropology* 37(5): 803–810.

Corona Nuñez, José
1955 *Tumba de El Arenal, Etzatlán, Jalisco*. Instituto Nacional de Antropología e Historia, Mexico City.

Dahlgren, Barbro
1990 *La Mixteca: Su Cultura e Historia Prehispánicas*. 4th ed. Imprenta Universitaria, Mexico City.

Demarest, Arthur A.
1992 Ideology in Ancient Maya Cultural Evolution: The Dynamics of Galactic Polities. In *Ideology and Pre-Columbian Civilizations*, edited by Arthur A. Demarest and Geoffrey W. Conrad, pp. 137–157. School of American Research Press, Santa Fe, NM.

Dietler, Michael, and Brian Hayden (editors)
2001 *Feasts: Archaeological and Ethnographic Perspectives on Food, Politics, and Power.* University of Alabama Press, Tuscaloosa.

Esparza López, Rodrigo
2015 La Tradición Teuchitlán. *Revista Occidente* 1: 1–13. http://www.mna.inah.gob.mx/contexto.html

Esparza López, Rodrigo, and Dolores Tenorio
2004 Las redes de intercambio de la obsidiana en la Tierra Caliente de Michoacán durante los periodos Epiclásico y Postclásico. In *Bienes estratégicos del antiguo occidente de México*, edited by Eduardo Williams, pp. 77–112. El Colegio de Michoacán, Zamora.

Esparza López, Rodrigo, and Phil C. Weigand
2008 *Proyecto Arqueológico Loma Alta, Teuchitlán, Jalisco: Redescubriendo la Tradición Teuchitlán.* Informe Técnico Temporada 2004–2006. Report submitted to the Instituto Nacional de Antropología e Historia, Mexico City.

Fargher, Lane F., and Richard E. Blanton
2012 Segmentación y acción colectiva: un acercamiento cultural-comparativo sobre la voz y el poder compartido en los Estados premodernos. In *El poder compartido: Ensayos sobre la arqueología de organizaciones políticas segmentarias y oligárquicas*, edited by Annick Daneels and Gerado Gutiérrez Mendoza, pp. 205–235. CIESAS/El Colegio de Michoacán, A.C., Mexico City/Zamora.

Fargher, Lane F., and Verenice Y. Heredia Espinoza (Editors)
2016 *Alternative Pathways to Complexity: A Collection of Essays on Architecture, Economics, Power, and Cross-Cultural Analysis in Honor of Richard E. Blanton.* University Press of Colorado, Boulder.

Fargher, Lane F., Verenice Y. Heredia Espinoza, and Richard E. Blanton
2011 Alternative Pathways to Power in late Postclassic Highland Mesoamerica. *Journal of Anthropological Archaeology* 30(3): 306–326.

Fox, John W.
1989 On the Rise and Fall of Tulan's and Maya Segmentary States. *American Anthropologist* 91(3): 656–681.

Fox, John W., Garrett W. Cook, Arlen F. Chase, and Diane Z. Chase
1996 Questions of Political and Economic Integration: Segmentary versus Centralized States among the Maya. *Current Anthropology* 37(5): 795–801.

Furst, Peter
1966 *Shaft Tombs, Shell Trumpets and Shamanism: A Culture-Historical Approach to Problems in West Mexican Archaeology.* PhD dissertation, Department of Anthropology, University of California, Los Angeles. University Microfilms, Ann Arbor.

Geertz, Clifford
1980 *Negara: The Theatre State in 19th Century Bali.* Princeton University Press, Princeton, NJ.

Heredia Espinoza, Verenice Y.
2011 Patrones de asentamiento and división regional de trabajo al norte del volcán de Tequila, Jalisco. In *Patrones de asentamiento y actividades de subsistencia en el*

occidente de México. Reconocimiento a la Dra. Helen Pollard, edited by Eduardo Williams and Phil C. Weigand, pp. 121–140. El Colegio de Michoacán, Zamora.

2016 Complexity without Centralization: Corporate Power in Postclassic Jalisco. *Alternative Pathways to Complexity: A Collection of Essays on Architecture, Economics, Power, and Cross-Cultural Analysis in Honor of Richard E. Blanton*, edited by Lane F. Fargher y Verenice Y. Heredia Espinoza, pp. 79–103. University Press of Colorado, Boulder.

2017 Long-term Regional Landscape Change in the Northern Tequila Region of Jalisco, Mexico. *Journal of Field Archaeology* 42(4): 298–311.

Heredia Espinoza, Verenice Y., Christopher S. Beekman, and Kirk Anderson

2018 *Proyecto Arqueológico en la Cuenca de la ex-Laguna de Magdalena, Jalisco (PAX)*. Report submitted to the Instituto Nacional de Anttropología e Historia, Mexico City.

Heredia Espinoza, Verenice Y., and Joshua D. Englehardt

2015 Simbolismo Pan-Mesoamericano en la Iconografía Cerámica de la Tradición Teuchitlán. *Travaux et Recherches dans les Amériques du Centre* 68: 9–34.

Herrejón Villicaña, Jorge

2008 Unidades habitacionales y estratificación social en la Tradición Teuchitlán. Unpublished MA thesis, El Colegio de Michoacán, La Piedad.

Jansen, Maarten

1990 The Search for History in Mixtec Codices. *Ancient Mesoamerica* 1(1): 99–112.

Johns, Catherine

2014 *Ceramic Activity Analysis of Navajas Circle 5 and the Need for Practice Theory in Unusual Monumental Architecture*. MA thesis, Department of Anthropology, University of Colorado Denver.

LeCount, Lisa J.

2001 Like Water for Chocolate: Feasting and Political Ritual among the Late Classic Maya at Xunantunich, Belize. *American Anthropologist* 103(4): 935–953.

Levine, Marc N.

2011 Negotiating Political Economy at Late Postclassic Tututepec (Yucu Dzaa), Oaxaca, Mexico. *American Anthropologist* 113(1): 22–39.

Long, Stanley V.

1966 *Archaeology of the Municipio of Etzatlán, Jalisco*. Unpublished PhD dissertation. Department of Anthropology, University of California, Los Angeles.

López Mestas C., Lorenza

2007 *Las Piedras Verdes en el Centro de Jalisco*. Foundation for the Advancement of Mesoamerican Studies, Crystal River, FL.

2011 *Ritualidad, Prestigio y Poder en el Centro de Jalisco Durante el Preclásico Tardío y Clásico Temprano. Un Acercamiento a la Cosmovisión e Ideología en el Occidente del México Prehispánico*. Unpublished dissertation, Departamento de Antropología Social, Centro de Investigaciones y Estudios Superiores en Antropología Social en Occidente, Guadalajara.

López Mestas C., Lorenza, and Jorge Ramos de la Vega

2006 Some Interpretations of the Huitzilapa Shaft Tomb. *Ancient Mesoamerica* 17(2): 271–281.

Mann, Michael

1984 The Autonomous Power of the State: Its Origins, Mechanisms, and Results. *European Journal of Sociology* 25(2): 185–213.

Marcus, Joyce, and Gary M. Feinman

1998 Introduction. In *Archaic States*, edited by Gary M. Feinman and Joyce Marcus, pp. 3–13. School of American Research, Santa Fe, NM.

Martin, Simon, and Nikolai Grube

2008 *Chronicle of Maya Kings and Queens: Deciphering the Dynasties of the Ancient Maya*. 2nd ed. Thames & Hudson, London.

Millhauser, John K.

2002 Malpaso Valley Obsidian Exchange. *Archaeology Southwest* 16(1): 9 (Special issue: La Quemada, a Monument on the Mesoamerican Frontier, edited by B. A. Nelson and E. C. Wells).

Ohnersorgen, Michael A., and Mark D. Varien

1996 Formal Architecture and Settlement Organization in Ancient West Mexico. *Ancient Mesoamerica* 7(1): 103–120.

Olay Barrientos, María de los Ángeles, and Ligia Sofía Sánchez Morton

2015 La Tradición Teuchitlán en Colima, notas para una discusión. *Revista Occidente,* INAH. Electronic document, http://www.mna.inah.gob.mx/contexto.html, accessed January 10, 2017.

Pastor, Rodolfo

1987 *Campesinos y Reformas: La Mixteca, 1700–1856*. El Colegio de México, Mexico City.

Pohl, John M. D.

2003a Creation Stories, Hero Cults, and Alliance Building: Confederacies of Central and Southern Mexico. In *The Postclassic Mesoamerican World*, edited by Michael E. Smith and Frances F. Berdan, pp. 61–66. University of Utah Press, Salt Lake City.

2003b Ritual and Iconographic Variability in Mixteca-Puebla Polychrome Pottery. In *The Postclassic Mesoamerican World*, edited by Michael E. Smith and Frances F. Berdan, pp. 201–206. University of Utah Press, Salt Lake City.

Pohl, John M.D., and Bruce E. Byland

1990 Mixtec Landscape Perception and Archaeological Settlement Patterns. *Ancient Mesoamerica* 1: 113–131.

Ramos de la Vega, Jorge, and Lorenza López Mestas C.

1996 Datos Preliminares sobre el Descubrimiento de una Tumba de Tiro en el Sitio de Huitzilapa, Jalisco. *Ancient Mesoamerica* 7(1): 121–134.

Scarborough, Vernon L.

1998 Ecology and Ritual: Water Management and the Maya. *Latin American Antiquity* 9(2):135–159.

Scarborough, Vernon L., Nicholas P. Dunning, Kenneth B. Tankersley, Christopher Carr, Eric Weaver, Liwy Grazioso, Brian Lane, John G. Jones, Palma Buttles, Fred Valdez, and David L. Lentz

2012 Water and Sustainable Land Use at the Ancient Tropical City of Tikal, Guatemala. *Proceedings of the National Academy of Sciences* 109(31): 12408–12413.

Schele, Linda, and Mary Ellen Miller
1986 *The Blood of Kings: Dynasty and Ritual in Maya Art*. Kimbell Art Museum, Fort Worth, TX.
Sharer, Robert J., with Loa P. Traxler
2006 *The Ancient Maya*. 6th ed. Stanford University Press, Stanford, CA.
Smith Márquez, Sean Montgomery
2009 El Caso Teuchitlán-Loma Alta: Un Ejemplo de Patrón de Asentamiento Mesoamericano. Unpublished MA thesis, El Colegio de Michoacán, La Piedad.
Southall, Aidan
1988 The Segmentary State in Africa and Asia. *Comparative Studies in Society and History* 30(1): 52–82.
2004 *Alur Society: A Study in Processes and Types of Domination*. LIT Verlag, Münster.
Spores, Ronald
1967 *The Mixtec Kings and Their People*. University of Oklahoma Press, Norman.
1974 Marital Alliances in the Political Integration of Mixtec Kingdoms. *American Anthropologist* 76(2): 297–311.
1984 *The Mixtecs in Ancient and Colonial Times*. University of Oklahoma Press, Norman.
2007 *Ñuu Ñudzahui La Mixteca de Oaxaca: La Evolución de la Cultura Mixteca desde los Primeros Pueblos Preclásicos hasta la Independencia*. Instituto Estatal de Educación Pública de Oaxaca, Oaxaca.
Spores, Ronald, and Andrew K. Balkansky
2013 *The Mixtecs of Oaxaca: Ancient Times to the Present*. University of Oklahoma Press, Norman.
Spores, Ronald, and Nelly M. Robles García
2007 A Prehispanic (Postclassic) Capital Center in Colonial Transition: Excavations at Yucundaa Pueblo Viejo de Teposcolula, Oaxaca. *Latin American Antiquity* 18(3): 333–353.
Stein, Burton
1980 *Peasant, State, and Society in Medieval South India*. Oxford University Press, Delhi.
Sumano Ortega, Kimberly
2016 Los Guachimontones, Jalisco: Hacia una Caracterización Sociopolítica desde la Arquitectura. Unpublished MA thesis, El Colegio de Michoacán, La Piedad.
Tambiah, Stanley J.
1977 The Galactic Polity: The Structure of Traditional Kingdoms in Southeast Asia. *Annals of the New York Academy of Sciences* 293(1): 69–97.
Tenorio, Dolores, Melania Jiménez Reyes, J. Rodrigo Esparza López, Thomas F. Calligaro and Luis Alfonso Grave Tirado
2015 The Obsidian of Southern Sinaloa: New Evidence of Aztatlán Networks through PIXE. *Journal of Archaeological Science Reports* 4: 106–110.
Terraciano, Kevin
2001 *The Mixtecs of Colonial Oaxaca: Ñudzahui History, Sixteenth through Eighteenth Centuries*. Stanford University Press, Stanford, CA.

Townsend, Richard F.
1998 Before Gods, before Kings. In *Ancient West Mexico: Art of the Unknown Past*, edited by Richard F. Townsend, pp. 107–135. Art Institute of Chicago, Chicago.
2000 Introducción: Renovando las investigaciones en el antiguo Occidente de México. In *El Antiguo Occidente de México: Arte y arqueología de un pasado desconocido*, edited by Richard F. Townsend, pp. 19–37. Art Institute of Chicago/ Secretaria de Cultura Gobierno de Jalisco/Tequila Sauza, Chicago/Guadalajara.

Trombold, Charles D., Jams F. Luhr, Toshiaki Hasenaka, and Michael D. Glascock
1993 Chemical Characteristics of Obsidian from Archaeological Sites in Western Mexico and the Tequila Source Area: Implications for Regional and Pan-Regional Interaction within the Northern Mesoamerican Periphery. *Ancient Mesoamerica* 4(2): 255–270.

Trujillo, Armando
2019 Aproximaciones Metodológicas para el Análisis del Territorio en la Tradición Teuchitlán. In *Nuevos Enfoques en la Arqueología de la Región de Tequila*, edited by Verenice Y. Heredia Espinoza, Joshua D. Englehardt, and Héctor J. Cardona Machado, pp. 55–82. El Colegio de Michoacán, Zamora.

Tyndall, Gregory
2008 Los Banquetes en Navajas, Jalisco. In *Tradición Teuchitlán*, edited by Christopher S. Beekman, Phil C. Weigand, and Rodrigo Esparza, pp. 251–273. El Colegio de Michoacán and Secretaría de Cultura del Estado de Jalisco, Zamora and Guadalajara.

Webster, David
1998 Warfare and Status Rivalry: Lowland Maya and Polynesian Comparisons. In *Archaic States*, edited by Gary M. Feinman and Joyce Marcus, pp. 311–351. School of American Research Press, Santa Fe, NM.

Weigand, Phil C.
1985 Evidence for Complex Societies during the Western Mesoamerican Classic Period. In *The Archaeology of West and Northwest Mesoamerica*, edited by Michael Foster and Phil C. Weigand, pp. 47–91. Westview Press, Boulder, CO.
1993 *Evolución de una Civilización Prehispánica*. El Colegio de Michoacán and Secretaría de Cultura del Estado de Jalisco, Zamora and Guadalajara.
1996 The Architecture of the Teuchitlan Tradition of the Occidente of Mesoamerica. *Ancient Mesoamerica* 7(1): 91–101.
2004 La Tradición Teuchitlán del Occidente de México. En *Tradiciones Arqueológicas*, edited by Efraín Cárdenas, pp. 217–241. El Colegio de Michoacán and Gobierno del Estado de Michoacán, Zamora and Morelia.
2008 The Teuchitlán Tradition and the Excavations at the Guachimontones de Teuchitlán, Jalisco. In *Urbanism in Mesoamerica*, edited by Alba Guadalupe Mastache, Robert Cobean, Ángel García Cook and Kenneth G. Hirth, pp. 557–592. Pennsylvania State University and Instituto Nacional de Antropología e Historia, State College and Mexico City.
2009 El Estado Segmentario en el Occidente de Mesoamérica. In *Las Sociedades Complejas del Occidente de México en el Mundo Mesoamericano: Homenaje al Dr. Phil*

C. Weigand, edited by Eduardo Williams, Lorenza López Mestas C., and Rodrigo Esparza, pp. 53–74. El Colegio de Michoacán, Zamora.

2010 El Estado Segmentario en el Occidente de Mesoamérica. *Ecumene* 1(1): 3–31.

Weigand, Phil C., and Christopher S. Beekman

1998 The Teuchitlán Tradition: Rise of a Statelike Society. In *Ancient West Mexico: Art and Archaeology of the Unknown Past*, edited by R. F. Townsend, pp. 35–51. Art Institute of Chicago, Chicago.

Weigand, P., and A. García de Weigand

2005 El Juego de Pelota Monumental de Los Guachimontones, Teuchitlán, Jalisco. En *El Antiguo Occidente de México: Nuevas Perspectivas sobre el Pasado Prehispánico*, edited by E. Williams, P. Weigand, L. López Mestas, and D. Grove, pp. 45–72. El Colegio de Michoacán, A.C., Zamora.

8

Cerro de Santiago

An Epiclassic Site within the World-System of Central-Northern Mexico

M. NICOLÁS CARETTA AND MANUEL DUEÑAS GARCÍA

In many ways, the state of Aguascalientes in Mexico is unknown territory in terms of knowing and understanding pre-Hispanic cultures. Studies conducted recently at the archaeological site of Cerro de Santiago show a population with characteristics that link them to their southern neighbors and show evidence of certain Mesoamerican canons in the creation of their material culture. Although the studies involve only field surveys and excavations, this contribution will present diagnostic cultural features and their implications for the dynamics of macroregional social interaction in Late Classic Mesoamerica.

Recent research conducted at the archaeological site of Cerro de Santiago, located north of the Río Verde Grande in the present-day state of Aguascalientes, Mexico (Figures 8.1 and 8.2), provides new data that not only illuminate its pre-Hispanic occupation, but also help us to understand the processes of interaction that took place throughout the expansive north-central border of Greater Mesoamerica during the Epiclassic period (circa AD 600–900) (Nicolás Caretta 2006, 2010, 2011a, 2011b, 2012, 2013, 2014a, 2014b; Puch Ku 2014; Silva 2015).

The study of northern Mesoamerica has focused mainly on understanding how the ancient monumental and ceremonial sites located in this portion of modern Mexico exhibit strong and clear signs of ties with their southern neighbors. Based on previous research, different scholars have used the term *fluctuant border* in an attempt to explain the appearance and subsequent abandonment of populations with Mesoamerican

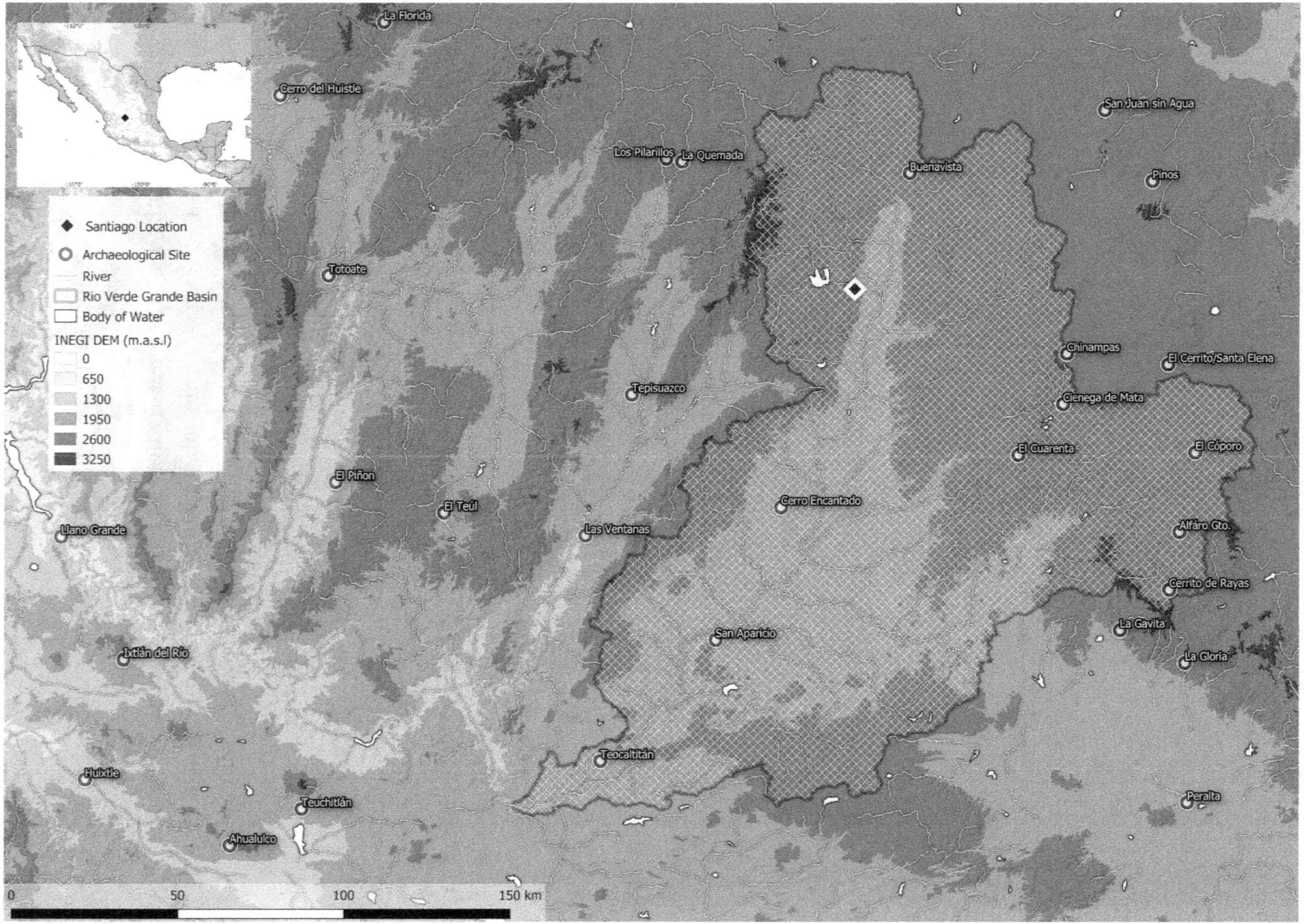

Figure 8.1. Location of archaeological sites in the Río Verde Basin (map by the authors).

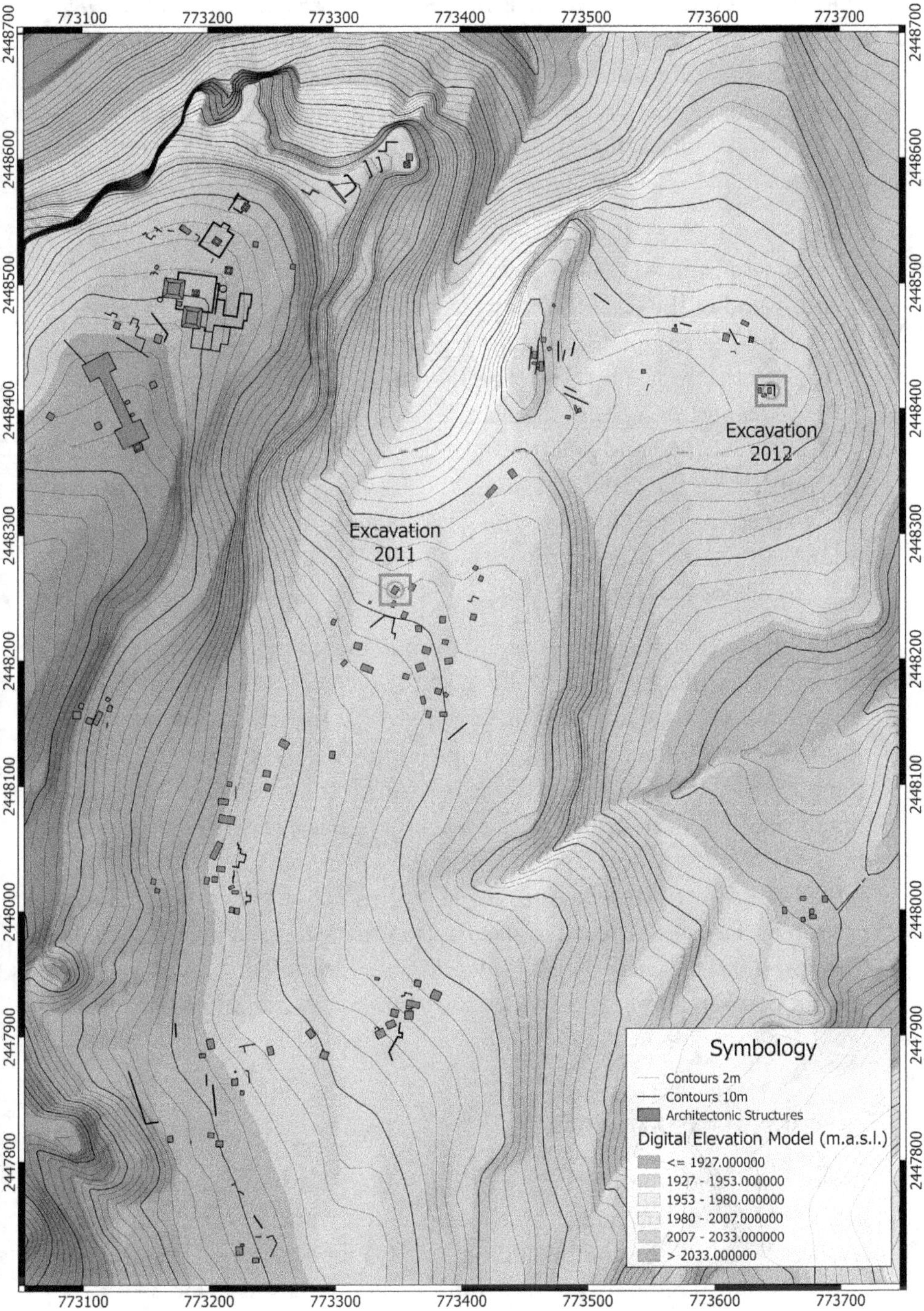

Figure 8.2. The archaeological site of Cerro de Santiago (map by the authors).

characteristics (compare Armillas 1964, 1969, 1987; Braniff 1965, 1974, 1989, 1994, 2001; Jiménez Betts 1989, 1995; Jiménez Betts and Darling 2000; Kelley et al. 1961; Kelley 1963, 1971, 1974, 1990).

References to the northern border of Mesoamerica describe a variety of processes of interaction that have been detected in at least five different periods:

At the end of the Formative and Early Classic periods, with dates ranging from AD 150 to 400.

At the end of the Classic, that is, AD 400–600, representing the rise of large ceremonial sites in the region.

Approximately AD 600–900, representing the largest expansion of Mesoamerican culture toward the north-central region

Between AD 900 and 1500, with the decline in the central part of the border, but the expansion along the Pacific coast toward the north connecting the southwestern United States with Mesoamerica

1520, when the occupation withdrew to the south along the Rio Grande de Santiago.

Undoubtedly, this progression was not simultaneous in all regions, making the research still more difficult to categorize. Nevertheless, when viewed progressively, it is notable that during the rise of the late Classic period in north-central Mesoamerica, communication was focused on the hydrographic region of the Lerma-Santiago Basin to the south. The Río Verde Grande Basin is located in what is now southern Zacatecas, most of the state of Aguascalientes, a small portion of northwestern Guanajuato and most of the highlands of Jalisco (Altos de Jalisco), which make up the southern portion of the basin. The cultural expansion that occurred during the Late Classic around the basin of the Río Verde Grande found no natural barriers to limit contact, and may be why strong similarities are noted among the sites (López Mestas et al. 1994: 280).

Theoretical Framework

The expansion of the northern cultural border early in the Late Classic period can be understood as an expansion of a world-system whose success was based on the combination of two main phenomena: the growing incorporation of different political units to the system and its inclusiveness. This expansion was associated with greater interconnectivity between elite regions that increased the demand for and the repertoire of available goods.

The world-system during the Late Classic period would have incorporated not only new territories, but also prior participants with new roles. The new system included the Basin of Mexico, which no longer formed a core with Teotihuacan at its head; rather, the periphery shifted toward a system of two competing political units (Jiménez Betts 1998; Pérez Cortés 2007; Solar 2002). As Jiménez Betts (2001: 163) argues:

> The Late Classic Mesoamerican world seems to have become integrated into an exchange system in which all regional systems that participated were favored. This interaction generated growth in nearly all areas of the regional systems. None of the regions or areas seemed to have played a dominant role, but it was a time of equipollent units within various areas of intertwined interregional interaction.

In this sense, the Late Classic period witnessed a new dynamic in the macroregional networks that were consolidated in prior periods, and in turn, formed the basis for subsequent networks (Jiménez Betts and Darling 2000: 5, 19, 22; Jiménez Betts 1998: 300).

From a world-systems perspective, human societies inevitably participate in networks that link them into a single economic and political system—a world-system—whose limits are conditioned by the technologies of transportation and communication at any given moment, but generally extend to connect different regions within a macroregion. It is possible to classify certain material traces from the networks in which societies participate. Chase-Dunn and Hall (1997) aimed to identify the components of a system and to focus attention on networks that comprise: information networks (IN), prestige-goods networks (PGN), political/military networks (PMN), and networks of bulk goods (NBG).

Bulk goods networks are usually the least geographically extensive, and their control is usually directed from a political-administrative unit (a chiefdom, or even a state) in the territory required to produce food and raw materials. The study of material culture both in urban and rural areas could suggest a tangible pattern that shows a tendency toward homogeneity of the material culture related to the central power.

Therefore, the analysis of the extension of bulk goods networks requires focusing on the design of materials in the territory beyond the central urban area, which may suggest the territorial integration of the interior. In the same way, as the boundaries of bulk goods networks are reached, the material patterns of these networks should show a decrease in density as the boundaries of these networks give way to a transition zone or border.

The political/military networks contrast with the material pattern of bulk goods networks in two ways. If there is a sustained-conflict relationship, the presence of defensive buildings or fortresses are indicative of a closed militarized border, and a transition zone would be present in between the bulk goods and military alliance networks. On the contrary, in the case of alliances between the core and independent political units within the political networks, these relationships are considerably more difficult to distinguish and characterize.

To see the differentiation between the center and periphery, the sustained relationship between elites should manifest itself through emulation of prestige material and symbols of status that could play a central role in the political economy within the network.

Local traditions would present evidence of integrating the central ideology, which the local elites would use to gain power within their communities. Because the political networks are nested within the prestige-goods and information networks, the imports, symbols of power, and ritual paraphernalia related to the core could be present within political/military networks.

The transition between the military and political networks and the prestige-goods networks could be more gradual if considerable distances existed between them. It will be a matter of overlapping different markers of interpolity prestige, which are the goods that show the communications between elites.

The last area is the transition to the territory of the information networks in which there is a decline in the presence of prestige goods. However, discrete elements occur in local contexts that would show modified versions or local interpretations of ideational elements related to the ideology and culture of the nuclear or center society. In information networks, the evidence of core-periphery differentiation would be characteristically discernible in ideas and symbols, in contrast to the portable objects in circulation between the networks of prestige-goods networks.

Comparative studies (Chase Dunn and Hall 1997) reveal that all world-systems exhibit cyclical change processes, the most important being the emergence and fall of large political units, and pulsations in the spatial extent and intensity of trade networks. "Rise and fall" correspond to changes in the centralization of political/military networks, a matter of relative size and distribution of power among political units that is defined by the term "cycle." David G. Anderson (1994) has used this term to describe the phenomenon among chiefdom-type societies in his study on the rise and fall of the Savannah Valley groups of the Mississippi River. In this study, the cycle is composed of a first phase in which chiefdom units extend control over neighbors and form a hierarchical administration of two levels, centralizing power regionally to eventually disintegrate and return to smaller and less-hierarchical political units.

All the world-systems in which hierarchical political units intervene experience a cycle in which some of these units grow in power and size and then decline (Frank and Gills 1993). Also, all systems, including the smallest and most egalitarian ones, exhibit cyclical expansions and contractions in their spatial extension and intensity in their exchange networks. This sequence of expansion and contraction is called "pulsation." Even small-scale and very egalitarian systems, although they do not have cycles of boom and bust, experience pulsations.

Chase-Dunn and Hall (1997) argue that the causal process for rise and demise differs in each system depending on the prevailing mode of accumulation. A huge difference between the emergence and fall of empires and the rise and fall of modern hegemonies is in the degree of centralization acquired within the nucleus.

The tributary systems alternate between structures of multiple competing cores (peer polities) and a system with a single core (or few cores). The modern system experiences rises and falls of hegemonies, but these never destroy the other core-states to form an imperial system of a single core, mainly because modern hegemonies seek a form of capitalist accumulation, and not tributary.

Rise and fall work differently among systems composed of chiefdoms because the institutions that facilitate the extraction of resources from distant groups are less vertical. Chiefdoms use hierarchical relations of kinship, control of rituals, and control of prestige goods more than other systems. These power techniques are highly dependent on ideological integration and consensus rules. However, states develop organizations to extract resources that chiefdoms do not have, for example: permanent armies. And states and empires in tributary systems are more dependent on their ability to mobilize troops over long distances than modern hegemonic states, since in modern times modes of production and financial control mechanisms have been developed, as have bureaucratic techniques to exert power that allow modern hegemonies to extract resources from distant places without the need to exert direct violence.

In short, waves of integration are a fundamental characteristic of world-systems. What we mean by integration is the increase of societies interacting economically and culturally, establishing increasingly extensive trade networks. These interaction networks are empirically determinable connections between different groups of people, which allows us to analyze the spatial aspect of subgroups within societies (especially the interaction between elites) as well as the ways in which their members are interconnected. The underlying idea in world-systems theory is that interaction is an agent of social change (Adams 1974), seen as a network that can bring about massive and rapid changes while motivating mobility and leadership in societies.

This classification is useful for showing that each type of network brought together the constituents of the system in a different way. When researching archaeological sites, we see the material remains of all the interaction networks that at some point in the past were working simultaneously. This enables us to get a picture of society with the dynamics of broad interaction. For example, networks that moved bulk goods occur in small geographic areas, while networks that moved luxury goods tended to link larger regions. At the same time, a political/military network may include more than one network of bulk goods; likewise, prestige-goods networks may encompass more than a single political/military network.

If we accept that more than one network may be operating simultaneously, then it is also possible to assume that the constituent units of the system had no clear boundaries, and that these variations in the flow of boundaries can indicate potential changes in the relationships between these units, making terms like "border" less appropriate in explaining the similarities and differences in the archaeological record between sites.

Archaeological Research in the Río Verde Grande Basin

The Río Verde Grande Basin connects the archaeological site of Cerro de Santiago with its neighbors (see Figure 8.1). Its most studied section includes the highlands of Jalisco (the Los Altos region), where there has been more formal archaeological research than in sections corresponding to Aguascalientes or southeastern Zacatecas.

One of the first sites studied in the region was the Cerro Encantado, near the present community of Teocaltiche, in the Los Altos region. The archaeologist Betty Bell excavated the site in 1970, and its 43 units revealed ceramics that would become diagnostic of the traits in this section of Mesoamerica, as would the "Cornudos" figurines found as an offering in a primary burial. Carbon dating shows an early occupation related to the shaft tombs and Bell also found ceramics related to the Chupícuaro culture (1974). When she noticed the overlap in this area of the two cultures, one of her conclusions was that it was most likely an earlier population that received influence from the two cultural epicenters of that period.

To support this idea of the native population receiving influences, Williams (1974) conducted surveys of the area in the northern section of the Los Altos region along the Zacatecas border, identifying 13 pre-Hispanic settlements of various sizes and complexities. There, he collected figurines and classified them, producing an interpretation that suggests that there are similarities in styles between these figurines and those of Chupícuaro. However, he concluded that the influence of that culture was modified by a local tradition, which merged with another from the northernmost basins, such as the Magdalena River in Nayarit, leading to its own tradition during the Late Formative period. We can see that from the very beginning of these explorations, the idea that local cultures had been influenced by their surrounding neighbors was firmly installed in archaeological considerations of the region.

The research conducted by Piña Chán and Taylor (1976) in the archaeological site of Cuarenta, in the municipality of Lagos de Moreno, suggested a link between this settlement and sites like La Quemada, Chalchihuites, and Teul, and they proposed that a cultural expansion from Zacatecas reached the site of El Cuarenta in Jalisco and was projected toward El Tunal Grande in San Luis Potosí, at sites such as Electra or Villa de Reyes.

In the 1980s, researchers Carolyn Baus and Sergio Sánchez set out to identify Caxcan, Coca, and Tecuexe settlements from sixteenth-century Spanish chronicles in the state of Jalisco, and conducted field survey and mapping in the area. They reported on materials recovered in the sites of Cerro Támara, Teocaltitán, and Tlacuitapán, located near the present populations of Jalostotilán, Teocatitlán, and Lagos de Moreno, northeast of Los Altos (Baus de Czitrom and Sánchez Correa 1995). The materials included elements such as "inverted rims," "ring base" ceramics, and certain types of figurines based on Williams's (1974) classification; types I and IV became diagnostic for the entire basin during the Late Classic period.

In the early 1990s, the Centro INAH Jalisco launched the Jalisco Highlands Archaeological Project, and conducted surveys of pre-Hispanic settlements. This project provided important data regarding the regional settlement pattern, types of settlements, architecture, and associated archaeological materials, the inverted rim and the ring base ceramic vessels (López Mestas et al. 1994). Three types of settlement patterns were identified. In the first, the sites are located on hilltops, either as part of a mountain system or on isolated hills, as in the case of Peñol de Chiquihuitillo (see Weigand and García de Weigand 1999). This settlement pattern is common in the southern part of Los Altos, at sites located by Castellón and Ramos de la Vega (1988; compare López Mestas et al. 1994; Valencia Cruz 1994), as well as certain sites in the northern part of the Río Verde Grande, such as Cerro de los Antiguos, El Tuiche, and Cerro de Chihuahua, near Nochistlán, Zacatecas.

Characteristic of these sites is the presence of a civic-ceremonial area normally represented by monumental architecture such as large platforms associated with other spaces and structures like plazas, courtyards, residential areas and housing, ball games, and shrines. In some cases, large platforms were surrounded by walls that defined and differentiated the various sectors of the settlements. The residential areas were associated with the terraces themselves, which were usually on the hillsides alongside agricultural activities (Castellanos 1994; López Mestas et al. 1994).

These preliminary studies propose several occupational phases ranging from AD 300 to 900 based on the orange-on-maroon group of ceramics, a local tradition. This ceramic group features pots with inverted rims, small pots with anthropomorphic representations on the rims, and annular-base bowls, which López Mestas et al. (1994) interpreted as indicative of a clear association with the Ixtépete–El Grillo phase in the Atemajac Valley (modern valley of Guadalajara), from AD 350 to 700. Elements of this phase include new forms such as the annular base, solid supports, punch marking, distinctive rims, pseudo-cloisonné as a decorative technique, and a diversified distribution of ceramic styles, which marks a new mode of interaction, sharing information between the elite groups that inhabited north-central Mesoamerica. Other types from this period include red-on-buff and red-on-orange; their presence is more useful in terms of the spatial information they provide, since these ceramic types are widely distributed throughout western Mexico from the Formative through the Late Classic periods. López Mestas et al. (1994) also mention a negative *tardío,* which they consider a more direct indicator of a temporal association between the highlands of Jalisco and Zacatecas from AD 600 to 900. This ceramic type is characterized by annular-base bowls, globular pots, and plates.

Based on the above information, the authors conclude that the central region of Los Altos shows human occupation from the Late Formative period (circa 200 BC–AD 300), as evidenced mainly by the polished ceramics. This occupational phase was related to the traditional shaft tombs from the Atemajac Valley, they argue, and there is a notable absence of the early negative-surface materials that characterize this period north of the Río Verde Grande. As mentioned above, the Classic period (AD 300–900) is represented by a local ceramic tradition (orange-on-maroon group polychromatic variants and negative *tardío*), while the Postclassic period is not even represented in this research, despite the historical references that mention the presence of groups such as Tecuexes and Cocas at the time of the Spanish contact (López Mestas et al. 1994: 287–288).

The Archaeological Site of Cerro de Santiago (PAS)

In the northern section of the Río Verde in the present-day state of Aguascalientes, INAH researchers have registered and worked on several sites.[1] In 2003, a rescue project focused on the use and protection of natural resources in the Santiago Canyon was carried out by Nicolás Caretta, Mario Pérez, and Jorge Martínez. Later, the PAS was established as an INAH research project and continues the exploration, identification, and topographical survey of the ceremonial area or zone "A," which has an approximate area of 28 hectares. In this area, 14 platforms, 12 mounds, 25 structures, 2 patios, 17 retaining walls, and a ballcourt were identified (Nicolás Caretta 2006).[2]

The archaeological site of Cerro de Santiago consists of a ceremonial center on top of a plateau and is comprised of several courtyards, mounds, and a ballcourt, with a residential zone at the foot of the plateau (see Figure 8.2). Recent research has revealed that it was occupied during the Epiclassic period, between AD 600–900 (Nicolás Caretta 2014a, 2014b; Puch Ku 2014; Silva 2015). This was determined by the correlation of ceramic objects compared to other sites in the north-central region of Mesoamerica.

Ceremonial Area

This area was the center of ritual life and specialized production for the settlement's inhabitants, and a reflection of its hierarchical organization. Here we have identified spaces designed for specific rituals, such as mounds and sunken courtyards that were common throughout the region and fundamental to the pre-Hispanic indigenous cosmovision; they also served as the residence of its leaders (see Figure 8.3).

Residential Area

The primary residential area of the site is composed of more than 190 elements including terraces, retaining walls, and residential structures,

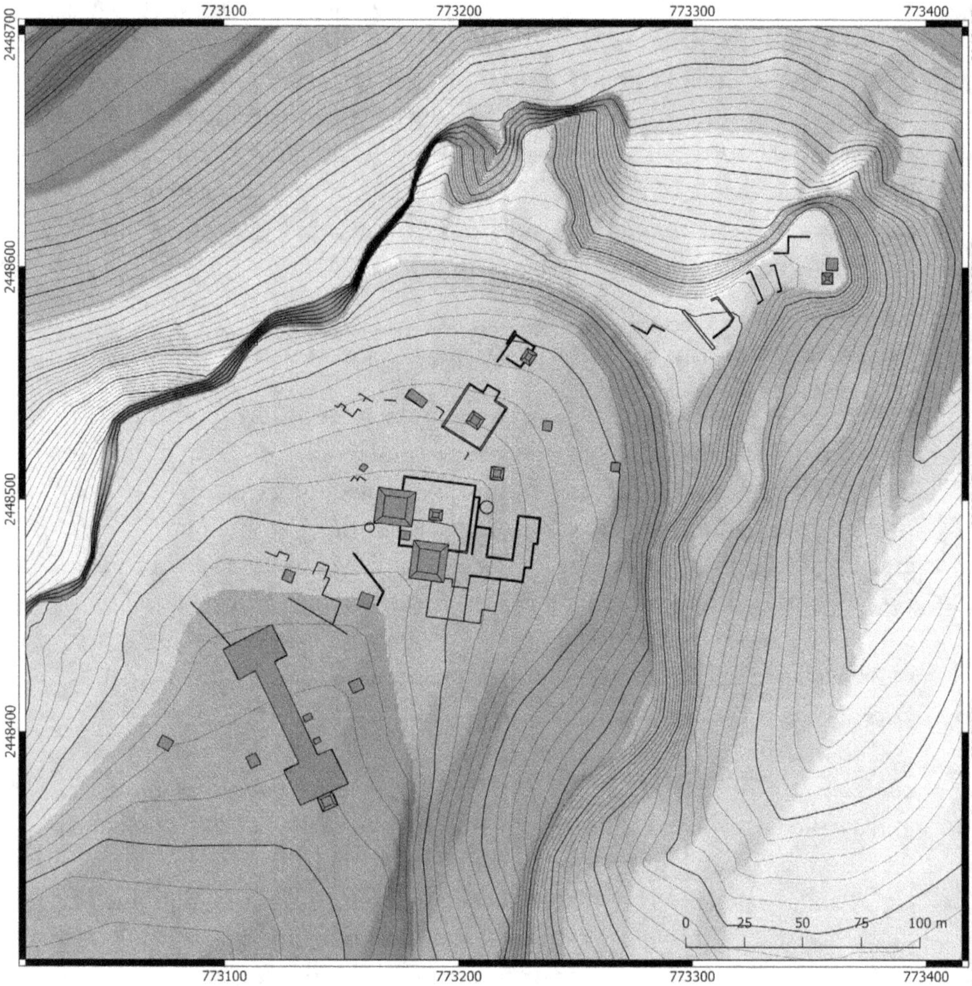

Figure 8.3. Map of the ceremonial area of Cerro de Santiago (map by the authors).

covering almost 20 hectares. It is divided into three parts from north to south. The northern part is located on a headland, the second part on a plateau with a lower altitude, and the third part is on the previous plateau, separated so that at some point it could have been the bed of a small stream that went down from the plateau into the valley. These latter areas appear to be composed of more compact housing units. However, excavations are required to establish the precise relationships between each of the foundations that were detected on the surface.

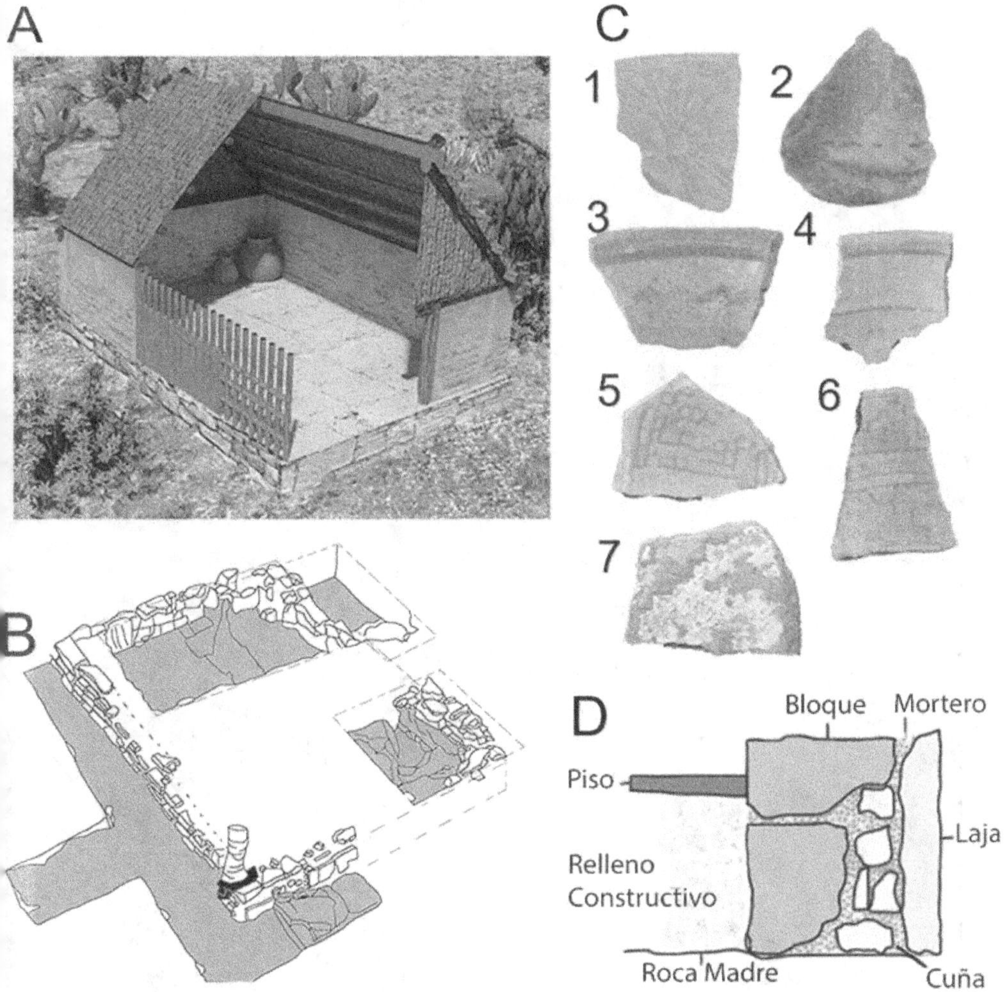

Figure 8.4. Excavation 53 and recovered material: *A*, reconstruction of the architectural structure. *B*, isometric view of the excavation process of structure 53. *C*, diagnostic materials found during the excavation of structure 53—1, 5, 6, 7 Pseudo-cloisonné; 2 Río Verde figurine; 3, 4 San Luis Polychrome. *D*, construction system of the platform (map by the authors).

Recent Preliminary Excavations

Two excavation units have been initiated at the archaeological site. The first was made on structure 53 and the second on the edge of a residential unit. The results of both operations and the materials recovered from them are presented below in Figures 8.4 and 8.5.

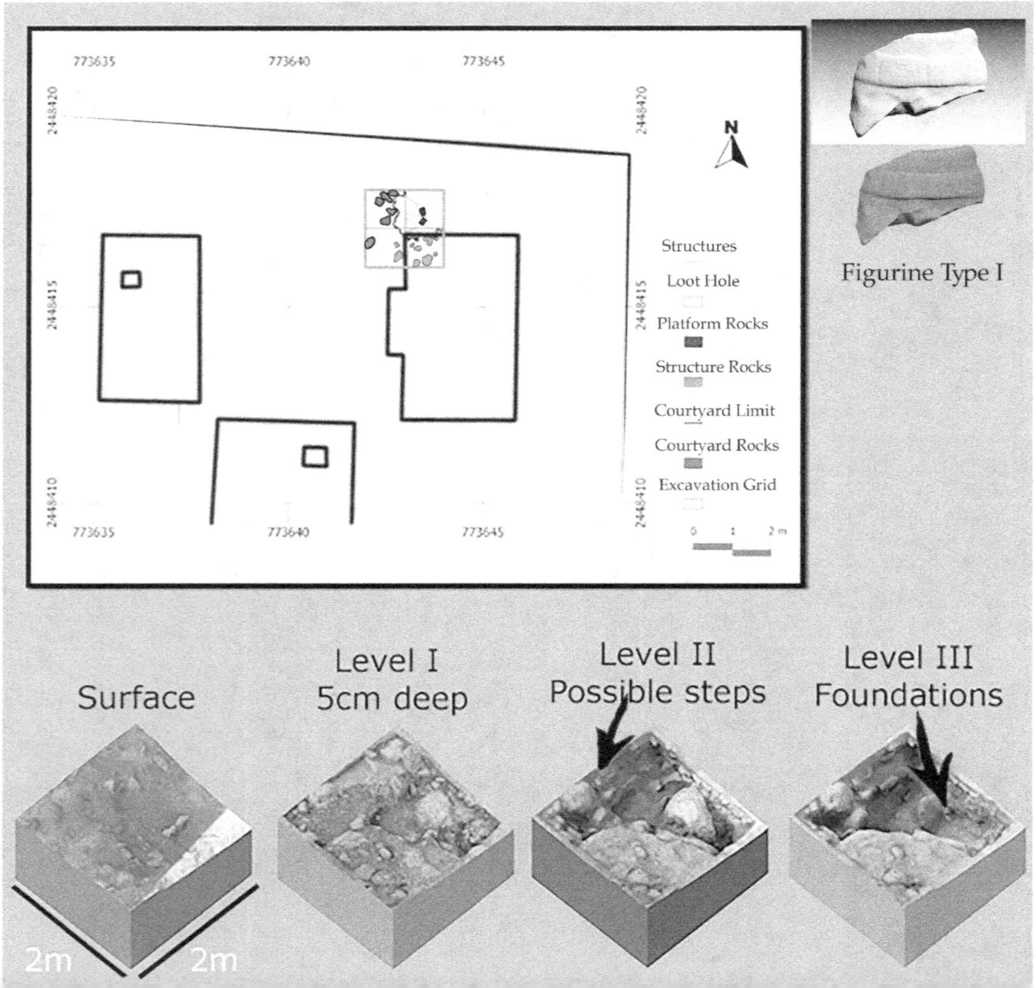

Figure 8.5. Housing unit excavation and recovered material

Ceramic Analysis

An analysis of the ceramics from Cerro de Santiago has allowed us to correlate a series of common archaeological features that clearly mark a zone of transition for the diagnostic materials from the different spheres, such as the Type I figurine, pseudo-cloisonné (northern sphere), the negative-decorated ring-shaped cups (Altos-Juchipila subsphere), San Luis Polychrome (San Luis valley subsphere) and the El Bajío sphere. Jiménez Betts and Darling (2000) identify these ceramic attributes as indicative of the Epiclassic period.

To date, the scale of analysis of the Cerro de Santiago ceramics has allowed us to understand the scope of the geographic distribution of the following diagnostic features: pseudo-cloisonné, San Luis Polychrome, Type I figurines, Río Verde figurines, and incised Atoyac, negative, and inverted edges. They reveal an extensive network of interaction between the sites in the Bajío, northwest and central-north regions of Mexico that permitted active participation within the Northern Sphere, including the regions of Los Altos–Juchipila, El Bajío, and the San Luis Valley (see Table 8.1, Figures 8.6–8.8).

From a broad perspective, it seems clear that pseudo-cloisonné ceramics reflect close communication between local and regional elites, which would result in the joint adoption of a successful ideological discourse that provided the elites with political support within their respective communities. This evidence suggests that a political-religious discourse underpinned the wide dispersion and acceptance of the ideological concepts associated with the use of these complex ritual ceramics (Holien 1977; Kelley 1974).

However, the red motifs outlined in white, which we have registered in smaller proportions in Santiago, have been reported at the sites in Buenavista and El Teúl, in Zacatecas. Peter Jiménez (personal communication, 2007; compare Pérez Cortés 2007) comments that this ceramic type is diagnostic and characteristic of the valley of Tlaltenango and the Bolaños Canyon during the Epiclassic period.

Certain authors (Jiménez Betts 2001; Jiménez Betts and Darling 2000: 180; Moguel and Sánchez 1988: 230; Ramos and López Mestas 1999: 258) have noted that the Northern Sphere intersects with the Bajío region and with the Los Altos–Juchipila sphere that extends slightly outside the limits of the Northern Sphere to the east, reaching Cerrito de Rayas, and also including La Gavia in Guanajuato. These two sites are important because they show the overlapping of the two spheres through the negative ceramics and

Table 8.1. Distribution of diagnostic ceramic features across north-central Mesoamerica

Diagnostic Features	Teul/Buenvista (Zac.)	Altos de Jalisco/ Cañón de Juchipila (Zac./Jal.)	El Cóporo/la Gloria/el Cobre (Gto.)	Cerrito de Rayas/La Gavia (Gto.)	El Ocote (Ags.)	Santiago (Ags.)	Cerro de en Medio (Ags.)	Montesita (Ags.)
Engraving			X		X	X	X	
Incised			X		X	X		X
Atoyac incised brown	X				X	X		
Raised white			X		X	X		
Red on cream		X	X		X	X		X
Red on cream		X					X	
Negative	X			X	X	X		
Negative red–black on cream	X	X				X		X
Pseudo-cloisonné	X	X	X		X	X	X	X
San Luis Polychrome	X	X		X		X		X
Inverted edge	X				X	X	X	
Ring-shaped base	X				X	X	X	X
Type I figurine	X		X			X		
Río Verde Figurine	X					X		

Source: Data from Baus de Czitrom and Sánchez (1995); Braniff (2001); Dueñas (2014, 2015, 2016); Jiménez (1992: 189–190, 2001), Jiménez and Darling (1992: 14–15, 18, 2000); Moguel and Sánchez (1988: 230); Nicolás (2006, 2010, 2011a, 2011b, 2012); Pelz and Jiménez (2007); Pelz et al. (2012, 2014); Porcayo (2001); Ramos and Sánchez (1988: 315).

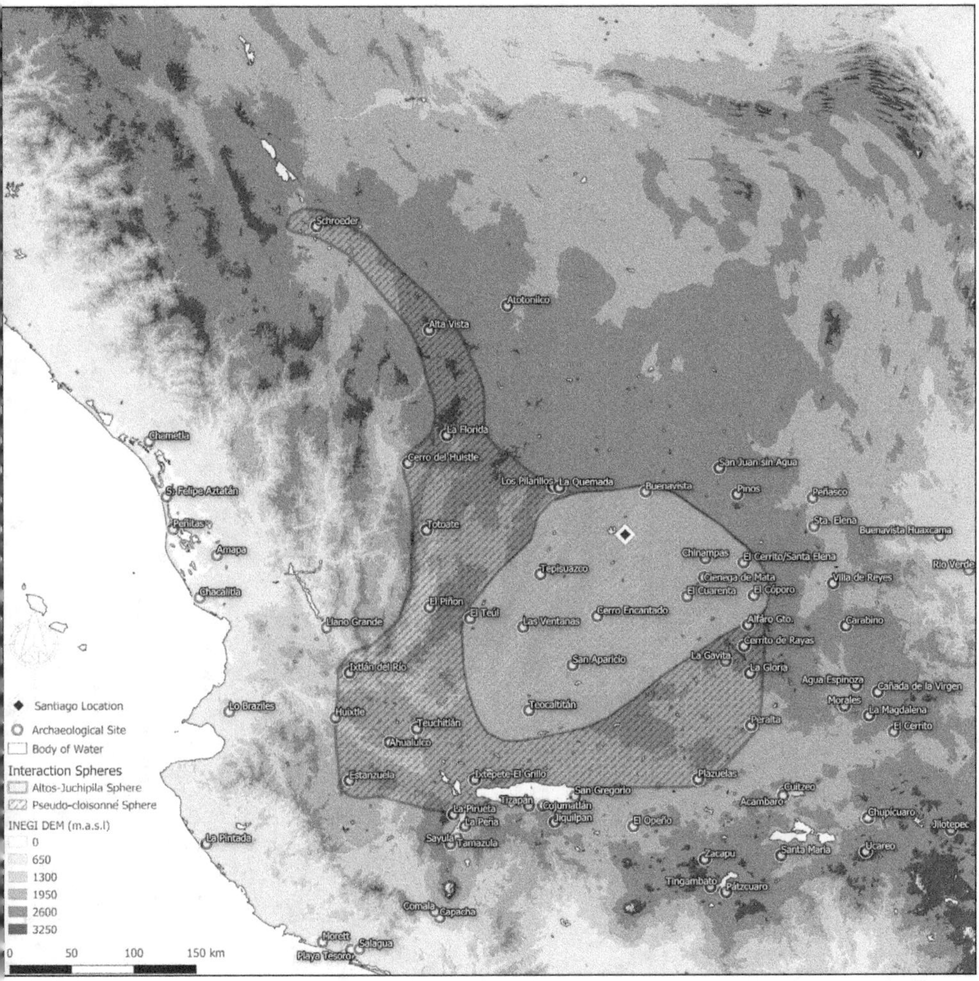

Figure 8.6. The Altos-Juchipila subsphere of ceramic interaction within north-central Mesoamerica (after Puch Ku 2014: 204).

the San Luis Polychrome, but they also represent an area of connection with the distributive limits of Garita, Cantinas, and Negro/Naranja types. Also, to the southeast, the Northern Sphere integrates the sites of El Cóporo, La Gloria, and El Cobre, in Guanajuato, which constitute the northwestern boundary of the Bajío Sphere (Jiménez and Darling 1992: 17), as well as the known distributional extent of Type I figurines, pseudo-cloisonné and negative (raised white, incised, punch-marked, and red/cream types (Braniff 2000: 40; Jiménez 1992: 189–190; Jiménez and Darling 1992: 14–15, 18).

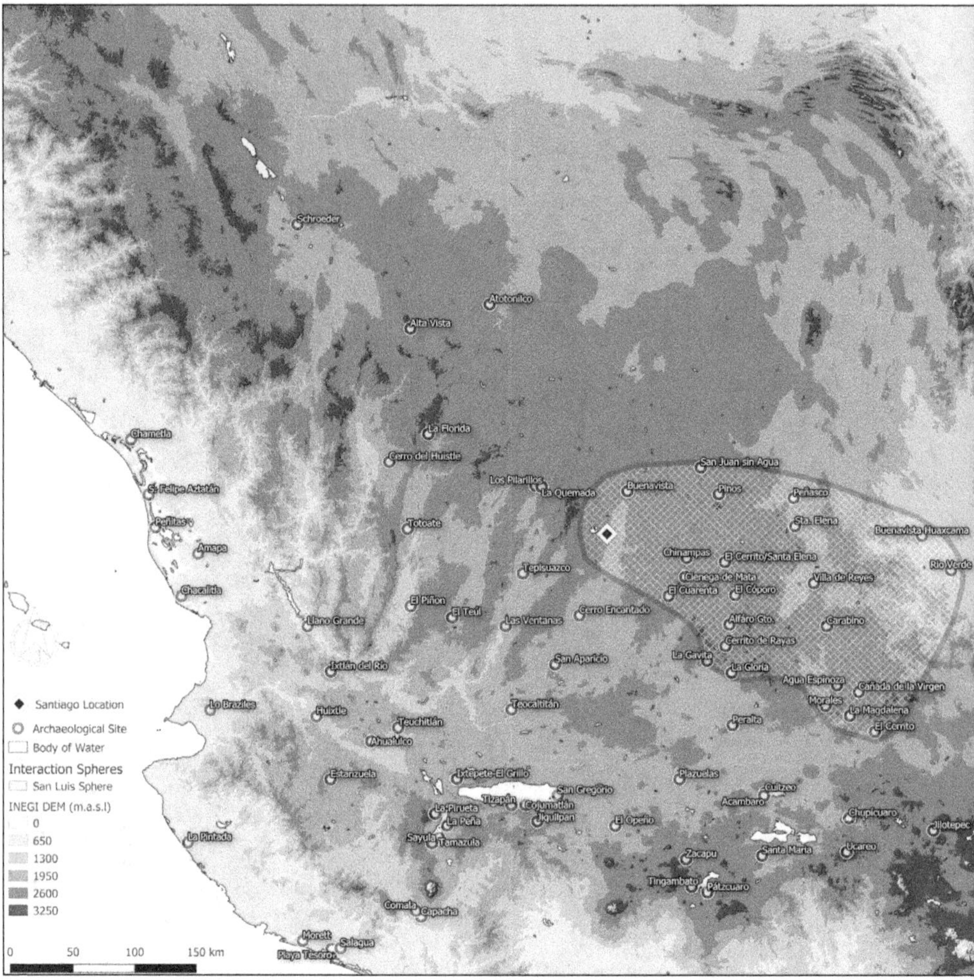

Figure 8.7. The San Luis sphere of ceramic interaction within north-central Mesoamerica (after Puch Ku 2014: 205).

Discussion: Evidence of Interaction at Cerro de Santiago

The archaeological site of Santiago is exemplary for understanding the political and economic climate of the Epiclassic (Nicolás Caretta and Kroefges 2014; see Figure 8.9). Its ceremonial area shows clear similarities with its southern neighbors in Los Altos and in the lowlands of El Bajío; namely, the sunken courtyard with two mounds that was an architectural space typical of those regions during the earlier Classic period between AD 200 and 400 (Silva 2015: 93). This may indicate that, on one hand, the site preserves

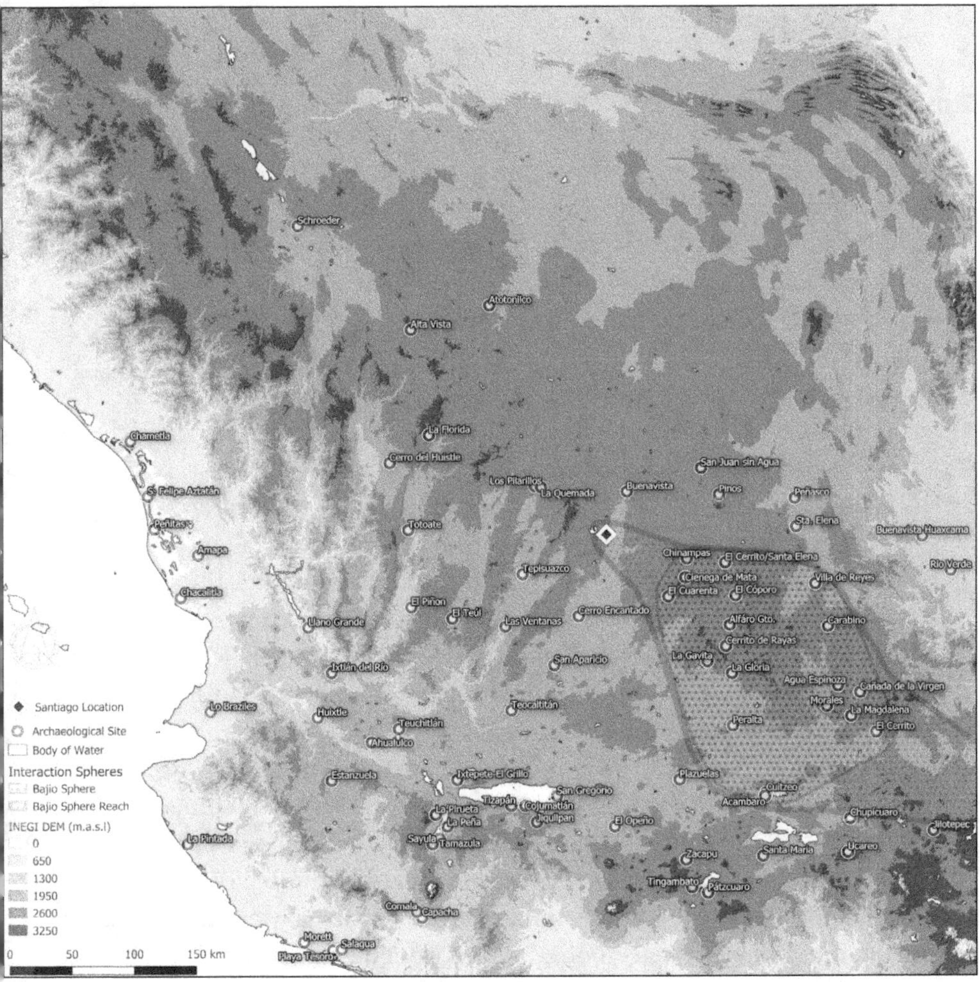

Figure 8.8. The Bajío complex sphere of ceramic interaction within north-central Mesoamerica (after Puch Ku 2014: 206).

architectural patterns possibly from the south, being connected to these since the Classic period. The incorporation of a main altar in the courtyard is also a phenomenon from the end of the Classic period in El Bajío (Cárdenas 1999), possibly due to the increased interaction that it had during this timewith the northern area and the Chalchihuites sites. There, the architectural pattern with the sunken courtyard and central altar is diagnostic of the Epiclassic period (AD 600–900). Based on this, Achimm Lelgemann (1997: 101–105) mentions that:

> The ceremonial architecture or housing, not only from La Quemada but also from other major sites of the northwestern outskirts of Mesoamerica, is characterized by a highly canonized group in its composition, that has been named (patio compound). . . . It is a group of buildings raised on platforms surrounding a sunken courtyard. This courtyard invariably has an altar in the center and in most cases a pyramid of small dimensions on one side. Generally, a room of considerable size forms part of this compound, while the other buildings that make up the compound are made up of several rooms.

During the Epiclassic period, the archaeological site of Santiago joined the prestige-goods trade networks, among which we can identify more clearly the pseudo-cloisonné ceramic and the Type I figurines.

In terms of the political domain, the basis of power is not only strength, but also legitimacy (Krueger 2008). Hence, large political entities almost always rely on ideological and religious foundations to legitimize themselves. Mesoamerica was no exception. In Mesoamerica, luxury goods often played an important role in the accumulation and retention of power by an elite through the controlled distribution of status symbols. Regionally, diagnostic materials considered as prestige goods include "pseudo-cloisonné pottery, type I figurines, and some architectural elements such as an enclosed courtyard with central altar, and shared items with other sites that indicate that there was an interregional interaction among its elites" (Jiménez Betts 2006: 38). In this light, the presence of pseudo-cloisonné ceramics and Type I figurines suggests that Cerro de Santiago was incorporated in a larger regional prestige-goods trade network during the Epiclassic period. Such elements are found throughout northern and western Mesoamerica at this time, primarily in the valley of Atemajac, the highlands of Jalisco, the Juchipila Canyon, the Malpaso Valley, and the area of Chalchihuites, in addition to the San Luis Valley (Puch Ku 2014: 209).

It is too early to jump to final conclusions; however, with the above information we can focus on different research lines. The few explorations at the site have revealed that a stage of greater effervescence and interaction took place. Its origin and decline will be evaluated in subsequent investigations, perhaps focused on the nuclear and ceremonial area of the site.

The origin of Cerro de Santiago can be found in societies among the Los Altos and El Bajío areas that, since the Formative period, displayed cultural associations through the shaft tomb and Chupícuaro traditions. Only in the archaeological site of Ocote hasmaterial has been found that indicates

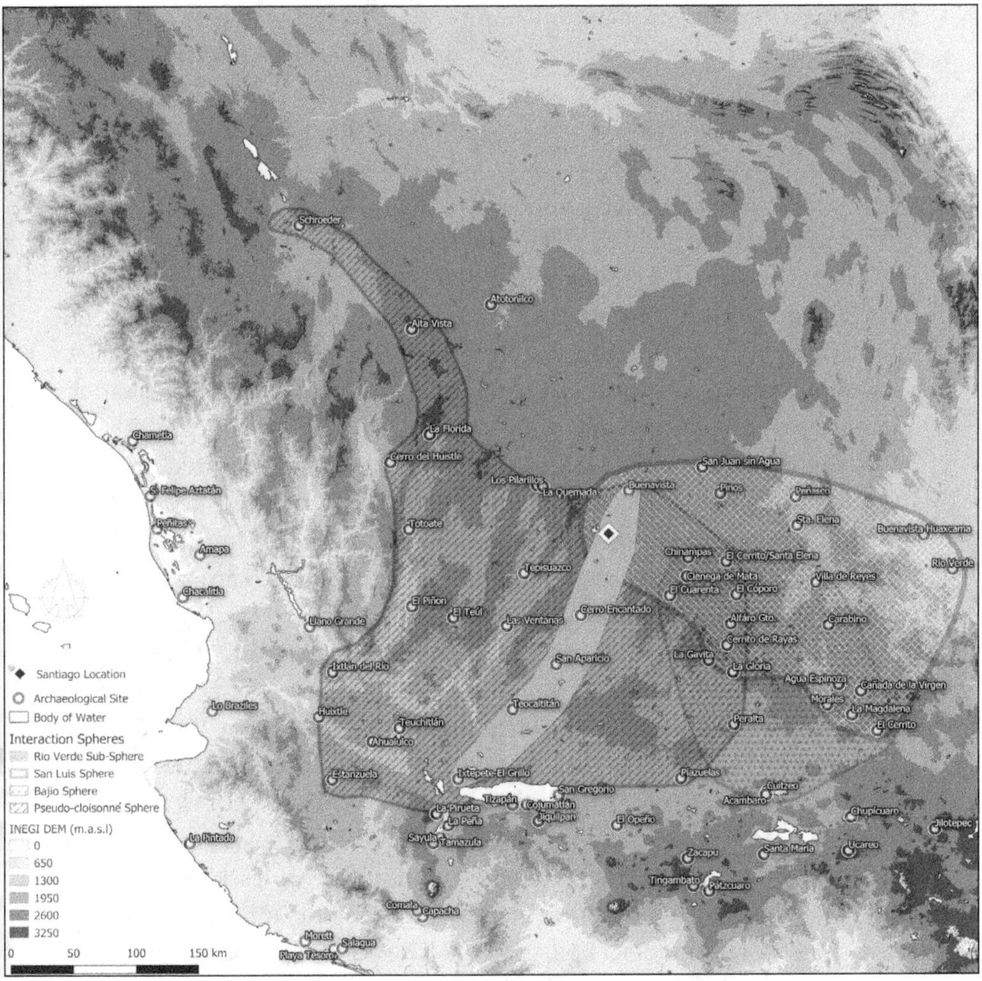

Figure 8.9. Classic and Epiclassic period interaction spheres in north-central Mesoamerica (map by the authors).

Chupícuaro origins, specifically the Morales phase. No shaft tombs have been detected in the region to date, but it would not be surprising if they were discovered in subsequent investigations.

The Late Formative and Early Classic periods on the northern border are characterized by an expansive stage with population development into new areas, which clearly demonstrates the increase of social differentiation (Beekman 2010: 21). The cultural traditions of El Bajío are an example of this complexity. They are characterized by an architectural pattern with a central element of sunken courtyards, while the valleys of Tequila, near the volcano, also developed a cultural complex with a distinctive architectural pattern (see the chapters by Beekman, and Sumano and Englehardt, this volume).

The ceremonial centers of the Teuchitlán tradition include two new architectural forms in the region, with similar social functions. The first of these are ballcourts, architectural devices widely spread throughout Mesoamerica, and whose possible role could be that of a mechanism to provide less-violent social skills among different groups, in addition to other ritual uses (Taladoire and Colsenet 1991). The second architectural form is the *guachimontón*, consisting of a platform around a circular stepped altar. A varying number of platform structures encircle these structures, creating a distinctive complex (Weigand 1985, 1996; compare Sumano and Englehardt, Fig. 5.2, this volume). Smaller versions have been reported north of the nuclear region of Los Guachimontones in the current Cañón de Bolaños, dated between AD 200 and 600 AD, and Tepizuasco in the southern Juchipila Canyon, suggesting a cultural and political dynamism (Teresa Cabrera, personal communication, 2016). It is almost certain that these architectural forms were not imposed, but rather were adopted by local elites seeking their association with agricultural rituals or opportunities to consolidate prestige within their own populations (Beekman 2010: 65).

In the Río Verde Basin, there is only one record of a structure similar to a *guachimontón*. This is located in the archaeological site of Cerro de en Medio, located 7 kilometers west of the Cerro de Santiago. However, it has no central altar, only the platform with rooms above, at least based on the surface survey. Similarly, more research in this architectural complex may account for its chronological position, which is essential to understanding the relationship between the Río Verde Basin and the areas surrounding the Tequila Volcano.

In contrast, the societies from El Bajío continued after the Chupícuaro tradition, in terms of work in ceramics, but the public architecture of these sites became diversified around a distinct construction principle, the "sunken courtyard." Sunken courtyard architecture consists of a platform that forms a banquette around a sunken space, and upon which there are rooms or pyramidal bases. The buildings on the platforms are accessed by a stairway, which distinguishes this architectural arrangement from other places. For example, in Teotihuacan, access to similar constructions uses wide alleys between adjacent buildings. Excavations at Cerrito de Rayas, San Bartolo Aguascalientes, and Plazuelas give clear examples of this type of sunken courtyard architecture (Cárdenas 1999: 160).

A major change in relation to the previous stage is the social complexity apparent in the archaeological settlements. Their hierarchical segmentation is evident in the distribution of the sites, with centers of power controlled by the elites (Jiménez Betts and Darling 2000). It is noteworthy that, to researchers such as Cárdenas (1999), the cultural area of El Bajío during the Classic period is outside of the realm of Teotihuacan influence, with almost no material links. Rather, the El Bajío situation suggests a set of communities that expanded from the Chupícuaro region, bringing a construction style with them to southern Guanajuato and northern Michoacán. The chronological process of this expansion is still unknown, since absolute dating is needed; however, the ceramic correlation dates this process from AD 200 to 600, squarely within the Classic period.

Cárdenas's (1999) spatial analysis suggests that El Bajío shared an architectonic tradition across different political units, which interacted differently with other areas. For example, settlements farther west have small *guachimontón* complexes, which appear to be intrusions (Beekman 2000; Weigand 2000). In contrast, the eastern part of El Bajío had contacts and connections with Central Mexico, confirming links with the city of Teotihuacan—suggested primarily by the appearance of sunken plaza architecture in the great city during the Tzacualli phase (AD 1–100) (Beekman 2010: 66). This region also shares certain ceramic types found at the site of Barrio de la Cruz in Querétaro, while in the rest of El Bajío, there are some types related to the Tlamimilolpa-Xolalpan phases (AD 200–600; Brambila and Velasco 1988; Crespo Oviedo 1998: 325–326, 330; Saint-Charles 1996: 148).

Most of these ceramics appear in burial contexts, which Filini and Cárdenas (2007) consider too insignificant to discuss in terms of domination, or even in terms of direct contact with Teotihuacan. Filini uses information from the Cuitzeo Basin and the theory of world-systems to propose that similar artifacts like the fine orange ceramics and Pachuca obsidian are scarce prestige goods imported by local elites, and for that reason it is common to find local copies with symbols carefully selected, since Teotihuacan iconography clearly was important to the validation of the autonomous local elites. Thus, although the importance of Teotihuacan during this period at a macroregional level is undeniable, especially as the core of a system that exported ideological terms and whose connection conferred prestige, there is little evidence of a direct Teotihuacan incursion in western and northern Mexico. The acculturation mechanism had to be less direct, but quite effective.

The explanation for changes throughout Mesoamerica during the period between AD 600 and 900 must be multifactorial. On one hand, an extended interpretation has to do with a possible political reorganization after the fall of Teotihuacan as the core of a macroregional system, which is very often conceptualized as the restructuring of the Mesoamerican world-system (Jiménez Betts 2006, 2007). The centralized political system in the Tequila valleys collapsed, and the use of *guachimontones* and shaft tombs stopped in the west (Beekman and Christensen 2003; Weigand 1990); simultaneously, multiple new political centers of various sizes emerged. Farther north, the ceremonial centers of Alta Vista and Quemada existed in splendor, but new ceremonial centers also appeared in more southern regions, such as El Grillo (Galván Villegas and Beekman 2000), Ixtépete (Galván Villegas 1975), Santa Cruz de Bárcenas (Weigand 1990), and La Higuerita (López Mestas and Montejano Esquivias 2003), in the central valleys of Jalisco. New towns were also founded in what is now the state of Michoacán during this period, including Tingambato (Piña Chán and Oi 1982), Urichu (Pollard and Cahue 1999), Guadalupe (Arnauld and Faugère-Kalfon 1998; Pereira 1999), Zaragoza (Fernández Villanueva 2004) and Jiquilpan (Noguera 1944) (compare Punzo et al., this volume). These sites emerged as new population centers in what has been described as the Lupe phase (Pollard 2008: 221–223).

These phenomena are connected to the reinforced production of prestige goods, as the multitude of relatively small political units (small compared to the previous stage, and possibly unstable) competed for legitimization in a highly volatile environment, increasing the demand for exotic materials and iconography and symbols of authority (Beekman and Christensen 2003: 145–149; Pollard and Cahue 1999).

Concluding Thoughts

Our early research at the archaeological site of Cerro de Santiago allows us to observe its integration in multiple spheres of interaction throughout western and northern Mesoamerica. Materials recovered suggest ties with the Los Altos–Juchipila subsphere, El Bajío, and the San Luis Valley. Comparative analysis of ceramics at sites—such as El Teul, Buenavista, Altos de Juchipila, and Juchipila Canyon in Zacatecas; El Cóporo, La Gloria, El Cobre, Cerrito de Rayas, and La Gavia in Guanajuato; and El Ocote in Aguascalientes—reveals many specific ceramic types that have been registered through research at Cerro de Santiago. These include red, brown, black, cream, and orange monochromes, incised Atoyac, Triana, and raised white, as well as red-on-buff bichromes, black-on-red, black-on-brown, red-on-red, and brown-on-gray negative polychromes. Further shared ceramic styles and types include outlined red motifs, Type I figurines, and Suchil, San Luis Polychrome, pseudo-cloisonné, and Río Verde types. These all display similarity to materials excavated at the Cerro Santiago site, which sheds light on the historical-cultural knowledge of the occupation period, trade relations, and cultural influences that occurred throughout the ceramic spheres that encompassed central-northern Mesoamerica, the Los Altos–Juchipila highlands and Tequila valleys, the San Luis Valley, and El Bajío during the Epiclassic period.

It is clear that the evidence of these objects and knowledge of pottery sheds light on the political, religious, and sociocultural ideas in relation to the cultural interaction that took place at the site and regional levels. There, direct or indirect participation permitted the development of networks and/or exchange routes where the similarities and differences of the ceramic types reflect the interregional cultural dynamics from southwestern (El Ocote) and northwestern (Cerro de Santiago) Aguascalientes to the sites in the Bajío, Los Altos–Juchipila, the Valley of San Luis, Tlaltenango, and Bolaños Canyon, among others.

The archaeological information to date seems to indicate that the Río Verde served as an important communication artery that, during the Epiclassic period, articulated a wide region that includes southeastern Zacatecas, Aguascalientes, Altos de Jalisco, and the Atemajac Valley, corresponding to the Los Altos–Juchipila subsphere proposed by Jiménez Betts and Darling (2000). Within this subsphere, it is possible to observe a whole range of ceramic traits that highlight one of the most interesting facets of an interaction dynamic that was both complex and interregional (Pérez Cortés 2007). What we see in this area during the Epiclassic is an integration of various cultural complexes, coming from different directions, transforming daily life and creating a cultural buffer zone in what is now North-Central Mexico.

The world-system frame allows us to understand the history and the nature of the cultural changes in Late Classic times. Our proposition is that change during this period was due to increased economic interaction and labor reorganization in the different political units that participated in such interaction. Following the same systemic logic, we can continue to research the same area but at different times, looking at the different societies that formed both in early times, when the first sedentary people settled, and in later periods, when the hunters and gatherers shared the territory.

Notes

1. Most of the research has been carried out by INAH-affiliated archaeologists led by Ana María Pelz, but important progress has also been achieved by a number of other researchers (see Nicolás Caretta 2006, 2014a, 2014b for useful overviews).

2. Cerro de Santiago has been the focus of two thesis investigations: the first by Elizabeth Puch Ku (2014), who focused on ceramic analysis and interaction during the Epiclassic period; and a second by Yesenia Silva (2015), whose investigations centered on settlement pattern analysis. Both works allow us to sketch the intensity of the cultural relations present at the site during the Epiclassic period.

References

Adams, Robert McC.

1974 The Mesopotamian Social Landscape: A View from the Frontier. *Bulletin of the American Schools of Oriental Research. Supplementary Studies* 20:1–20.

Anderson, David G.

1994 *The Savannah River Chiefdoms: Political Change in the Late Prehistoric Southeast.* University of Alabama Press, Tuscaloosa.

Armillas, Pedro

1964 Condiciones Ambientales y Movimientos de Pueblos en la Frontera Septentrional de Mesoamérica. In *Homenaje a Fernando Márquez Miranda, Publicaciones del Seminario de Estudios Americanistas y el Seminario de Antropología Americana*, pp. 62–82. Universidades de Madrid y Sevilla, Madrid.

1969 The Arid Frontier of the Mexican Civilization. *Transactions of the New York Academy of Sciences* 31(6): 97–704.

1987 Chichimecas y Esquimales: La Frontera Norte de Mesoamérica. In *La Aventura Intelectual de Pedro Armillas: Visión Antropológica de la Historia de América*, edited by José Luis Rojas, pp. 35–66. El Colegio de Michoacán, Zamora.

Arnauld, Charlotte, and Brigitte Faugère-Kalfon

1998 Evolución de la Ocupación Humana en el Centro-Norte de Michoacán (Proyecto Michoacán, CEMCA) y la Emergencia del Estado Tarasco. In *Génesis: Culturas y Espacios en Michoacán*, coordinated by Véronique Darras, pp. 13–34. Centro de Estudios Mexicanos y Centro Americanos, Mexico City.

Baus de Czitrom, Carolyn, and Sergio Sánchez Correa

1995 Arqueología en la Región Tecuexe. In *Arqueología del Norte y del Occidente de México, Homenaje al Dr. J. Charles Kelley*, edited by Barbro Dahlgren and Ma. de los Dolores Soto de Arechavaleta, pp. 267–283. UNAM, IIA, Mexico City.

Beekman, Christopher S.

2000 The Correspondence of Regional Patterns and Local Strategies in Formative to Classic Period West Mexico. *Journal of Anthropological Archaeology* 19(4): 385–412.

2010 Recent Research in Western Mexican Archaeology. *Journal of Archaeological Research* 18(1): 14–109.

Beekman, Christopher S., and Alexander F. Christensen
2003 Controlling for Doubt and Uncertainty through Multiple Lines of Evidence: A New Look at the Mesoamerican Nahua Migrations. *Journal of Archaeological Method and Theory* 10(2): 111–164.
Bell, Betty
1974 Excavations at el Cerro Encantado, Jalisco. In *The Archaeology of West Mexico*, edited by Betty Bell, pp. 147–167. Sociedades Estatales Avanzadas del Occidente de México, Ajijic, Jalisco.
Brambila, Rosa, and Manuel Velasco
1988 Materiales de La Negreta y la expansión de Teotihuacan al norte. In *Primera Reunión sobre las Sociedades Prehispánicas en el Centro Occidente de México*, pp. 287–297. Centro Regional de Querétaro, INAH, Querétaro.
Braniff Cornejo, Beatriz
1965 Culturas del Occidente y Marginales. *Novena Conferencia del 31 de Marzo en auditorio del Museo Nacional de Antropología, ciclo de conferencias sobre Antropología en México*. Manuscript on file, Biblioteca Juan Comas, UNAM, Mexico City.
1974 Oscilación de la Frontera Septentrional Mesoamericana. In *The Archaeology of West Mexico*, edited by Betty Bell, pp. 40–50. Sociedades Estatales Avanzadas del Occidente de México, Ajijic, Jalisco.
1989 Oscilación de la Frontera Norte Mesoamericana: Un Nuevo Ensayo. *Arqueología* 1: 99–114.
1994 La Frontera Septentrional de Mesoamérica. In *Historia antigua de México, Volumen 1, El México antiguo, sus áreas culturales, los orígenes y el horizonte Preclásico*, coordinated by Linda Manzanilla and Leonardo López Luján, pp. 114–137. Porrúa ediciones, Mexico City.
2000 A Summary of the Archaeology of North-Central Mesoamerica: Guanajuato, Querétaro, and San Luis Potosí. In *Greater Mesoamerica: The Archaeology of West and Northwest Mexico*, edited by Michael S. Foster and Shirley Gorenstein, pp. 35–42. University of Utah Press, Salt Lake City.
2001 La Colonización Mesoamericana en la Gran Chichimeca: La Tradición del Golfo y la Tradición Chupícuaro-Tolteca. In *La Gran Chichimeca: El Lugar de las Rocas Secas*, coordinated by Beatriz Braniff, pp. 83–112. CONACULTA, Jaca Books, Mexico City.
Cárdenas García, Efraín
1999 *El Bajío en el Clásico: Análisis Regional y Organización Política*. El Colegio de Michoacán, Zamora.
Castellanos, Eloy
1994 *Informe final de la identificación y catalogación de sitios arqueológicos en el estado de Aguascalientes, Instituto Nacional de Antropología e Historia*. Report submitted to the Instituto Nacional de Antropología e Historia, Mexico City.
Castellón Huerta, Blas Román, and Jorge Ramos De La Vega
1988 *Informe de las excavaciones en el sitio Cerrito de Moctezuma, Tepatitlán, Jalisco*. Report submitted to the Instituto Nacional de Antropología e Historia, Mexico City.

Chase-Dunn, Christopher, and Thomas D. Hall

1997 *Rise and Demise: Comparing World-Systems.* Westview Press, Boulder, CO.

Crespo Oviedo, Ana María

1998 La Expansión de la Frontera Norte (y la Cronología Oficial para Teotihuacán). In *Los Ritmos de Cambio en Teotihuacán: Reflexiones y Discusiones de su Cronología,* edited by Rosa Brambila and Rubén Cabrera, pp. 323–334. Instituto Nacional de Antropología e Historia, Mexico City.

Dueñas García, Manuel de Jesús

2014 Registro Arqueológico en 3D mediante la Fotogrametría de Rango Corto. Unpublished undergraduate thesis, Autonomous University of San Luis Potosí, San Luis Potosí.

2015 *Informe verano 2015, Proyecto arqueológico Cerro de En medio, Municipio de San José de Gracia, Aguascalientes.* Report submitted to the Instituto Nacional de Antropología e Historia, Mexico City.

2016 *Informe 2016, Proyecto arqueológico Cerro de En medio, Municipio de San José de Gracia, Aguascalientes.* Report submitted to the Instituto Nacional de Antropología e Historia, Mexico City.

Fernández Villanueva, Eugenia

2004 Evidencias de una Tradición Mesoamericana en Zaragoza. In *Tradiciones Arqueológicas,* edited by Efraín Cárdenas García, pp. 291–306. El Colegio de Michoacán, Zamora.

Filini, Agapi, and Efraín Cárdenas García

2007 El Bajío, la Cuenca de Cuitzeo y el Estado Teotihuacano: Un Estudio de Relaciones y Antagonismos. In *Dinámicas Culturales entre el Occidente, el Centro-Norte y la Cuenca de México, del Preclásico al Epiclásico,* edited by Brigitte Faugère-Kalfon, pp. 137–156. Centro de Estudios Mexicanos y Centro Americanos, El Colegio de Michoacán, Zamora.

Frank, Andre Gunder, and Barry K. Gills

1993 World System Economic Cycles and Hegemonial Shift to Europe 100 BC to 1500 AD. *Journal of European Economic History* 22(1): 155.

Galván Villegas, Luis Javier

1975 Informe preliminar de las exploraciones efectuadas en la zona arqueológica de "'El Ixtépete," Jalisco, durante el mes de mayo de 1973. *Balance y perspectiva. XIII mesa redonda, 1,* pp. 395–410. INAH, Mexico City.

Galván Villegas, Luis Javier, and Christopher S. Beekman

2000 El Grillo. In *The Archaeology of Ancient Mexico and Central America: An Encyclopaedia,* edited by Susan Toby Evans and David Webster, pp. 299–300. Routledge, New York.

Holien, Thomas

1977 *Mesoamerican Pseudo-Cloisonne and other Decorative Investments.* PhD dissertation, Department of Anthropology, Southern Illinois University. University Microfilms, Ann Arbor.

Jiménez Betts, Peter

1989 Perspectivas sobre la Arqueología de Zacatecas: Una Visión Periférica. *Arqueología* 5: 7–50.

1992 Una Red de Interacción del Noroeste de Mesoamérica. In *Origen y Desarrollo de la Civilización en el Occidente de México: Homenaje a Pedro Armillas y Ángel Palerm*, edited by Brigitte Boehm de Lameiras and Phil C. Weigand, pp. 177–204. El Colegio de Michoacán, Zamora.

1995 Algunas Observaciones sobre la Dinámica Cultural de la Arqueología de Zacatecas. In *Arqueología del Norte y del Occidente de México, Homenaje al Dr. J. Charles Kelley*, edited by Barbro Dahlgren and Ma. de los Dolores Soto de Arechavaleta, pp. 35–66. UNAM, IIA, Mexico City.

1998 Áreas de Interacción del Noroeste Mesoamericano: Consideraciones y Tiestos. In *Arqueología, Historia, y Medio Ambiente: Perspectivas Regionales. Actas del IV Coloquio de Occidentalistas*, edited by Ricardo Ávila, Jean-Pierre Emphoux, Luis Gómez Gastélum, Susana Ramírez, Otto Schöndube, and Francisco Valdez, pp. 295–303. ORSTOM, Universidad de Guadalajara, Guadalajara and Paris.

2001 Alcances de la interacción entre el Occidente y el Noroeste de Mesoamérica en el Epiclásico. Lecture given during the CEMCA Archaeology Workshop, Mexico City, March 30.

2006 La Problemática de Coyotlatelco vista desde el Noroccidente de Mesoamérica. In *El Fenómeno Coyotlatelco en el Centro de México: Tiempo, Espacio y Significado. Memoria del Primer Seminario-Taller sobre Problemáticas Regionales*, edited by Laura Solar Valverde, pp. 375–392. INAH-CNA, Mexico City.

2007 El Noroeste en el Epiclásico y sus vínculos con el Bajío. Lecture given during the Panel Discussion: *La Cerámica del Bajío y Regiones Aledañas en el Epiclásico: Cronología e Interacciones*. Mexico City, October 15–16.

Jiménez Betts, Peter, and Andrew Darling

1992 Cultural Dynamics on the North-Central Periphery of Mesoamerica: Zacatecas and Northern Jalisco in the Middle Classic Period. Paper presented at the conference "The Culture Dynamics of West and Northwest Mexico," SISA Mesa Redonda, Phoenix, Arizona.

2000 Archaeology of Southern Zacatecas. The Malpaso, Juchipila, and Valparaiso-Bolaños Valleys. In *Greater Mesoamerica: The Archaeology of West and Northwest Mexico*, edited by Michael S. Foster and Shirley Gorenstein, pp. 155–180. University of Utah Press, Salt Lake City.

Kelley, J. Charles

1963 Northern Frontier of Mesoamerica. First Annual Report: August 15, 1961, submitted to the National Science Foundation, Alexandria, VA.

1971 Archaeology of the Northern Frontier: Zacatecas and Durango. In *Handbook of Middle American Indians, Vol. 11, Archaeology of Northern Mesoamerica*, edited by Robert Wauchope, Gordon F. Ekholm, and Ignacio Bernal, pp. 668–801. University of Texas, Austin.

1974 Speculations on the Culture History of Northwestern Mesoamerica. In *The Archaeology of West Mexico*, edited by Betty Bell, pp. 19–39. Sociedades Estatales Avanzadas del Occidente de México, Ajijic, Jalisco.

1990 Evaluación del Concepto de Mesoamérica desde la Frontera Noroccidental. In *La Validez Teórica del Concepto Mesoamérica*, edited by Antonio Guzmán and

Lourdes Martínez, pp. 145–197. XIX Mesa Redonda de la Sociedad Mexicana de Antropología, Colección Científica. INAH, Mexico City.

Kelley, J. Charles, Walter Taylor, and Pedro Armillas

1961 Archaeological and Ecological Investigation of the North-Central Frontier of Mesoamerica and the Relationships of the Cultures of Central Mesoamerica, the Gran Chichimeca, and the American Southwest. Application to the National Science Foundation, 1961. Southern Illinois University, Carbondale.

Krueger, Michal

2008 Pasado, Presente y Futuro de la Economía de Bienes de Prestigio como Modelo Interpretativo en Arqueología. *Revista d'Arqueología del Ponent* 18: 7–29.

Lelgemann, Achim

1996 Estudio Introductorio: Balance de la Arqueología de Zacatecas. In *Descripción de las Ruinas de Asentamientos Aztecas durante su Migración al Valle de México, a través del Actual Estado Libre de Zacatecas*, edited and translated by Carl de Berghes. Colección Joyas Bibliográficas Zacatecanas II. State Government of Zacatecas, Autonomous University of Zacatecas and Centro Bancario del Estado de Zacatecas, Zacatecas.

1997 Orientaciones Astronómicas y el Sistema de Medida en La Quemada, Zacatecas, México. *Indiana* 14: 99–125.

2000 Proyecto Ciudadela de La Quemada, Zacatecas. Report submitted to the Instituto Nacional de Antropología e Historia, Mexico City.

López Mestas C., Lorenza, and Marisol Montejano Esquivias

2003 Investigaciones arqueológicas en la Higuerita, Tala. *Revista del Seminario de Historia Mexicana* IV(1): 11–53.

López Mestas C., Lorenza, Jorge Ramos de La Vega, and Carlos Santos Rodríguez

1994 Sitios y Materiales: Avances del Proyecto Altos de Jalisco. In *Contribuciones a la Arqueología y Etnohistoria del Occidente de México*, edited by Eduardo Williams, pp. 245–267. El Colegio de Michoacán, Zamora.

Moguel Cos, María Antonieta, and Sergio A. Sánchez Correa

1988 Guanajuato y Noreste de Michoacán: Algunas Apreciaciones Cerámicas. In *Primer Reunión sobre las Sociedades Prehispánicas en el Centro Occidente de México*, pp. 223–235. Cuaderno de Trabajo No. 1. Centro INAH Querétaro, INAH, Mexico City.

Nicolás Caretta, M.

2006 *Proyecto Arqueológico Santiago, Municipio de Pabellón de Arteaga, Aguascalientes. Propuesta de Investigación*. Coordinación de Ciencias Sociales y Humanidades, Universidad Autónoma de San Luis Potosí, Instituto Nacional de Antropología e Historia, Centro INAH–Aguascalientes, Aguascalientes.

2010 *Informe técnico de prospección arqueológica y análisis cerámico del Proyecto Arqueológico Santiago, Municipio de Pabellón de Arteaga, Aguascalientes*. Report submitted to the Instituto Nacional de Antropología e Historia, Mexico City.

2011a *Informe técnico de prospección arqueológica y análisis de materiales del Proyecto Arqueológico Santiago, Municipio de Pabellón de Arteaga, Aguascalientes*. Report submitted to the Instituto Nacional de Antropología e Historia, Mexico City.

2011b *Informe técnico de excavación arqueológica y análisis de materiales del Proyecto*

Arqueológico Santiago, Municipio de Pabellón de Arteaga, Aguascalientes. Report submitted to the Instituto Nacional de Antropología e Historia, Mexico City.

2012 *Informe técnico de prospección arqueológica y análisis de materiales del Proyecto Arqueológico Santiago, Municipio de Pabellón de Arteaga, Aguascalientes*. Report submitted to the Instituto Nacional de Antropología e Historia, Mexico City.

2013 *Informe técnico de excavación y análisis de materiales. Santiago 2012–2013. Estructura 53*. Report submitted to the Instituto Nacional de Antropología e Historia, Mexico City.

2014a *Informe técnico de prospección y análisis de materiales 2013–2014. Proyecto arqueológico Santiago*. Report submitted to the Instituto Nacional de Antropología e Historia, Mexico City.

2014b *Informe técnico de excavación y análisis de materiales. Santiago 2013–2014*. Report submitted to the Instituto Nacional de Antropología e Historia, Mexico City.

Nicolás Caretta, M., and Peter Kroefges

2014 Ni todos Nómadas ni todos Sedentarios: El Sitio de Cerro de Santiago, Aguascalientes. In *Building Transnational Archaeologies: The 11th Southwest Symposium, Hermosillo, Sonora*, edited by Elisa Villalpando and Randall H. McGuire, pp. 23–33. State Museum Archeological Series 209. Arizona State Museum, Tucson.

Noguera, Eduardo

1944 Exploraciones en Jiquilpan. *Anales del Museo Michoacano* (2da época) 3: 37–52.

Pelz, Marín Ana María, and Jorge Luis Jiménez Meza

2007 Arqueología en Aguascalientes. El Ocote. In *Paleontología, Arqueología y Demografía. La reinvención de la memoria. Ensayos para una nueva historia de Aguascalientes*, coordinated by Víctor Manuel González Esparza, pp. 83–108. Instituto Cultural de Aguascalientes, Aguascalientes.

Pelz Marín, Ana María, Niklas Schulze, and Gilberto Pérez

2012 *Primer Informe Técnico Parcial Proyecto Arqueológico La Montesita, Temporada 2011*. Report submitted to the Instituto Nacional de Antropología e Historia, Mexico City.

2014 *Segundo Informe Técnico Parcial Proyecto Arqueológico La Montesita, Temporada 2013*. Report submitted to the Instituto Nacional de Antropología e Historia, Mexico City.

Pereira, Grégory

1999 *Potrero de Guadalupe: anthropologie funéraire d'une communauté pré-tarasque du nord du Michoacán, Mexique*. Centro de Estudios Mexicanos y Centro Americanos, Mexico City.

Pérez Cortés, Enrique

2007 *La región del río Verde Grande y el sitio arqueológico de Buenavista. Una aproximación a la dinámica de interacción interregional del sureste de Zacatecas durante el epiclásico*. Unpublished undergraduate thesis, Anthropology Academic Unit, Autonomous University of Zacatecas, Zacatecas.

Piña Chán, Román, and Kuniaki Oi

1982 *Exploraciones Arqueológicas en Tingambato, Michoacán*. INAH, Mexico City.

Piña Chán, Román, and Joan Taylor

1976 Cortas Excavaciones en El Cuarenta, Jalisco. *Boletín INAH* 1: 221–223.

Pollard, Helen Perlstein

2008 A Model of the Emergence of the Tarascan State. *Ancient Mesoamerica* 19: 217–230.

Pollard, Helen P., and Laura Cahue

1999 Mortuary Patterns of Regional Elites in the Lake Patzcuaro Basin of Western Mexico. *Latin American Antiquity* 10(3): 259–280.

Porcayo Michelini, Antonio

2001 La Cerámica Prehispánica de Aguascalientes. Ponencia presentada en el CEMCA. Unpublished manuscript in possession of the author.

Puch Ku, Elizabeth

2014 *Análisis cerámico e interacción cultural en el sitio cerró Santiago, Aguascalientes durante el Epiclásico*. Unpublished MA thesis, Faculty of Philosophy and Letters, Institute for Philological Research, UNAM, Mexico City.

Ramos de la Vega, Jorge, and Lorenza López Mestas C.

1999 Materiales Cerámicos en la Región Alteña de Jalisco. In *Arqueología y Etnohistoria de la Región del Lerma*, edited by Eduardo Williams and Phil C. Weigand, pp. 279–296. El Colegio de Michoacán, Zamora.

Ramos de la Vega, Jorge, and Sergio Vázquez Sánchez Correa

1988 Rescate e Investigación en "Cerrito de Rayas," León, Guanajuato. In *Primer Reunión sobre las Sociedades Prehispánicas en el Centro Occidente de México*, pp. 307–319. Cuaderno de Trabajo No. 1. Centro INAH Querétaro, INAH, Mexico City.

Saint-Charles, Juan Carlos

1996 El Reflejo del Poder Teotihuacano en el sur de Guanajuato y Querétaro. In *Tiempo y Territorio en Arqueología. El Centro-Norte de México*, edited by Ana Maria Crespo and Carlos Viramontes, pp. 143–169. INAH, Mexico City.

Silva Rivera, Yesenia

2015 Patrón de Asentamiento del Sitio Arqueológico Santiago, Aguascalientes. Unpublished undergraduate thesis, Autonomous University of San Luis Potosí, San Luis Potosi.

Solar Valverde, Laura

2002 *Interacción interregional en Mesoamérica. Una aproximación a la dinámica del Epiclásico*. Unpublished undergraduate thesis, ENAH, Mexico City.

Taladoire, Eric, and Benoit Colsenet

1991 "Bois Ton Sang, Beaumanoir": The Political and Conflictual Aspects of the Ballgame in the Northern Chiapas Area. In *The Mesoamerican Ballgame*, edited by Vernon L. Scarborough and David R. Wilcox, pp. 161–174. University of Arizona Press, Tucson.

Valencia Cruz, Daniel

1994 *Informe Técnico Final del Proyecto: Identificación, Catalogación y Conservación de Sitios con Pintura Rupestre en el Estado de Aguascalientes*. Report submitted to the Instituto Nacional de Antropología e Historia, Mexico City.

Weigand, Phil C.

1985 Evidence for Complex Societies during the Western Mesoamerican Classic Period. In *The Archaeology of West and Northwest México*, edited by Michael Foster and Phil C. Weigand, pp. 47–91. Westview Press, Boulder, CO.

1990 Discontinuity: The Collapse of the Teuchitlán Tradition and the Early Postclassic Cultures of Western Mesoamerica. In *Mesoamerica y Norte de Mexico, Siglo IX–XII, Vol. I*, coordinated by Federica Sodi Miranda, Wigberto Jiménez Moreno, and George E Stuart, pp. 215–222. Instituto Nacional de Antropologia e Historia, Mexico City.

1996 La Evolución y Ocaso de un Núcleo de Civilización: La Tradición Teuchitlán y la Arqueología de Jalisco. In *Las Cuencas del Occidente de México: época Prehispánica*, edited by Eduardo Williams and Phil C. Weigand, pp. 185–246. El Colegio de Michoacán, Zamora.

2000 The Evolution and Decline of a Core of Civilization: The Teuchitlán Tradition and the Archaeology of Jalisco. In *Greater Mesoamerica: The Archaeology of West and Northwest Mexico*, edited by Michael S. Foster and Shirley Gorenstein, pp. 43–58. University of Utah Press, Salt Lake City.

Weigand Phil C., and Acelia García de Weigand

1999 Arqueología de los Altos de Jalisco: El Peñol del Chiquihuitillo y su Contexto Regional. In *Arqueología y Etnohistoria: La Región del Lerma*, edited by Eduardo Williams and Phil C. Weigand, pp. 269–285. El Colegio de Michoacán, Zamora.

Williams, Glyn

1974 External Influences and the Upper Río Verde Drainage Basin at Los Altos, West Mexico. In *Mesoamerican Archaeology: New Approaches*, edited by Norman Hammond, pp. 21–50. University of Texas Press, Austin.

9

Weaving Our Life

The Economy and Ideology of Cotton in Postclassic West Mexico

MICHAEL MATHIOWETZ

Introduction to the Economy and Ideology of Cotton in Postclassic Mesoamerica

The Postclassic period (AD 900–1521) in Mesoamerica marked an era of significant cultural reorganization; this led to expansive economic systems and interregional trade that bound together disparate societies into an international world-system. For the far-flung regions of northwestern Mesoamerica, scholars have paid little attention over the past 40 forty years to the important Aztatlán culture, to the extent that it is one of the least-understood archaeological cultures of ancient Mesoamerica. The Aztatlán region has been positioned as a critical link when modeling the Postclassic-period "Mesoamericanization" of West Mexico as well as for conceptualizing the nature of interaction between Mesoamerican societies and Pueblo cultures of the U.S. Southwest. It is problematic, however, that research in this broad region has progressed relatively slowly (until recently), as only a handful of scholars actively work in this far-western region to provide the necessary new data to address questions on pre-Hispanic interregional interaction.

This chapter examines the dramatic social transformations that characterized the Postclassic period in the Aztatlán region of far west Mexico within the context of the heightened connectivity that characterized this era in Mesoamerica. In particular, it is proposed that these changes relate to the onset of a new and nonlocal political-religious solar and rain complex focused upon the Mesoamerican solar deity Xochipilli and Flower World complex. This religious complex was centered in the Pacific-coastal Aztatlán heartland where cotton spinning, weaving, and finished textiles played a central economic and ideological role in newly burgeoning macroregional economic networks. In this chapter, I examine the evidence of diverse weaving technologies and related artifacts derived not from my own archaeological field research at a particular site—the perspective of many of the authors in this volume—but from a more synthetic approach focused upon data derived from previous systematic excavations and surface surveys at a multitude of Aztatlán sites. In addition, a rich database of archaeological literature compiled over the past 80 years and analyses of extant museum collections derived from past field research informs the interpretation offered herein. My particular area of expertise and fieldwork focuses upon cultural dynamics of the Aztatlán core zone, the highlands of Nayarit, and northern Jalisco where I have conducted field research since 2005. In conjunction with archaeological analyses of cotton and weaving practices in the context of Aztatlán religion, the examination of similar ethnographically documented practices by presumed and known descendant cultural groups in West Mexico allows for new and more nuanced interpretations of patterned material remains in the archaeological record. Comparative archaeological and ethnohistoric data from adjoining regions that exhibit more distant cultural connections complement this work.

A case study on cotton industries and ideology can shed light on the nature of pre-Hispanic social, political, economic, and religious organization in the little-understood Aztatlán world. This research adds to a diverse body of extant work on cotton production, distribution, and consumption practices in the pre-Hispanic New World on a range of topics of interest, including: the origins of cotton domestication; cotton industry technologies; specialization and craft production; feasting and gift exchange; market exchange; gendered weaving activities; clothing styles and regalia; and the ideological and religious significance tied to conceptual metaphors of weaving and cotton, among other themes (Brumfiel 1991; Carpenter et al. 2012; Hamann 1997; Hendon 2006; G. McCafferty and S. McCafferty 2015; S. McCafferty and G. McCafferty 1991, 1994; Milbrath 2000; Pohl 1994; Schaafsma 2013; Smith and Hirth 1988; Sullivan 1982; Taube 1983; and others).

Connecting West Mexico's Archaeological Past with the Ethnographic Present

The study of religion, ideology, and belief systems from the material remains of archaeological cultures in the Americas dates over a century, but with the development of processual archaeology during the middle twentieth century, a primary focus upon prehistoric adaptations to the environment dominated the field. Systematic approaches to the archaeological study of religion and ritual in the past few decades have accelerated and resulted in a number of recent publications (for example, Fowles 2013; Hall 1997; Hodder 1989, 2010; Inomata and Coben 2006; Insoll 2004; Pauketat 2013; VanPool et al. 2006; Whitley and Hays-Gilpin 2008). In more humanistic approaches to the archaeology of Mesoamerica, the incorporation of ethnohistoric texts, ethnographic data of descendant indigenous cultures, and linguistic studies has been central in providing the cultural context for interpretive breakthroughs by archaeologists, art historians, and epigraphers in the study of religion, symbolism, political and economic organization, the decipherment of pre-Hispanic hieroglyphic texts, and insights into many aspects of a multitude of cultures. The question of how to study past religions and ideas is key.

Conceptual Metaphor Theory and the Study of Religion

Ideologies are aspects of symbol systems within which power is negotiated, and since ideas cannot be measured objectively, they must be studied "from the inside." As such, the principles of meaning of symbol systems and ideologies must be historically generated—although archaeologists are often theoretically ill-equipped for this task (Hodder and Hutson 2003: 88). Ideologies are "the framework within which, from a particular standpoint, resources are given value, inequalities are defined and power is legitimated. Ideas are themselves the 'real' resources used in the negotiation of power; and material resources are themselves part of the ideological apparatus" (Hodder and Hutson 2003: 88).

One of the concerns of archaeologists in recent decades is the search for meaning in the material culture of past societies (for example, Hodder 1987a, 1989; Tilley 1990, 1993). Likewise, scholars in recent decades have advocated for the resurgence of historical approaches in archaeology (for example, Deetz 1988; Hodder 1987b; Lekson 2008; Whiteley 2004), which in the United States increased after 1990 with the passage of the North American Graves Protection and Repatriation Act (NAGPRA). For current historical analyses of religion in antiquity, one useful tool for discerning cultural meaning in material culture and continuity or discontinuity in cosmological tenets and religious beliefs and practices is conceptual metaphor theory. This theoretical approach has become more prevalent in various regional studies with the understanding that metaphors—acting as mnemonic devices for shared cultural knowledge rooted in analogy—are basic elements of human expression and discourse in speech, narrative, ritual, art, and other forms of material culture, including the production, form, and use of material objects (for example, Culley 2006, 2008; Ortman 2000; Sekaquaptewa and Washburn 2004; Tilley 1999). A conceptual metaphor is an "image-based, nonlinguistic phenomenon that is expressed in material culture as well as language" (Ortman 2000: 613), a fact that positions analyses of metaphor as an empirical research method and tool useful in conjunction with art historical techniques.

Conceptual metaphor theory has roots in the cognitive sciences as a cognitive-processual approach. It is at once a scientific endeavor in the definition of metaphors as systematic and predictable, and humanistic in the metaphorical expression of ideas and values (Culley 2006: ii) in a way that "operationalizes conceptual metaphor theory as an ethnographically bound research method that moves us toward a science of meaning" (Culley 2006: 11–12). Ortman (1998: 183) noted that "there are systematic relationships between the system of metaphors that motivated ancient behavior and the patterns in practice that have left material traces in the archaeological record." The challenge then is to recognize the system and interplay of complex metaphors that underpin the cosmology of a particular archaeological culture, which is identified, when possible, through the analysis of ethnographic cultures that are presumed or have been demonstrated to be descendants of that culture. Ethnographic patterns in the material culture—whether in symbolism, architecture, colors, artifacts, or even activities that use the material objects in which metaphors are also encoded—can be defined. In other words, "it is the ethnographic context that evidences the repeated use of specific metaphors and the coherence of metaphorical reasoning" among cultures, wherein "[it] is then the ethnographic context that actually predicts prehistoric metaphorical reasoning and that contextualizes prehistoric materials within a specific logic" (Culley 2006: 85).

With an understanding of the systems of intersecting metaphors evident within past societies and living descendant peoples, one can begin to discern from the material culture and material patterns of behavior when certain metaphors and complexes of metaphors are present in a society or region, and when they are not. Thus, through conceptual metaphors one can trace historical developments, continuities, and disjunctions of metaphors through time and across space—such as those that occurred during the significant social transformations in West Mexico which distinguish the Classic period from the Postclassic period when the Aztatlán tradition emerged. Here, new patterns in material culture (and attendant metaphors) became apparent among societies within the region as well as in adjoining areas among people with whom they were clearly in contact. For the Aztatlán region, these extraregional contacts include Pueblo cultures of the U.S. Southwest and northern Mexico as well as societies in highland central and southern Mesoamerica (see Mathiowetz 2011). Ideally, discerning shared metaphors and elucidating the nature of interactions require detailed knowledge of the archaeology, ethnohistory, ethnography, religion, art, and symbolism of multiple cultures, of multiple time periods and in multiple regions, rather than expertise of one single culture or region.

Previous Studies of Religion in Far West Mexico

The study of religion among pre-Hispanic West Mexican societies is comparatively recent. Most seminal studies of West Mexican art and symbolism were simply descriptions of icons, design layouts, and deities; this included limited comparisons to stylistic parallels in iconography with other regions, particularly highland Mexico (for example, Bell 1960; Ekholm 1942; Furst and Scott 1975; Schöndube Baumbach 1974). Early systematic studies of shaft tomb sculptures were largely descriptive and primarily focused upon literal interpretations of secular activities and scenes of daily life (for example, von Winning 1974). While some scholars cautioned against interpreting the religious meaning of shaft tomb art (for example, Taylor 1970), among the earliest examples of the use of ethnographic data to interpret archaeological phenomena beginning in the 1960s was Peter Furst's (1974) analyses of shaft tomb sculpture. His research characterized the Huichol religion in terms of concepts of shamanism and psychotropic plant use that largely derived from his work with a single Huichol consultant and drew upon direct comparisons to other unrelated societies on a worldwide scale. Weigand (1985: 150–153) and others roundly critiqued Furst's interpretations of the Huichol religion while Fikes (1993) detailed how popular fads tied to Carlos Castaneda influenced the shamanic interpretations of shaft tomb art. Pickering and Beekman (2016: 9–10) recently summarized this history of research and the two divergent approaches of Von Winning and Furst.

One of the main issues involved with using the ethnography of modern indigenous groups of the Gran Nayar to interpret the archaeological patterns and symbolism of shaft tomb cultures is the presumed cultural continuity between earlier shaft tomb cultural formations and modern (and largely composite) indigenous groups in the region, despite being separated by a long time gap. Concerns with the presumption of continuity are further exacerbated by the fact that significant cultural disjunctions in the broader region clearly developed during the Postclassic period by AD 900, and these intervening disjunctions are rarely taken into account. To be fair, the Postclassic-period social formations, symbolism, and religious beliefs and ritual practices of the Aztatlán culture have only recently come into focus, and scholars' conceptions were much less developed during the time of Furst's initial work. In my estimation, much clearer linear connections are evident between late pre-Hispanic Aztatlán societies and indigenous groups in the Gran Nayar today. West Mexican archaeologists in recent decades continue to rely on the ethnographic record for interpretations (for example, Beekman 2003; Mathiowetz 2011; Vidal Aldana and Gómez Ambríz 2017; Williams 1990, 1991).

Aztatlán iconographic and religious connections to highland Mexico were considered by some archaeologists in the middle twentieth century to be so clearly evident that migration was considered to be a viable explanation for these similarities. Today, those who specialize in Postclassic Mesoamerican art, religion, and symbolism, including the Aztatlán region, continue to find close parallels between the ethnographic cultures and the archaeological record of this region in comparison to highland Mexican cultures and religion, particularly in the Aztatlán core zone (for example, Mathiowetz 2011, 2019a, 2019b; Mathiowetz et al. 2015; Pohl 2012). With increasing knowledge of Aztatlán religion and regional social and political dynamics, scholars of earlier cultural traditions in the broader region can begin to incorporate the results of these analyses into their interpretations so as to gauge continuities or disjunctions. While the above summary is by no means comprehensive, it is a point of departure for the analysis of beliefs and worldview in the Aztatlán region, which often have no clear regional antecedents.

In considering Postclassic West Mexican religion in relation to conceptual metaphors, I aim to illustrate the point that ideology and religion matter in archaeological interpretations. While such a statement is simple enough, ideas and beliefs often have been dismissed or downplayed in some regards as esoteric, unknowable, and/or relatively unimportant. However, religion is multilayered and often intersects with politics, economies, social organization, architecture, gender roles, and other domains; it is often manifest in seemingly mundane material culture and expressed in the performance of everyday actions. With a clearer understanding of how religion and ideology often underpin basic economic activities among ancient societies—such as weaving industries in the Aztatlán world—the discovery of a simple spindle whorl at a site and the recognition of the underlying ideological significance and conceptual metaphors imbued in that object and its use can then provide clues to assess the cosmology and worldview of the community; the nature and order of the natural and social worlds; the role of individuals in negotiating that world; and the continuity or discontinuity of ideas through time. One can then gauge how weaving as an economic act simultaneously reflected the individuals' agency in their role in the recreation and sustenance of the world while contributing toward the legitimization (or resistance) to an ideology that sustained or reinforced social hierarchies.

The Aztatlán Culture in the Postclassic Mesoamerican World

Following the collapse and reorganization of Classic-period polities and regional sociopolitical systems such as Teotihuacan and Monte Albán in highland central Mexico and centers across the Classic Maya world, new and smaller polities and interaction spheres emerged in the ensuing Epiclassic and Postclassic periods. The Postclassic International World in Mesoamerica was characterized by: population growth; a proliferation of small polities; an expansive international connectivity and increased volume of long-distance trade between cultures in disparate regions; a greater diversity of trade goods; commercialization of the economy; widespread evidence of metallurgical technologies (see Mountjoy et al., this volume); and the emergence of new forms of writing and iconography and new patterns of stylistic interaction in the form of the Postclassic International Style and symbol set that encoded a shared Flower World ideology (Smith and Berdan [eds.] 2003; Smith and Berdan 2003: 6–8; Taube 2010). In this new international economy, cotton was a key commodity (Berdan et al. 2003). At the onset of the Postclassic by AD 900, the Aztatlán culture—with its heartland and core zone centered on the Nayarit and southern Sinaloa coastal plain—reflected many of these changes. Aztatlán polities emerged in far west Mexico partly as a result of the extensive sociopolitical reorganization that occurred following the collapse of the highland Epiclassic sites of Alta Vista and La Quemada in Zacatecas.

After this change, far west Mexico became integrated into the Postclassic-period world-system and newly expansive economic and intellectual interaction spheres. In a recent study on pre-Hispanic West Mexican cultural developments, Christopher Beekman (2010: 73) noted that "the collapse of the interior was undoubtedly related to the rise of Pacific Coast [Aztatlán] communities" at the onset of the Postclassic period. Nicolás Caretta and Dueñas García (this volume) detail aspects of the dissolution and transformation of Classic and Epiclassic systems in north-central Mexico in Aguascalientes, an area that lies directly east of the Aztatlán center of El Teúl in southeastern Zacatecas, and north of Aztatlán sites around Lake Chapala on the border of Jalisco and Michoacán in the broad region along the Lerma-Santiago basin. Prior to the rise of the Aztatlán system, cultural developments in this area had reflected earlier connections to adjoining regions in highland Mexico within the Late Classic Mesoamerican world-system, which in some ways helped to set the framework for later Toltec influence in the rise of the Aztatlán tradition during the Early Postclassic period (see Jiménez Betts 2017). It is important to note that although the Xochipilli and Flower World complex discussed below does appear to have precursors in Classic and Epiclassic cultures around the Basin of Mexico and Gulf Coast, Xochipilli does not appear to have been present in the Classic and Epiclassic systems in north-central Mexico (see Turner 2016: 75–81), nor is it evident in the Teuchitlán tradition of Jalisco (see chapters by Heredia Espinoza; Beekman; and Sumano Ortega and Englehardt, this volume). Perhaps the best evidence we have for very limited Flower World imagery in the Bajío region is the Teotihuacan-related butterfly depictions and images of the *xiuhcoatl*-related Teotihuacan War Serpent found at the Epiclassic site of Plazuelas in Guanajuato (Turner 2016: 132–136). However, Plazuelas imagery does not appear to have a direct linear relation to the Toltec- and Mixtec-related development of the Postclassic Xochipilli and Flower World complex in the Aztatlán region.

Following the Epiclassic and the demise of what Jiménez Betts (2017) calls the Inland Northern Network (INN), the subsequent cultural reorganization and social, political, and religious developments in the emerging Aztatlán heartland on the fertile coastal alluvial floodplains involved the consolidation of populations from more widely dispersed and smaller Classic-period settlements into large and highly nucleated civic-ceremonial centers; these centers were constructed around the Marismas Nacionales estuary system and along every major river drainage along the coast from northern Sinaloa to central Jalisco (Mountjoy 2000). They were strategically located to take advantage of fertile soils and marine, estuarine, and riverine resources and to provide easy access to transportation corridors. Other Aztatlán sites of a range of sizes and ethnic and linguistic affiliations in the Nayarit, Jalisco, Sinaloa, Durango, and Zacatecas highlands formed component parts of a vast interlinked political, economic, and ideological interaction sphere across much of northwest Mesoamerica between AD 850/900–1350 (Foster 1999; Kelley 1986, 1995, 2000; Mountjoy 2000; Publ 1986, 1990).

The formation of large, independent Aztatlán civic-ceremonial centers on the coast within apparently defined provinces indicates that these forms of sociopolitical organization fit into the "city-state model" characterized by the pan-Mesoamerican Postclassic period rise of small, politically autonomous but interdependent polities comprised of a single capital city or town with surrounding territories of farmland and a hierarchy of smaller settlements (see Smith 2003). The numerous *señoríos* or *cacicazgos* organized within densely populated and multiethnic native provinces along the Sinaloa and Nayarit coastal plain as described in contact-period documents—including Cinaloa, Petatlan, Culiacán, Piaxtla, Chametla, Aztatlán, and Sentispac—may represent similar forms of territorial boundaries that likely extended into the Postclassic period (see Carpenter 1996: fig. 3.2). Some contact-period provinces may correspond to pre-Hispanic provinces of archaeologically known large Aztatlán centers, such as the Aztatlán sites of San Felipe Aztatán in the Aztatlán province, and Sentispac in the Sentispac province. Michael Ohnersorgen (personal communication, 2008) suggested that Tzapotzingo—a *señoríos* located north of San Blas, Nayarit (see Anguiano 1992: map 9)—may correspond to the large Aztatlán site of Chacalilla.

The development of a system of interconnected Aztatlán polities certainly derived in some ways from local, indigenous developments and social networks that existed in the Classic period as sites from this period—albeit much smaller, simpler, more dispersed, and with very different material culture and symbolism—often underlie Postclassic sites, yet these earlier influences are poorly understood. However, transformations in the Early Postclassic period in West Mexico in some ways were closely tied into developments in the Toltec world, both through highland corridors of interaction through the Bajío, and through Pacific-coastal networks extending to Central America and perhaps beyond. Conceptualizations of direct political control, however, have been replaced with ideas tied to ceremonial influence and religious structures partly focused upon what Jiménez Betts (2017) characterized as a Tlaloc-related ceremonialism (or "Ceremonial Subcomplex"). Censers, or *incensarios*, linked to this subcomplex are thought to have been shared along a highland corridor and appear to be manifest at the far southeastern Aztatlán site of La Peña in the Sayula Basin around AD 1050 following an earlier influx of Aztatlán coastal core zone material culture (see Ramírez and Cárdenas 2006). Helen Pollard (2015: 109) contends that Early Postclassic Toltec ceremonial influence tied to this subcomplex is present in elite burials at the Early Postclassic Pátzcuaro Basin site of Urichu in northern Michoacán, while Jadot (2016: 538) indicates similar Toltec connections at the site of El Palacio in the Zacapu Basin in northern Michoacán (adjacent to Lake Chapala). Taken together, this evidence indicates that Toltec connections to West Mexico through highland corridors were established in the realm of political, prestige goods, and information networks. For further synthesis of these developments, see Jiménez Betts (2017).

Finally, the presence of Early Postclassic Toltec-affiliated Tohil Plumbate ceramics at Aztatlán sites—manufactured on the Pacific coast of Guatemala near the Chiapas border (Neff 1989a, 1989b; Neff and Bishop 1988)—followed by evident interaction with later Nahua-Mixteca polities also signified the onset, intensification, and significant expansion of highly developed coastal economic and information networks (Mathiowetz 2011; Pohl 2012, 2016). It is noteworthy that neither the Flower World–related Xochipilli and Tlahuizcalpantecuhtli solar and Venus complexes (Mathiowetz 2011, 2018a, 2019a, 2019b; Mathiowetz et al. 2015) that dominated the Aztatlán core zone and broader region have been identified by scholars in those adjoining highland regions to the east of Lake Chapala—a fact that indicates a coastal information network also played a significant role at this time. In sum, both highland and coastal corridors from central and southern Mexico appear to have contributed in various ways toward the development of Aztatlán political and religious ideologies, a subject of further debate.

Xochipilli Worship in the Aztatlán Region of Postclassic West Mexico

In order to understand the intensive production and consumption of cotton that occurred after AD 900 in the Aztatlán world, one must first set the context for this discussion by examining the origin and distribution of worship of the young solar deity Xochipilli/Piltzintli in this region. Much like the Gran Nayar region today (Jaúregui and Neurath 1998) and the former Aztatlán core zone and broader coastal region at contact in the early sixteenth century (Anguiano 1992: map 9), the pre-Hispanic Aztatlán region likely was comprised of multiple interlinked ethnic and cultural groups bound by a set of shared political, economic, social, and religious practices, but distinguished by local and regional variations and other cultural differences. Elsewhere I have argued that the onset of the Postclassic period in far west Mexico was marked by the widespread adoption of the decidedly nonlocal Xochipilli solar complex that crosscut or transcended these ethnic and linguistic differences. This solar-oriented, political-religious complex was most intense in the Aztatlán core zone, and the transformative effect that the adoption of this complex had on the social, political, and economic organization of the macroregion and on expansive interregional interactions has been amply demonstrated (Mathiowetz 2011, 2018a, 2019a, 2019b; Mathiowetz et al. 2015). The widespread worship of Xochipilli at this time provides insight into the simultaneous florescence of intensive cotton production and use across this region.

Xochipilli—also known in highland central Mexico by the name Piltzintli or Piltzintecuhtli, Seven Flower, Macuilxochitl, Centeotl, and other names—is a young solar deity whose attributes mark him as the god of dawn and the patron of flowers, love and sex, generation (maize), music, dance, art and crafts, painting, song, games, feasting, and the soul, among other related domains (Mathiowetz 2011; Pohl 1994; Renteria 2018; Taube 2004, 2005). The Xochipilli religious complex—which is closely linked to the Postclassic Flower World solar realm—has roots at Classic-period Teotihuacan (Taube 2006). While evidence for this deity's presence has recently been proposed for various Epiclassic sites in highland Mexico, the Gulf Coast, and the Maya world (Turner 2016: 75–81), the Xochipilli complex seems to have become most pronounced during the Postclassic period. During the Early- to Late Postclassic in highland central Mexico, Xochipilli was evident among the Toltec and the Aztec and was known by a variety of names among the Mixtec, Zapotec, and Eastern Nahua of Puebla and Oaxaca.

In far west Mexico, Xochipilli worship first became evident at the onset of the Postclassic period. Elsewhere I have described in more depth that the evidence for Xochipilli worship in this region is apparent in archaeological portrayals, toponyms, ethnohistorical accounts, and ethnographic information (see Mathiowetz 2011, 2019a, 2019b). Here, due to space constraints,

I will only summarize these data. Early contact-period accounts from the southern Sinaloa and Nayarit coastal plain (that is, the former Aztatlán heartland) primarily identify this region as the territory of the Totorame, also known as the Pinome or Pinonuquia. The Totorame territory overlaps with the Chametla, Aztatlán, and Sentispac indigenous provinces first encountered between 1530 and 1531 during the Nuño de Guzmán entrada (Carpenter 1996: 86–91). Regarding language, the former Aztatlán heartland in the early historic period was quite diverse; for example, in the far northern area of Nayarit around the community of Acaponeta, Fray Alonso Ponce reported in 1587—some 50 fifty years after the conquest and about two centuries after the end of the Aztatlán tradition in the Middle Postclassic period—that there existed "seven languages or seven differences of languages," including Pinome (Totorame) (in Sauer 1934: 7). Totorame is thought to have been a lowland dialect of the highland Cora whose political center is at the Mesa del Nayar, and there appears to have been a close cultural affinity between groups in the two regions (Sauer 1934: 14). Based upon data from the 1587 *visita* of Fray Alonso Ponce, Carl Sauer (1934: 7–8) indicated that the Totorame (Pinome) territory and language extended from coastal southern Sinaloa through the Nayarit lowlands and southeastward into the highlands around Tepic. At contact, the Totorame population in the Aztatlán core zone was estimated at 100,000 (Sauer 1935: 5). The Totorame are important for the present discussion because the extent of their cultural territory largely overlaps with the pre-Hispanic Aztatlán heartland, and they (and their highland Cora relatives) are known to have worshipped the solar deity Piltzintli (Xochipilli) at contact, much like their Aztatlán ancestors.

Piltzintli/Xochipilli was widely worshipped by Totorame, Tzayahueca, and Tepehuan people in southern Sinaloa, Nayarit, and southern Durango (Anguiano 1992: chart 7). In the northern Totorame region of southern Sinaloa, a travertine effigy vessel portraying Xochipilli was recovered from the site of Siqueros near Mazatlán (Figure 9.1). The most detailed account of Piltzintli worship among the Cora is the account of Fray Antonio Arias de Saavedra, who in 1673 described the worship of Piltzintli in the annual ceremonial and agricultural cycle and noted a lineage of Cora Sun Kings—living representations or intermediaries of Piltzintli—based at the Mesa del Nayar (McCarty and Matson 1975). Contact-period documents indicate the worship of Xochipilli in the town of Acaponeta in northern Nayarit, with toponyms bearing his name known from Ahuacatlán in southeastern Nayarit. The Ortelius Map of AD 1579 depicts two towns named Suchipila,

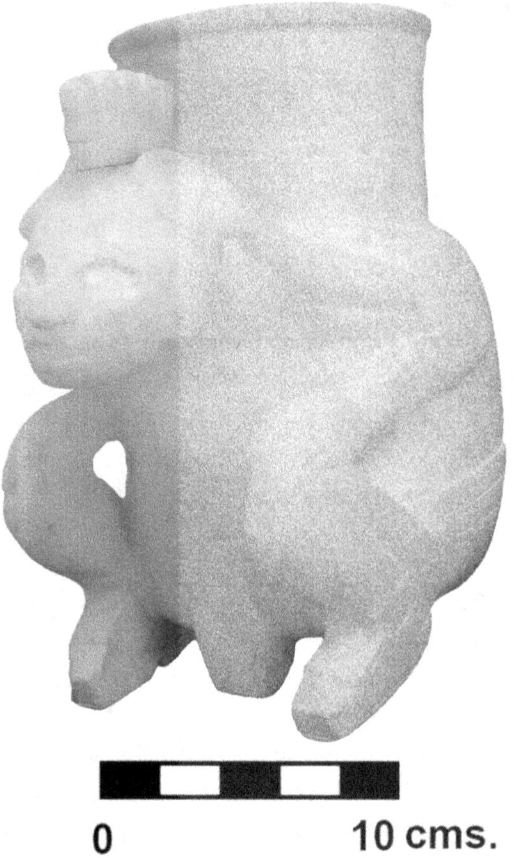

Figure 9.1. Travertine effigy vessel in the form of the young solar deity, Xochipilli. Siqueros, southern Sinaloa (Museo Arqueológico de Mazatlán, Sinaloa; photo by author).

one located in southern Zacatecas (modern Juchipila) and the other in Jalisco to the west of modern Guadalajara. Thus, in the early historic period, Xochipilli/Piltzintli worship was prevalent among multiple ethnic and linguistic groups in the Aztatlán lowland core zone and in surrounding highland regions of Nayarit, southern Durango, northern Jalisco, and southern Zacatecas—an area that essentially overlaps with a broad section of the southern segment of the Aztatlán interaction sphere (see Mathiowetz 2011) (Figure 9.2). The extent of the Xochipilli complex in the Aztatlán region surely will be revised to encompass a broader geographical territory as new data come to light.

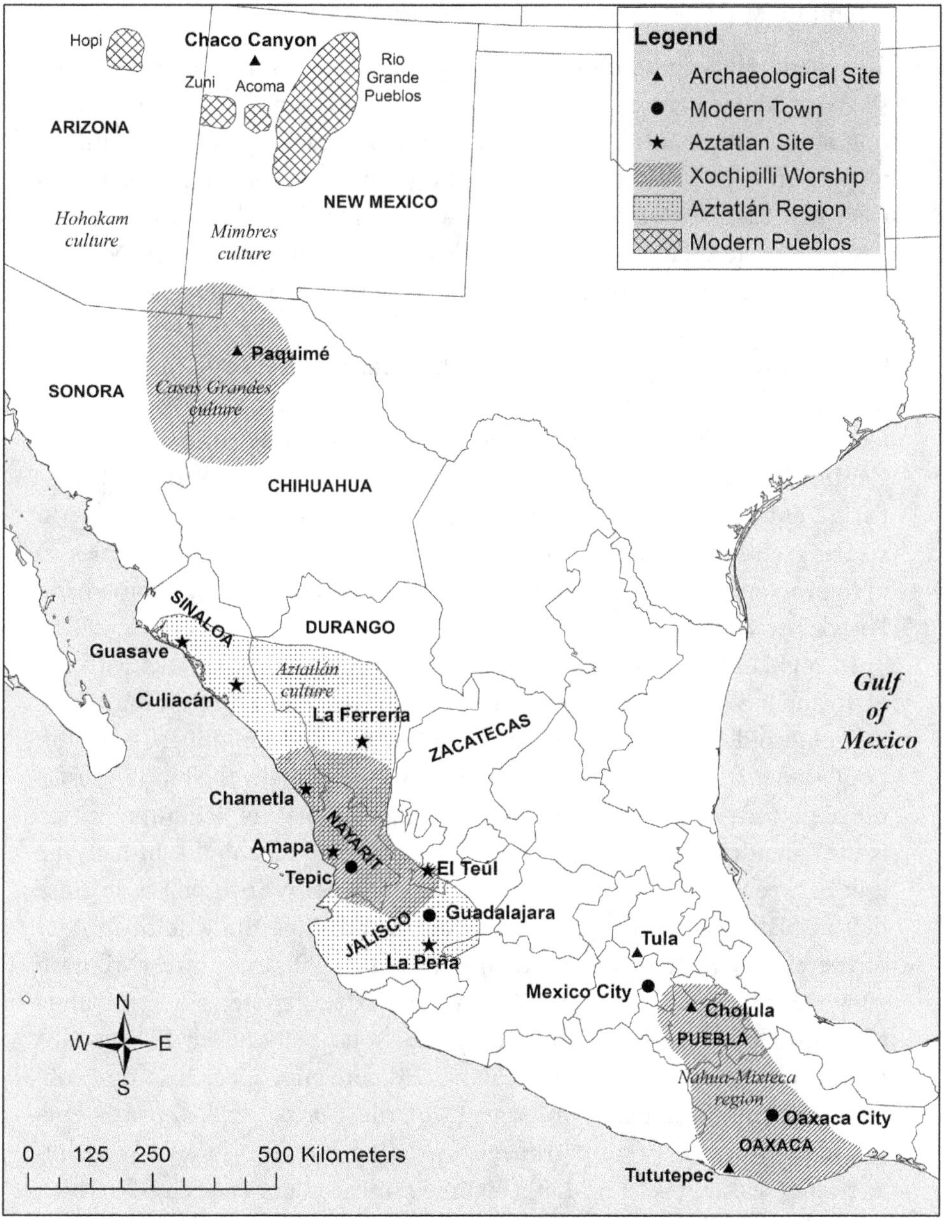

Figure 9.2. Currently known regional distribution of Xochipilli worship in Postclassic West Mexico as evident from the archaeological record and ethnohistorical documents, in relation to other zones of Xochipilli worship (map by Daniel Pierce).

Xochipilli and Cotton/Weaving Industries in Postclassic Mesoamerica

The Xochipilli and cotton/weaving complex that flourished in Postclassic West Mexico were surely closely intertwined, and the extensive ethnographic documentation of the ideology of weaving and cotton linked to the sun and rain among descendent groups in the Gran Nayar makes clear that this ideology and worldview continues into the present day, as described below. However, this complex did not originate in West Mexico. The template for this complex—and this deity in particular—has roots in a very similar ideological complex focused on this deity in highland central Mexico and Oaxaca. Because of the clear post–AD 900 material and ideological connections between these disparate regions—as well as the fact that this solar deity in West Mexico originates in Postclassic highland Central Mexico and/or Oaxaca and Puebla with Classic- and Epiclassic-period antecedents—it is instructive to examine how Xochipilli relates to weaving, cotton, and textile exchange and to the Flower World complex so as to provide a context for comparison to similar cultural developments in Postclassic West Mexico.

In highland Mexico, the solar deity Xochipilli has close links with the earth and lunar deity Xochiquetzal, who is the patroness of spinning, weaving, embroidery, crafts, artisans, love, flowers, and childbirth. Sahagún (1950–1982: 2: 36) noted that every 260 days—essentially the human gestational period—women honored Xochiquetzal on the day "7 Flower," which is the calendrical name of the deity Seven Flower–Xochipilli. In fact, the link between Xochiquetzal and the solar floral deity Xochipilli is so close that in historical documents she is considered to be the wife or mother of the closely related deities Xochipilli and/or Piltzintecuhtli (Milbrath 2000: 32–33; Nicholson 1971: 417). In the Mixtec/Zapotec/Eastern Nahua regions of Oaxaca and Puebla, the young solar deity Seven Flower–Xochipilli was the patron god of weaving and elite craft specialization (Pohl 1994). Xochipilli appears on page 15 of the *Codex Vindobonensis* surrounded by objects of his patronage, including garments known as *xicollis* and other textiles (Pohl 1994: 9) (Figure 9.3). The major feast days in which Seven Flower–Xochipilli was celebrated involved elite palace feasting networks and the gift exchange of woven garments and other objects (Pohl 1994). Similarly, during the Aztec feast of Xochilhuitl ("Feast of Flowers") in which Xochipilli was feted, "the seamstresses celebrated a feast. They fasted . . . [in order] to weave textiles well" (Sahagún 1950–1982, Book 2: 35–36, Book 4: 7).

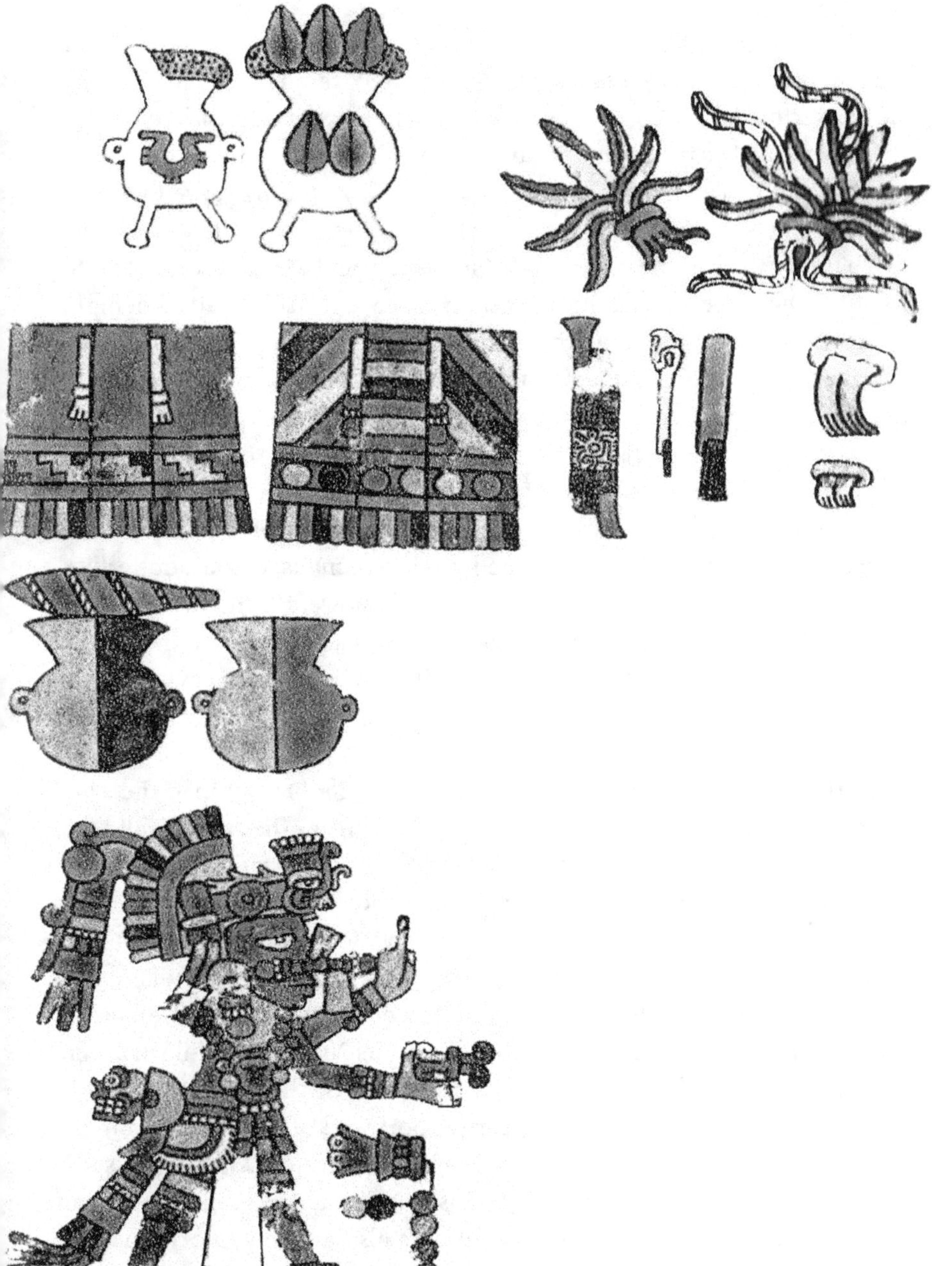

Figure 9.3. Seven Flower–Xochipilli as patron of craftworkers. Depiction of Seven Flower–Xochipilli with crafts and objects of his patronage including textiles, paintbrushes, stone-working tools, pulque, and cacao (*Codex Vindobonensis* 15; drawing courtesy John M. D. Pohl).

An ideology of weaving in relation to Xochipilli and the Flower World is evident in carved-bone weaving implements recovered from Late Postclassic Mixtec burials in Tomb 7 at Monte Albán, Oaxaca—considered one of the richest burials in ancient Mesoamerica. Within this tomb, a number of spinning and weaving implements were found, including carved weaving battens, small combs and picks, spindle whorls, and spinning bowls used as a base for the spindle (Caso 1969; McCafferty and McCafferty 1994: 148). Many of the objects found in the tomb reference the solar deity Xochipilli (Caso 1969). The link between the bone weaving battens and the Flower World ritual complex centered upon Xochipilli is evident in the finely carved codex-style imagery on these objects. Caso (1969: 191–193, fig. 185) noted that Seven Flower–Xochipilli appears in a birth scene on a carved weaving batten. More recently, Karl Taube (2010: 171) concluded that these bone weaving implements are heavily ornamented with Flower World symbolism, including solar rays and the heads of humans, deities, butterflies, eagles, and other creatures that emerge from flowers: "The heads on the carved bones emerge from or are placed on blossoms as celestial denizens of the solar Flower World" (Figure 9.4). These Postclassic weaving implements clearly reflect a close association between the act of weaving and the Xochipilli-oriented Flower World.

In the Valley of Mexico, Elizabeth Brumfiel (2007) examined the symbolism of 96 Postclassic-period spindle whorls from the site of Xaltocan. Here she found that pre-Aztec motifs on spindle whorls from domestic household contexts and burials, beginning in the Early Postclassic and leading up to the Aztec conquest of the site in 1430, focused upon a prominent solar theme including quadripartite divisions, solar rays, the solar cycle, and flowers; the latter likely alluded to the *tonalli* (solar life force) and to the patroness of weaving Xochiquetzal (Brumfiel 2007: 99). While Brumfiel (2007: fig. 8) identified one spindle whorl as bearing the head of the crocodile Cipactli, the head instead is clearly a portrayal of a butterfly—a motif that again substantiates the presence of a Flower World theme tied to spinning and weaving at the Early to Late Postclassic highland central Mexican site of Xaltocan. I suggest that these motifs also allude to Xochiquetzal's consort Xochipilli, the god of flowers and patron of crafts, including weaving and textiles. In fact, Franco (1961: 236–237, plate 18) indicated that Aztec-era spindle whorls and spindle whorl molds in the Valley of Mexico bear representations of butterflies and flowers, and he noted that butterfly imagery often is associated with Xochipilli—a clear allusion to the Flower

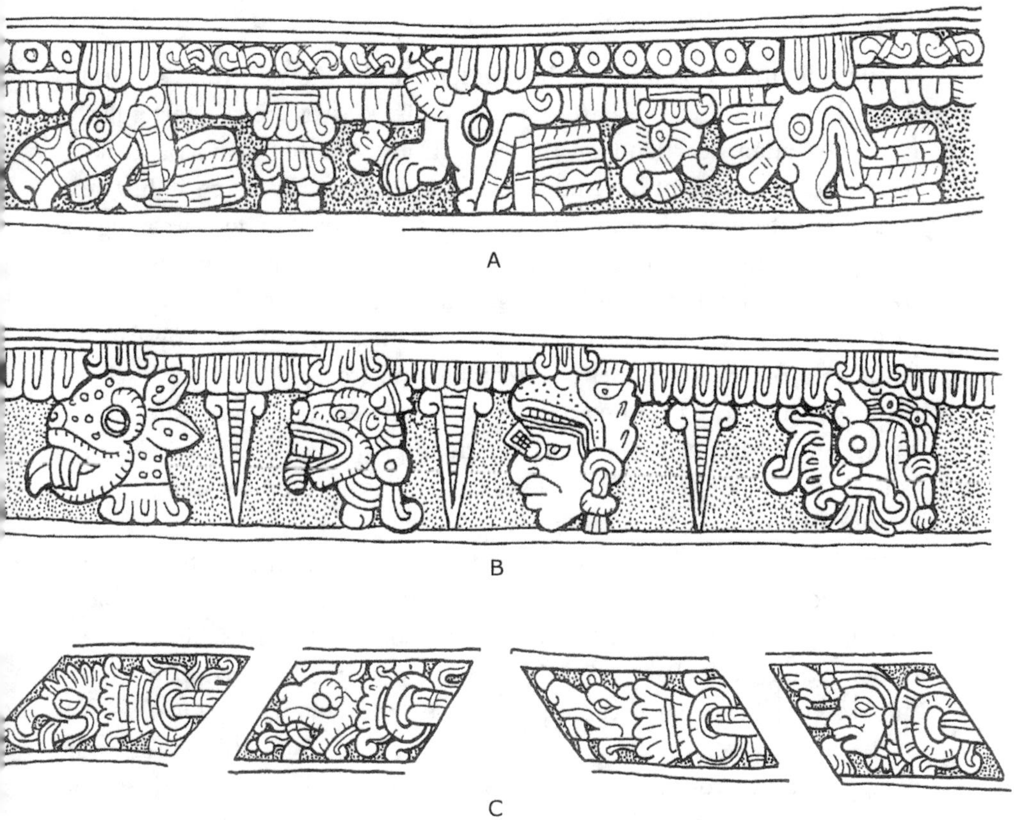

Figure 9.4. Late Postclassic Mixtec carved bone weaving tools with Flower World motifs from Tomb 7, Monte Albán: *A*, Creatures emerging from flowers in skyband; note butterfly head (second from right) central figure holding flower (from Taube 2010: fig. 22a). *B*, Heads attached to blossoms in skyband; note solar rays and the deity Tlaloc (from Taube 2010: fig. 22b). *C*, Human and animals emerging from flowers (from Taube 2010: fig. 22c) (drawings by Karl Taube, used with his permission).

World. Brumfiel (2007) emphasized the role of women and their domestic spinning and weaving crafts as being oriented toward the solar-centered reproduction of the cosmos and the production of cotton textiles imbued with solar energy. In sum, the data indicate that a Flower World–oriented cotton and weaving complex centered upon Xochipilli flourished in parts of highland central Mexico and Oaxaca—and likely West Mexico—during the Postclassic period.

The Archaeology of Cotton in the Aztatlán Region

While Xochipilli worship and the ideology of the Flower World began to flourish at the onset the Early Postclassic in West Mexico—particularly in the Aztatlán heartland on the Nayarit and southern Sinaloa coastal plain—so, too, did cotton cultivation, spinning, and weaving industries. In addition to more direct evidence of cotton cultivation and weaving in the New World and elsewhere—such as cotton thread and finished textile fragments—a number of additional supplementary forms of archaeological material evidence for cotton can complement these more fragile, perishable items. Supplementary evidence can include macroscopic and microscopic plant parts including cotton seeds, cotton seed beaters, spindle whorls and shafts, loom parts, weaving tools, textile imprints in clay, and even symbolic representations of cotton or cotton textiles in ceramics, murals, rock art, and other forms (Kent 1957: 462–463).

In West Mexico, Postclassic-period cotton textile fragments found in association with copper bells were discovered in an urn burial by a local resident in Chametla, Sinaloa, in 1961 with some of the bells still attached to the cotton cloth (Pang 1971). It might be presumed that copper bells—the presence of which became pronounced in the archaeological record of the Postclassic period—were commonly strung on cotton cords and/or appended to cotton textiles. At Guasave in northern Sinaloa, several fragments of cotton cloth were recovered in association with copper bells and the cotton threads that bound them—the copper apparently serving as the means of preservation (Ekholm 1942: 115). One textile fragment was affixed to a copper plate from Amapa, Nayarit (Meighan 1976: 114, plate 109d). Textile impressions of fine threads—likely cotton rather than the coarser *maguey*—were recovered in hardened mud at Cojumatlán, Michoacán (Lister 1949: 85) while fabric impressions were reported on ceramic sherds at Peñitas, Nayarit (Bordaz 1964: 106).

Aside from the textile fragments or seeds and plant parts, spindle whorls represent the most common material means by which to discern evidence for cotton production and weaving in the archaeological record, along with more rare weaving implements like bone picks, combs, and copper needles. Beginning with Sauer and Brand's (1932) pioneering survey along the Pacific Coast, scholars have consistently noted the presence of spindle whorls, both ceramic and stone, in Aztatlán contexts. According to Michael Foster (1999: 158), a "marker of the Early Aztatlán tradition is a distinctive round, incised spindle whorl decorated with incised circles." While

Classic-period spindle whorls are present to some extent, the heightened ubiquity of modeled and incised ceramic spindle whorls at large and small Aztatlán sites—particularly along the humid coastal lowlands of Nayarit and southern Sinaloa where intensive agriculture was possible—suggests the emergence of a well-developed and widespread cotton industry in the Postclassic period. Joseph Mountjoy (2000: 96) noted: "Judging from the relative abundance of spindle whorls in Aztatlán deposits at Amapa and other sites to the north in Sinaloa, cotton was probably grown in many of the coastal valleys in these areas. Cotton was probably spun locally into yarn for weaving cloth most likely used by the elites." In the present day, cotton (*Gossypium hirsutum*) grows wild and is cultivated by Huichol in the sierra (Grimes 1980: 270), which may have been the case at highland Aztatlán sites as well.

At present there has been no systematic study of Aztatlán spindle whorls or other weaving tools; however, an examination of collections of spindle whorls from Amapa, Nayarit, and other sites indicates that they were manufactured in a range of sizes and weights. This variation may indicate the presence of multiple weaving industries focused on different fibers, including *maguey*, as noted in other regions of Mesoamerica. For example, in a prominent study of Aztec-era spindle whorls from the Teotihuacan Valley, Mary Parsons (1972) noted a bimodal distribution pattern whereby the small whorls likely were used for cotton spinning, while the large whorls functioned within *maguey* fiber (*ixtle*) industries. Subsequent scholars have added to the literature on this subject in various regions of Mesoamerica (Brumfiel 2007; Carpenter et al. 2012; Smith and Hirth 1988).

Spindle whorls are present in West Mexico from the Formative through Classic periods but appear to be much more prevalent during the Postclassic period. While no systematic study has been conducted on pre-Aztatlán spindle whorls, the variation in sizes of Classic-period whorls has been observed to be markedly less distinct than the variability that characterizes Aztatlán spindle whorls in the Aztatlán core zone; collectively, this reflects heightened economic specialization and a marked social division of labor (Mauricio Garduño Ambriz, personal communication, 2018). Stanley Long (1966: 55, 59, fig. 184) recovered two clay spindle whorls from Tomb 3 at the Late Formative shaft tomb site of El Arenal, Jalisco, and two clay spindle whorls from Tomb 2 at the site of Santa Maria near Etzatlán, Jalisco. Verenice Heredia (personal communication, 2016) has characterized as "rare" the presence of spindle whorls at Late Formative to Early/Middle Classic Teuchitlán sites in highland Jalisco. Ben Nelson (personal communication, 2016) indicated that spindle whorls were "very uncommon" at the Epiclassic site of La Quemada in Zacatecas. For the pre-Aztatlán region in the far west, rare Classic period examples—often just a few in number—were reported from sites in northern Jalisco, at Amapa and in the San Blas area of Nayarit, and at Chametla in southern Sinaloa (Mountjoy 1995: 78–79), among others. Garduño Ambriz (personal communication, 2016) noted that Late Classic/Epiclassic spindle whorls are prominent on the northwest coastal plain of Nayarit; however, further documentation is needed. In the broader region, the relative rarity of spindle whorls prior to the Postclassic period indicates that while cotton and *maguey* fiber production was practiced in various areas earlier in time, there is not sufficient evidence to suggest intensive industries for cotton textile production.

The widespread presence of Postclassic-period spindle whorls suggests that the coast of West Mexico was a major producer and exporter of raw and woven cotton; it likely was produced locally in certain coastal regions to be exported and spun elsewhere in nearby or distant areas (Mountjoy 1995: 77–80). For the Postclassic era (beginning from north to south), at the site of Guasave—a large burial mound located in far northern Sinaloa—36 spindle whorls were recovered, most of which were found interred beside individuals (Ekholm 1942: 87–88, fig. 17). Sauer and Brand (1932: 27–29, fig. 4) reported spindle whorls at sites near the Río Piaxtla on the coastal plain between Culiacán and Mazatlán. They also reported spindle whorls from sites in the Sierra de Tacuichamona and San Lorenzo Valley regions located south of Culiacán. A total of 278 whorls were recovered from Culiacán (Kelly 1945: 137–141, Table 14). Incised spindle whorls recovered at Culiacán resembled those from Guasave (Foster 1999: 152). Spindle whorls at Chametla, Sinaloa, were affiliated with Aztatlán contexts in Late Chametla II and El Taste–Mazatlán assemblages (Kelly 1938: 52–53, fig. 24). Spindle whorls were reported from sites in the Marismas Nacionales in southern Sinaloa and northern Nayarit (Sweetman 1974: 68–69) and from Juana Gómez, a site located in the foothills near Escuinapa, Sinaloa. Gill (1974) reported on the results of burial analyses from excavations at three sites (Panales, Chalpa, and Tecualilla) in the Teacapan Estuary, a section of the larger Marismas Nacionales located in southern Sinaloa and northern Nayarit (see map in Gill 1974: fig. 1) where female burials sometimes were interred with spindle whorls (Gill 1974: 88, 94–96, and tables 1–5).

At the site of Amapa along the central coastal plain of Nayarit, Meighan (1976: 77–81) recovered 164 spindle whorls—only a few of which were in Classic contexts—where roughly half were stone (some in the form of pottery vessels) while the rest were ceramic (Figure 9.5). Spindle whorls also were reported at other Aztatlán archaeological sites along the coastal plain of Nayarit, including at sites north of Amapa such as Coamiles and San Felipe Aztatán (Garduño Ambriz 2007; personal communication, 2009) and at Chacalilla (Ohnersorgen 2007: 41, fig. 4.13), a large center located to the south of Amapa. A private collection housed at the Casa-Museo Vladimir Cora in the town of Acaponeta in northern Nayarit (near the major site of San Felipe Aztatán) contains roughly 100 spindle whorls. In southeastern Nayarit, Gifford (1950: 233, figs. 19a–19l) reported 12 spindle whorls from the community around the site of Ixtlán del Río. One spindle whorl fragment was excavated at Structure 7 at the nearby site of La Pitayera near the modern town of Ahuacatlán (personal observation, 2015).

In highland Durango, spindle whorls from the Sinaloa coast—likely from sites such as Guasave and Culiacán—were recovered from Late Aztatlán–period Guadiana Chalchihuites sites in the highlands of Durango (Kelley 1986: 88). At the Schroeder site (now called La Ferrería), Kelley and colleagues (Kelley 1971, 1990; Kelley and Winters 1960) recovered a number of intrusive objects from the Pacific coast, including large, incised spindle whorls. At Cañon del Molino in Durango, Ganot and Peschard (1995: 158; 1997: fig. 56) reported Aztatlán spindle whorls resembling those from the west coast. Spindle whorls of Pacific coast origin also were recovered from Navacoyán in the Guadiana Valley (Kelley 1986: 88).

Joseph Mountjoy (1982, 1990, 1995: 79) noted spindle whorls in the Tomatlán River Valley of coastal Jalisco, and at Ixtapa in the Banderas Valley of Jalisco, with recent reports of spindle whorls from Maito along the northwest Jalisco coast (see Mountjoy et al., this volume). In highland Jalisco, Meighan and Foote (1968: 126–132) reported 41 spindle whorls from Tizapán el Alto, 2 of which were described as having a "flower-shaped design." Of the 41 recovered whorls at Tizapán el Alto, 7 were found in three burials, 4 with a female, 1 with a male, and 2 from the burial of a person of an unknown sex (Meighan and Foote 1968: 45–47, table 2). Lister (1949: 63–67) reported roughly 97 spindle whorls from the nearby site of Cojumatlán, Michoacán, 8 of which display a "four-petaled floral" motif. Located west of Tizapán el Alto and Cojumatlán, the site of La Peña in the Sayula Basin of Jalisco (south of modern Guadalajara) contained an Aztatlán component with spindle whorls (Ramírez de Swartz et al. 2005: 313).

Coastal lowland and highland Aztatlán communities may have engaged in diverse weaving economies centered upon both cotton and *maguey* fiber production, the latter of which is a highland resource that is absent

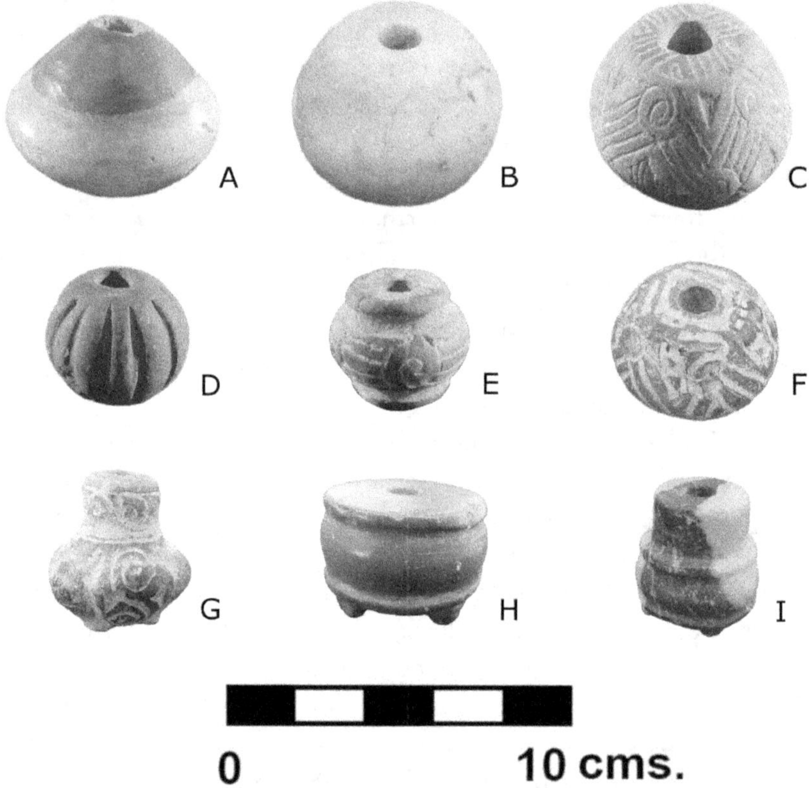

Figure 9.5. Ceramic and stone spindle whorls from Amapa (246) and Penitas (184), Nayarit (images courtesy of Fowler Museum at UCLA Archaeology Collections Facility and the Instituto Nacional de Antropología e Historia). *A*, Ceramic (Classic period), Catalog No. 246-2780; *B*, Stone; Catalog No. 246-2164; *C*, Ceramic, Catalog No. 246-2334; *D*, Ceramic, Catalog No. 246-2619; *E*, Ceramic, Catalog No. 246-221; *F*, Ceramic, Catalog No. 246-1615; *G*, Ceramic tripod jar effigy, Catalog No. 246-430; *H*, Stone tetrapod jar effigy, Catalog No. 184-110d; *I*, Stone tetrapod jar effigy, Catalog No. 184-339h (photos by author).

from the humid coastal lowlands. For example, Lawrence Feldman (1978: 142, table 16 and fig. 21) noted that the products in contact-period market towns (*tiangues*) in Nayarit each reflected economic specializations within particular ecological zones. The 1524 expedition by Francisco Cortés, the *alcalde mayor* of Colima, to subjugate towns between the Ameca and Santiago rivers—along with tax assessment documents from over two decades later—provides a view into local economies that indicate *maguey* specialization in highland communities of southeastern Nayarit, such as Ixtlán del Río, Mexpan, and Tetitlan (Feldman 1978: 142, table 16), among other highland specializations. Many market towns in the region offered cotton and/or finished textiles, and these will be described in more detail below.

Along with spindle whorls, archaeological evidence for cotton and weaving industries and related technologies includes the presence of bone awls and combs, copper needles for sewing and embroidery, and artifacts linked to textile decorative practices including ceramic stamps with geometric and naturalistic designs. Byron Hamann (1997: 157, fig. 5) illustrated modern bone-weaving picks in Mexico and noted that these tools functioned in tasks that required more delicate work with individual threads and small areas of the cloth. In a study of weaving implements from sites in Postclassic Nicaragua, Sharisse and Geoffrey McCafferty (2008: 150–154) found that rarely preserved bone implements such as small- and medium-sized bone awls served as weaving tools. Similar small- to medium-sized bone awls and combs—probable weaving tools—have been recovered at some Aztatlán sites, including in Nayarit at Amapa (Meighan 1976: 124 and plates 112a–112m) and Peñitas (Bordaz 1964: 130); in Sinaloa at Culiacán (Kelly 1945: 141–142, figs. 71b–71i); in highland northern Jalisco at Tizapán el Alto (Meighan and Foote 1968: 154–155 and plates 23a–23g and 23m) and coastal Jalisco at the Arroyo Piedras Azules site (see Mountjoy et al., this volume); and in Durango at Cañón del Molino (Ganot and Peschard 1995: 166) (Figures 9.6–9.7). Copper needles have been recovered from Aztatlán sites including: Cojumatlán in Michoacán (Lister 1949: 72, figs. 32a–32b); La Peña and Tizapán el Alto in Jalisco (García and Liot 2006: 404, photo 228; Meighan and Foote 1968: 135–136, plates 22f–22g); Amapa (Meighan 1976: 112, plates 99a–99f, 100a–100h, and 100k), Coamiles (Garduño Ambriz 2011: 33, 2012: 43–44), and Peñitas (Bordaz 1964: 126–129, plate 44) in Nayarit; and Cañón del Molino in Durango (Ganot and Peschard 1995: 161) (Figure 9.8).

Ceramic stamps in cylindrical and rectangular flat form (and other forms) also are present to varying degrees at sites in the Aztatlán region and appear to be limited to Postclassic contexts. Meighan (1976: 154–156) noted that ceramic stamps are absent in Amapa-phase (AD 500–750) contexts but appear during the Postclassic-period Cerritos (AD 900–1100) and Ixcuintla (AD 1100–1350) phases in several forms with decorative variations. In Nayarit, 38 stamps were reported from Amapa (Meighan 1976: 87–90) (Figure 9.9). Jacques Bordaz (1964: 113–114, plate 29) reported fragments of 5 ceramic stamps from the kilning area at Peñitas. One fragment was found at Coamiles (Garduño Ambriz 2013: photo 60) and 2 fragments

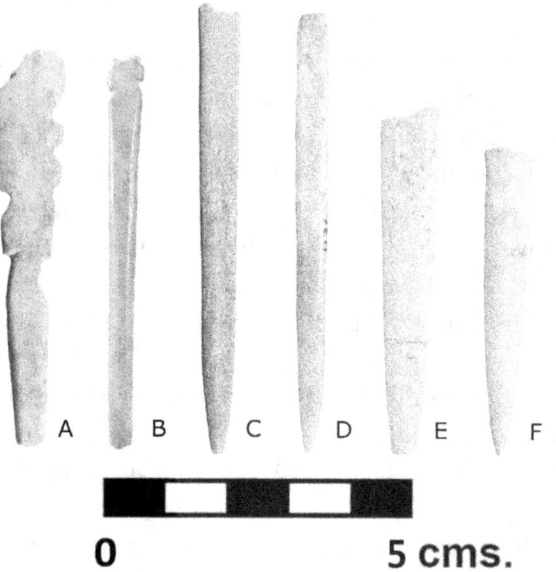

Figure 9.6. Bone picks from the Aztatlán site of Amapa, Nayarit (images courtesy of Fowler Museum at UCLA Archaeology Collections Facility and the Instituto Nacional de Antropología e Historia. *A*, Catalog No. 246-979; *B*, Catalog No. 246-400; *C*, Catalog No. 246-574; *D*, Catalog No. 246-1136; *E*, Catalog No. 246-2256; *F*, Catalog No. 246-1172 (photos by author).

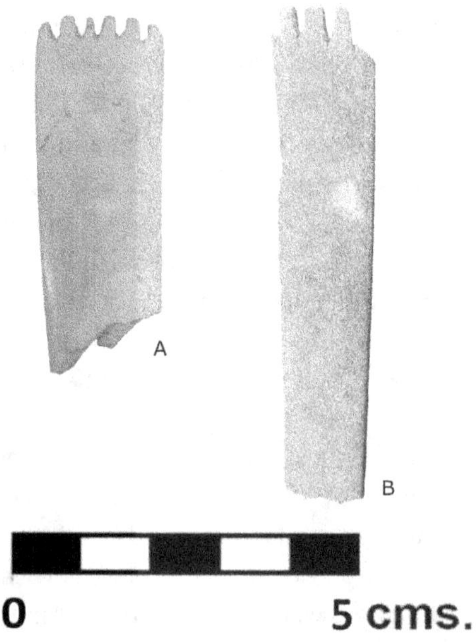

Figure 9.7. Bone weaving forks/combs from Amapa, Nayarit (images courtesy of Fowler Museum at UCLA Archaeology Collections Facility and the Instituto Nacional de Antropología e Historia). *A*, Catalog No. 246-979; *B*, Catalog No. 246-574 (photos by author).

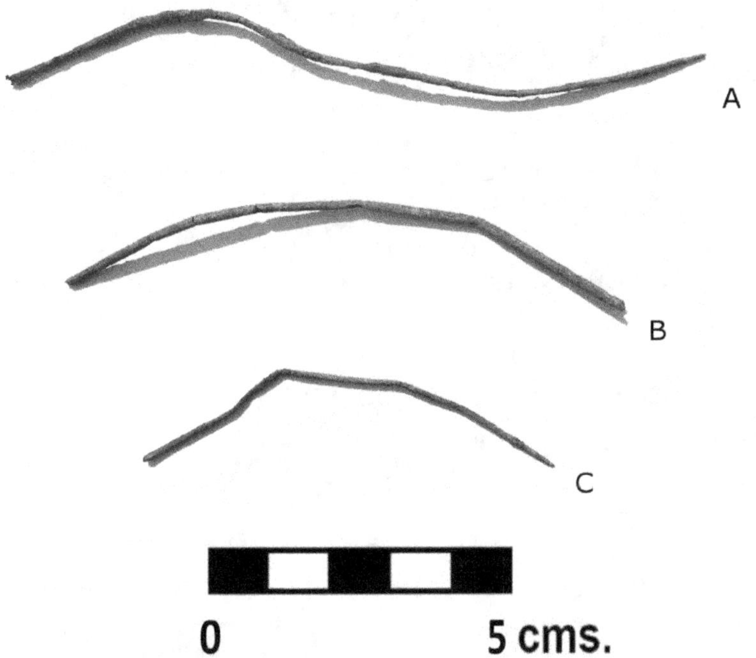

Figure 9.8. Copper needles, Coamiles, Nayarit (modified from image courtesy of Mauricio Garduño Ambriz). *A*, Cerritos phase (AD 850/900–1100); *B*, Cerritos phase (AD 850/900–1100); C, Ixcuintla phase (AD 1100–1350).

were recovered at Chacalilla (Ohnersorgen 2007: fig. 4.14). One roller stamp with a depiction of a feathered serpent was reported from Ixtlán del Río (Gifford 1950: 233). Ceramic stamps have been recovered at Aztatlán sites in Sinaloa including 12 cylinder stamps at Culiacán (Kelly 1945: 126, fig. 60), 2 examples (1 cylinder, 1 flat) from Chametla (Kelly 1938: 48–49), and 1 cylinder stamp at Guasave (Ekholm 1942: 88, 90, fig. 17dd). Three flat stamps were reported from Tizapán el Alto in Jalisco (Meighan and Foote 1968: 132–133, fig. 40) and 1 cylindrical stamp was recovered at La Peña in Jalisco (Ramírez and Cárdenas 2006: 361). One stamp fragment was reported from Cojumatlán, Michoacán (Lister 1949: 57). Although not as widespread as spindle whorls, the appearance of stamps signified new techniques for decorating textiles in the Postclassic.

With regard to the function of ceramic stamps, John Pohl (1994: 8, fig. 8) noted that embroidered and stamped geometric and naturalistic designs on Mixtec textiles in Postclassic Oaxaca likely served as mnemonic devices for reciting community, clan, and family lineage histories and religious stories,

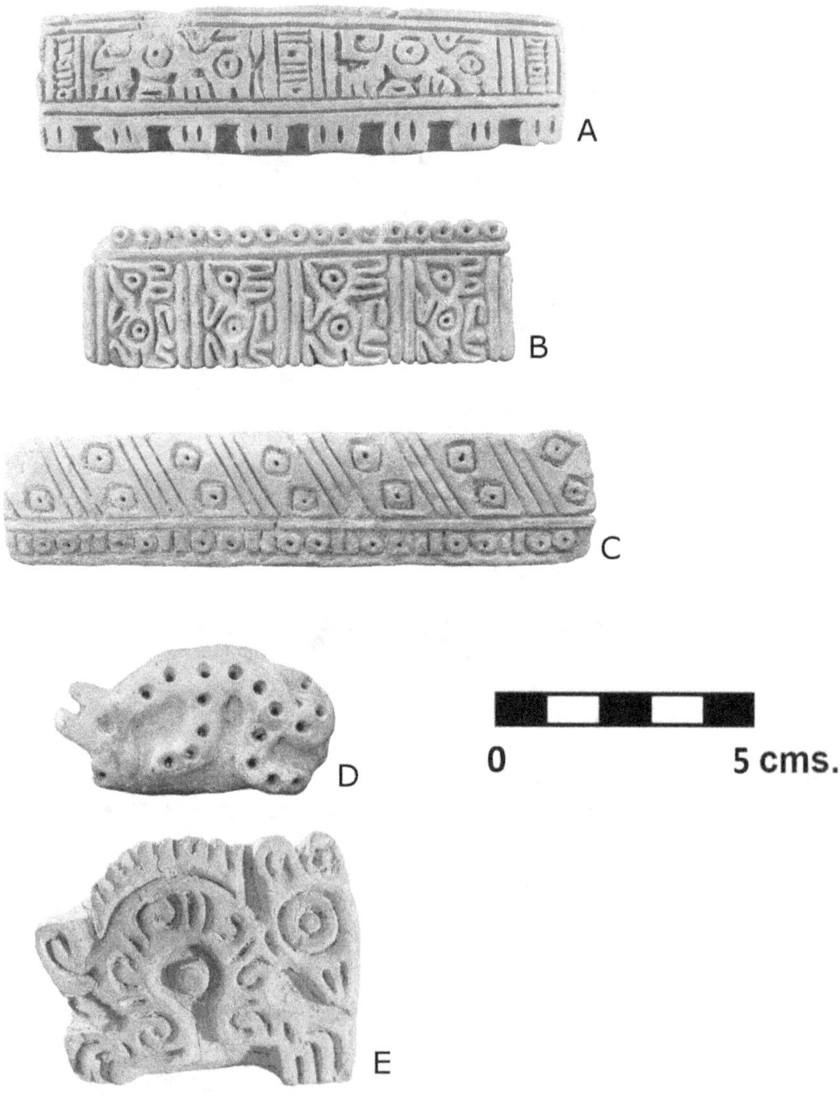

Figure 9.9. Ceramic stamps from Amapa, Nayarit (images courtesy of Fowler Museum at UCLA Archaeology Collections Facility and the Instituto Nacional de Antropología e Historia). *A*, Catalog No. 246-2727; *B*, Catalog No. 246-2460; *C*, Catalog No. 246-2459; *D*, Catalog No. 246-2461; *E*, Catalog No. 246-2708; *F*, Catalog No. 246-2708 (photos by author).

similar to Mixteca-Puebla–style designs that were incorporated into polychrome ceramics. These shared icons also are present in Mixtec weaving battens found in Tomb 7 at Monte Albán that portray codex-style imagery, as indicated above. Pohl (1994) noted that design patterns on spindle whorls often replicated designs on clay stamps, which suggests a degree of mutuality in symbolic content on spindle whorls, stamps, bone weaving implements, and ceramics (see Enciso 1971, 1980). The overlap in symbolic sets on different weaving implements and material culture suggests that shared designs formed part of an interrelated complex tied not just to economic activities, but to ethnic identity, religion, cosmology, and worldview, as was likely the case in the Aztatlán world.

Mountjoy (1995) proposed that the expansion of the Aztatlán system was tied to colonization efforts into resource-rich zones, such as areas with an abundance of obsidian and cacao. In addition to Spondylus found along the coast near the Arroyo Piedras Azules in northwest Jalisco, evidence for cotton production at the site (for example, spindle whorls, bone and ceramic combs, bone picks, and shell species important for cloth dyeing) indicates that cotton industries may have been an important factor in colonization efforts by elites from the Aztatlán core zone around Amapa—an assertion substantiated by the presence of ritually significant fine polychrome ceramics at the site originating from the Amapa region (see Mountjoy et al., this volume). With the ideology of cotton closely linked to Xochipilli and the presence of foreign ceramics of types typically found in the Aztatlán core zone where Xochipilli worship flourished (Mathiowetz 2011), it is possible that the worship of Xochipilli was spread to various regions through the establishment of enclaves for economic and ideological colonization. This assessment indicates that the Xochipilli complex likely was present at the site of Arroyo Piedras Azules, a subject that deserves further attention. Furthermore, this proposal should be considered where we see the influx of core coastal material culture into highland sites in other areas, for example, in Jalisco (Sayula Basin), Zacatecas (El Teúl), and Durango (Cañón del Molino).

In other parts of pre-Hispanic western Mesoamerica outside of the Aztatlán region, colonial enclaves have been reported in the late-pre-Hispanic Ucareo Valley of Michoacán on the eastern Tarascan frontier. Here, core-zone elites in the Lake Pátzcuaro Basin sought to enrich their lineages and political influence through military campaigns, including raids. This involved the establishment of distinct enclaves, settlements, or colonization of regions within broader multilingual and multiethnic territories

(Hernández and Healan 2008). For Tarascans, late-pre-Hispanic enclaves included those near the economically valuable Ucareo-Zinapecuaro obsidian source. While military campaigns are not clearly evident, similar political processes may have played a role in parts of the Aztatlán world. Furthermore, while Tarascans established defensive garrisons along the Aztec and Tarascan borderlands, no such systematic defensive measures (at least to this degree) appear to have been taken along the Aztatlán frontiers—although defensive sites on the far eastern Aztatlán border with the expanding Tarascan empire, around the Sayula Basin, are relevant to this topic. Finally, the use of LiDAR in an effort to identify elite and commoner residences may be useful; it can provide further insight into architectural groupings and hierarchically organized elite and commoner space within and around Aztatlán urban centers, along with more refined understanding of socioeconomic organization. These efforts could then shed more light on those core-zone Aztatlán elites who participated in colonization efforts.

The Ethnohistory of Cotton in the Aztatlán Region

Historical accounts from Nayarit provide insight into contact-period economic patterns that may help define pre-Hispanic economies. As noted above, the expedition of Francisco Cortés in southern Nayarit between the Río Ameca and Río Santiago in 1524 and the subsequent reports on tax assessments of subject towns provide an understanding of contact-period cotton cultivation and consumption patterns, which can produce more insight into local and regional pre-Hispanic economies (Feldman 1978). These reports describe market towns in the lowlands and highlands that reflected zones of economic specialization where people would travel to exchange their goods. Cotton production was focused in the Lower Piedmont regions (Feldman 1978: 141–142, table 16). Many of the towns in this region were found to have some cotton or finished textiles, and the zone of cotton production between San Blas and Jaltemba Bay to the south of Matanchen Bay appears to have been especially intense. Likewise, in 1673 Antonio Arias de Saavedra (McCarty and Matson 1975: 214) noted that cotton crops were "abundant" in the fertile alluvial plains in the region where the Río Grande de Santiago meets the sea on the Nayarit coastal plain. In the above-mentioned tax assessments, most of the sixteenth-century towns described with finished cotton products (*mantas* or *panizuelos*) are located in the Nayarit highlands (10 of 12 towns mentioned). It may be the case that lowland coastal communities produced the bulk of raw cotton (spun or unspun) that was traded to highland communities for weaving into textiles (Mountjoy 1995: 78), likely with meaningful woven or stamped designs that reflected local ethnic or community identities. Further north, the manufacture and trade of cotton textiles was noted by Francisco de Ibarra as he passed through the town of San Juan de Río, Durango (north of the city of Durango), in 1563. Here he observed that local indigenous people—possibly Tepehuanes—had traded goods for woven cotton clothing from the Topia region, located in the Durango highlands to the east of Culiacán (Kelley 2000: 150; Obregon 1924). Cotton also was important as a form of tribute. Chroniclers in the Nuño de Guzmán expedition in the 1530s reported that Sentispac in northern Nayarit was ruled by a cacique named Ocelotl who had over 20,000 subjects in the surrounding region, including the highland Cora and Tepehuan who paid tribute in the form of gold, silver, honey, and cotton (Tello 1858–1866: 2: 348).

The close connection between cotton and the solar deity Piltzintli/Xochipilli was evident in the historic period. The Franciscan friar Antonio Arias de Saavedra reported in AD 1673 that the Cora at the Mesa del Nayar offered cotton to Piltzintli and to deceased ancestor-rulers in the shrine at the ranchería of Tzacaimutta, at the Cora stronghold on the Mesa del Nayar (McCarty and Matson 1975: 205). Arias de Saavedra also noted that one local indigenous man made a pilgrimage to the Mesa del Nayar with offerings of cotton and other items in order to petition the sun (Nayarit/Piltzintli) for rain (McCarty and Matson 1975: 215). Even today, offerings of cotton and flowers are made to the remains of the last Cora Sun King in the church at the Mesa del Nayar (Aldana and Madrigal 2007: 110–111).

The Ideology of Cotton in the Ethnography of the Gran Nayar

The link between Piltzintli/Xochipilli solar worship and cotton spinning and weaving as rain- and cloud-making activities continues into the present day. The most comprehensive studies of contemporary indigenous weaving industries of the Huichol in the Gran Nayar are those of Stacy Schaefer (1989, 1993, 1996, 2002, 2015). Schaefer noted that every component of the process of spinning and weaving is comprised of multiple layers of metaphors tied to the dawn and diurnal path of the sun; to birth and death in the human life cycle; to breath, rain- and cloud-making and the ancestral rain spirits in the alternating rainy and dry seasons; to the planting of seeds and the flowering and fruiting of plants on the landscape; to the annual ceremonial cycle; and to the peyote pilgrimage to the flowery lands of Wirikuta where the sun was first born, among many other overlapping themes. These insights are important, particularly as the formation of many aspects of the contemporary indigenous worldview in the Gran Nayar appear to have taken form in the Postclassic period. Due to space constraints, these topics will be discussed only briefly.

Among the Huichol and Cora, weaving is closely linked with women from birth to death. At the time when a young Cora girl is introduced to the ceremonial *mitote* cycle, she is given a spindle with black- and white-colored yarn that represents the darkened clouds to symbolize her role as a weaver (Coyle 2001: 53, 108). Thus, as she spins and weaves the cotton thread, she symbolically creates the ancestral clouds and rain. A girl's weaving implements remain important throughout her life. One Huichol funerary rite reported in 1981 indicated that the deceased woman was interred in her grave with a number of objects, including brazilwood battens from her backstrap loom along with thread and spun wool (Fikes 2011: 175). Likewise, the Franciscan friar Antonio Arias de Saavedra reported in AD 1673 that upon death, Cora women were interred with objects they used in life, such as *metates* and spindles (McCarty and Matson 1975: 207). Interments of spindle whorls with women were later noted by Ortega (1754[1944]: 28). The historic-period interment of personal spindle whorls and weaving implements with women recalls the burials with spindle whorls at Aztatlán sites in the Teacapan Estuary as previously noted, which indicates great time depth for ritual practices centered upon women and weaving over the past millenium.

For the Huichol, the loom is a microcosm of the world. The component parts—including the sticks, beams, pickup sticks, heddle sticks, and rod sticks that are perpendicular to the yarn—represent landmarks that the sun passes on his daily journey and that form a stairway or ladder along the sun's pathway extending from the east to the west (Schaefer 1989: 181). The loom and weaving embody the symbolism inherent in the cyclical birth and death of an individual and the ascent and descent of the sun each day. As Schaefer (1989: 192) noted:

> . . . with reference to the loom as a model of the sun's path, the top beam on the loom represents the rising of the sun in the east, its movement down the loom sticks until it sets in the west, and its travel through the underworld and reappearance once again at the top of the loom. Mirroring this imagery, at birth, the arrival of the soul of an individual is like the sun rising in the east, in Virikuta. Following

> the sun, the soul travels down the warp of the loom along its life path, and at death it reaches the west, the Pacific Ocean, and then goes on to the sky above Virikuta, completing the cycle of the sun, the warp, and life in the Huichol world.

She further concluded that women are active participants in creating the rain, moving the sun along his path, and recreating the mythical time of the journey of the ancestors (peyote pilgrims) to witness the first dawn: "When weaving, the weaver metaphorically gives birth to the sun and helps it along its path. She brings the rains, plants the seeds, and helps the crops grow. She recreates mythical time by weaving the pilgrims into the path to Virikuta," the sacred floral land of dawn (Schaefer 1989: 192–193). As each textile progresses on the loom, Huichol weavers, in essence, create the flowering landscape in their woven designs, much like when embroidering. One Huichol noted (in Valadez 2010: 80):

> When women embroider they recreate the embroidery that appears on the garments of Grandmother Growth whose skirts are made of all the flowers that are blossoming in the fields. When the flowers blossom, the skirts of Our Mothers are filled with flowers of many colors, like the beautiful embroidered outfits we wear. When we make and wear our embroidered outfits, we also blossom. Everything in the universe began when it was embroidered onto the skirts of the goddesses. When Father Sun saw the designs he gave names to everything and brought them to life. That is why all of the embroidery designs are sacred, and why we bring designs to the altar, so that we will have more life, more luck, more abundance.

In the simple acts of spinning, weaving, and embroidering, Huichol women enact the annual ceremonial cycle that creates the Flower World realm in overlapping mythological and real time.

In a similar way, Cora women who spin the spindle in effect symbolically enact the rotation of the ceremonial *mitote* cycle that is centered upon the annual and seasonal cycle of the sun (see Mathiowetz 2011: 386–465). On a larger scale, Cora incorporate into their ceremonial dances a tall *chanaka* pole—an 8-foot-tall symbolic spindle whorl and shaft that symbolizes the world and represents the circular seasonal ceremonial cycle (Coyle 2001: 57, 113). In their creative acts of spinning and weaving, indigenous people of the Gran Nayar spin and weave into existence a realm of beauty in the blossoming landscape as a daily lived experience brought to life by the ancestral clouds and rain and the vivifying power of the sun, much as their ancestors likely did across the Aztatlán world during the Postclassic period.

Discussion: The Economy and Ideology of Cotton in the Postclassic Aztatlán World

The Postclassic-period Aztatlán world was marked by the simultaneous florescence of a new politico-religious complex centered upon the solar deity Xochipilli/Piltzintli as well as a widespread ritual economy centered upon cotton cultivation, the manufacture of textiles, and the distribution of both raw and finished goods through trade, tribute, and markets. According to Joseph Mountjoy (2000: 98), this pan-regional exchange and the similarity of material culture among Aztatlán sites across the region, "all support the idea that there was some rather uniform system of political, economic, and religious organization that linked these coastal valleys, probably through some sort of political alliance mechanism, such as elite marriage." While textiles in Mesoamerica had significant political and economic importance as items of trade and tribute, they also were important vehicles for transferring artistic styles and religious beliefs and practices (Taube 2010: 150).

The proposed system of Aztatlán elite marriage alliances—which undoubtedly would have connected lowland and highland Aztatlán sites—likely was facilitated through ritual feasting that may have included the exchange of fine textiles. These items would have represented the material manifestation of a shared solar- and rain-oriented ideology and cosmological order that was created by individuals on a daily basis through the act of cotton spinning and weaving. The proposed pan-regional multiethnic political, economic, and religious system that linked civic-ceremonial centers across the Aztatlán region during the Postlassic period was probably oriented around a new form of worship of the solar deity Xochipilli/Piltzintli and the Flower World complex, much like the politico-religious system described by Pohl (1994, 2003a, 2003b) for the Mixtec/Eastern Nahua/Zapotec region of Puebla and Oaxaca—the region where this complex in West Mexico likely originated. Among Aztatlán societies where community or polity leaders used this ideology to position themselves as the living sun god on earth or intermediary between the Sun and the community—such as was known for the historic-period lineage of Cora Sun Kings who were the living embodiment of the young solar deity Piltzintli (Xochipilli)—women's daily domestic cotton-spinning and weaving duties not only served to create the world anew, but they also may have unwittingly served to perpetuate, legitimize, and naturalize social hierarchies in the Aztatlán world that were rooted in a ruler's identity as the living sun within this new solar-oriented worldview and religion. Thus, as Aztatlán women likely symbolically guided and stabilized the sun in his diurnal journey through their weaving obligations, they simultaneously encouraged, guided, and stabilized the ruler as the living sun god in his maintenance of the community. In so doing, they reinforced the social hierarchy. By extension, the threat of chaos in the natural and social order could have been seen as a potential consequence of not fulfilling individual work obligations—a means by which elites may have exploited commoners within their polities. I suspect that the practice of "working for the sun" that is embedded in various industries—with attendant conceptual metaphors imbued within the related material culture and activities—was common in many parts of pre-Hispanic Mesoamerica and the Pueblo world where solar worship (particularly the Flower World complex) was manifest (see Mathiowetz 2011).

Scholars in the past decade have begun to explore the nature of Mesoamerican ritual economies and the ways that ritual practices and worldview (as materialized in objects) helped to structure economic behaviors (Wells and Davis-Salazar 2007). In the Aztatlán world of Postclassic West Mexico, the onset of the Xochipilli-oriented Flower World complex coincided with intensive cotton production and consumption. It may well be the case that it was this newly introduced religion that drove the expansive regional economic and social transformations. At this time, the production of cotton and the transformation of this raw material through spinning and weaving into finished textiles was not simply quotidian labor. These activities formed a fundamental component of a regional ritual economy and political ideology centered upon the solar-oriented Flower World—heavily imbued with metaphors centered upon birth, life, the ancestors as clouds and rain, the reproduction of the cosmos, and the ritualized actions required for maintaining the diurnal pathway of the sun—which bound together local and disparate Aztatlán sites into expansive economic and intellectual interaction networks that extended into highland central Mexico and the U.S. Southwest.

On a broader scale, Flower World symbolism long predates the Postclassic period in Mesoamerica among various (but not all) Formative, Classic, and Epiclassic cultures (see Taube 2006, 2010; Turner 2016). However, it remains to be understood how ritual economies that involved the intensification of cotton spinning, weaving, and textile production may have operated (if at all) among other cultures within the ideology of the Flower World complex. Scholars have noted cotton-industry intensification at Classic Teotihuacan (Carballo 2013: 126–128), along the Gulf Coast in Veracruz in the Classic to Postclassic (Stark et al. 1998), on the Pacific coast of Oaxaca in the Early Postclassic (King 2011), and at Early Postclassic Chichén Itzá in Yucatán (Ardren et al. 2010), among others—all regions where evidence of the Flower World has been documented. The possible interrelationship of intensified cotton industries and the presence of the Flower World is a topic that deserves further investigation.

Furthermore, the florescence of a solar and Flower World–related cotton industry in Postclassic West Mexico by AD 900—which was centered upon conceptual metaphors linking cotton to ancestors as clouds and rain—sheds light on Mesoamerican and U.S. Southwestern connections, as it is clear that this ideology long predates the very similar fourteenth-century katsina rain spirit complex found among Pueblo cultures of the U.S. Southwest (see Schaafsma 2000). While clear connections are evident between the Aztatlán culture of West Mexico and cultures in highland central Mexico and Oaxaca by AD 900, new questions arise as to whether there are historical connections between then-contemporary Southwestern and West Mexican cotton industries and their affiliated ideological complexes. Recent archaeological evidence, including the presence of scarlet macaws and cacao at Chaco Canyon, New Mexico, and other ritual items, indicates that Mesoamerican influence among Pueblo cultures became pronounced after AD 900 (Crown and Hurst 2009; Mathiowetz 2019a; Watson et al. 2015). Mesoamerican connections continued at least into the fifteenth century with clear Aztatlán influence at the important archaeological site of Paquimé in northern Chihuahua, Mexico (Mathiowetz 2011, 2018a, 2019b), and by extension in the Pueblo world in general (Mathiowetz 2018b; Mathiowetz et al. 2015).

With this influence in mind, future research on cotton industries in greater Mesoamerica and the U.S. Southwest should address the following questions: (1) How did the expansive Postclassic West Mexican economy and Flower World–related ideology of cotton influence the appearance of cotton cultivation, loom weaving, and cotton textile symbolism in rock art on the Colorado Plateau around AD 1100 (Schaafsma 2013; Teague 1998: 180–183)? (2) How did cotton weaving and textiles in West Mexico relate to new Mesoamerican-influenced textile designs in the late twelfth to early thirteenth century and garment styles—such as the kilt—that appear both in the Medio period (AD 1200–1450) Casas Grandes culture of northern Mexico and at the onset of the Pueblo IV period in the Southwest after AD 1300 (Webster et al. 2006; Webster 2007)? and (3) How did the Aztatlán ideology of cotton as clouds, rain, and ancestors between AD 900 and 1350 influence the development of the Puebloan katsina complex—which has the same focus on cotton, clouds, and ancestral rain spirits—that first became evident in the Mimbres region of southwest New Mexico after AD 1000, in the Casas Grandes culture by AD 1200, and in rock art, kiva murals, and ceramics across the Pueblo world after AD 1325 (Mathiowetz 2011: 593–596; Schaafsma 2000)?

In sum, with the arrival of the Xochipilli and Flower World complex in the Postclassic Aztatlán world, the act of spinning and weaving in far west Mexico—whether for household consumption, tribute, or gift and market exchange—was not simply a daily domestic task of women. Instead, these crucial activities positioned women as active participants in the symbolic and literal creation of life itself by enacting cosmological order, ensuring the creation and daily transit of the sun across the sky, and facilitating the arrival of the ancestral clouds and rain to Aztatlán communities.

References

Aldana, Guillermo, and Áurea Madrigal

2007 *Eco de Montaña: Espejo de Coras y Huicholes*. Cerezanegra, Mexico City.

Anguiano, Maria

1992 *Nayarit: Costa y Altiplanicie en el Momento del Contacto*. Universidad Nacional Autónoma de México, Mexico City.

Ardren, Traci, T. Kam Manahan, Julie Kay Wesp, and Alejandra Alonso

2010 Cloth Production and Economic Intensification in the Area Surrounding Chichén Itzá. *Latin American Antiquity* 21(3): 274–289.

Beekman, Christopher S.

2003 Agricultural Pole Rituals and Rulership in Late Formative Central Jalisco. *Ancient Mesoamerica* 14(2): 299–318.

2010 Recent Research in West Mexican Archaeology. *Journal of Archaeological Research* 18(1): 41–109.

Bell, Betty B.

1960 *Analysis of Ceramic Style: A West Mexican Collection*. Unpublished PhD dissertation, Department of Anthropology, University of California, Los Angeles.

Berdan, Frances F., Marylin A. Masson, Janine Gasco, and Michael E. Smith

2003 An International Economy. In *The Postclassic Mesoamerican World*, edited by Michael E. Smith and Frances F. Berdan, pp. 96–108. University of Utah Press, Salt Lake City.

Bordaz, Jacques

1964 *Pre-Columbian Ceramic Kilns at Peñitas, a Post-Classic Site in Coastal Nayarit, Mexico.* Unpublished PhD dissertation, Department of Anthropology, University of California, Los Angeles.

Brumfiel, Elisabeth M.

1991 Weaving and Cooking: Women's Production in Aztec Mexico. In *Engendering Archaeology: Women and Prehistory*, edited by Joan M. Gero and Margaret W. Conkey, pp. 224–251. Basil Blackwell, Oxford.

2007 Solar Disks and Solar Cycles: Spindle Whorls and the Dawn of Solar Art in Post-classic Mexico. *Treballs d'Arqueologia* 13: 91–113.

Carballo, David M.

2013 The Social Organization of Craft Production and Interregional Exchange at Teotihuacan. In *Merchants, Markets, and Exchange in the Pre-Columbian World*, edited by Kenneth G. Hirth and Joanne Pillsbury, pp. 113–140. Dumbarton Oaks, Washington, DC.

Carpenter, John P.

1996 *El Ombligo en la Labor: Differentiation, Interaction and Integration in Prehispanic Sinaloa, Mexico.* Unpublished PhD dissertation, Department of Anthropology, University of Arizona, Tucson.

Carpenter, Lacey B., Gary M. Feinman, and Linda M. Nicholas

2012 Spindle Whorls from El Palmillo: Economic Implications. *Latin American Antiquity* 23(4): 381–400.

Caso, Alfonso

1969 *El tesoro de Monte Albán.* Instituto Nacional de Antropología e Historia, Mexico City.

Coyle, Philip E.

2001 *From Flowers to Ash: Náyari History, Politics, and Violence.* University of Arizona Press, Tucson.

Crown, Patricia L., and W. Jeffrey Hurst

2009 Evidence of Cacao Use in the Prehispanic American Southwest. *Proceedings of the National Academy of Sciences of the United States of America* 106(7): 2110–2113.

Culley, Elisabeth V.

2006 *The Meaning of Metaphor: A Cognitive Approach to Prehistoric Ideation.* Unpublished MA thesis, Department of Anthropology, Northern Arizona University, Flagstaff.

2008 Supernatural Metaphors and Belief in the Past: Defining an Archaeology of Religion. In *Belief in the Past: Theoretical Approaches to the Archaeology of Religion*, edited by David S. Whitley and Kelley Hays-Gilpin, pp. 67–83. Left Coast Press, Walnut Creek, CA.

Deetz, James

1988 History and Archaeological Theory: Walter Taylor Revisited. *American Antiquity* 53(1): 13–22.

Ekholm, Gordon F.

1942 *Excavations at Guasave, Sinaloa, Mexico.* Anthropological Papers of the American Museum of Natural History, Vol. 38, Part II: 23–139. New York.

Enciso, Jorge
1971 *Designs from Pre-Columbian Mexico*. Dover, New York.
1980 *Sellos del Antiguo Mexico*. Editorial Innovación, Mexico City.
Feldman, Lawrence H.
1978 Postscript to the Santa Cruz Complex: A Trial Survey of Specialization and Trade in the Period of Initial Contact. In *Mesoamerican Communication Routes and Cultural Contacts*, edited by Thomas A. Lee, Jr., and Carlos Navarrete, pp. 141–144. Papers of the New World Archaeological Foundation, No. 40, Brigham Young University, Provo, UT.
Fikes, Jay C.
1993 *Carlos Castaneda, Academic Opportunism, and the Psychedelic Sixties*. Millenia Press, Victoria, BC.
2011 *Unknown Huichol: Shamans and Immortals, Allies against Chaos*. AltaMira Press, Lanham, MD.
Foster, Michael S.
1999 The Aztatlán Tradition of West and Northwest Mexico and Casas Grandes: Speculations on the Medio Period Florescence. In *The Casas Grandes World*, edited by Curtis F. Schaafsma and Carroll L. Riley, pp. 149–163. University of Utah Press, Salt Lake City.
Fowles, Severin M.
2013 *An Archaeology of Doings: Secularism and the Study of Pueblo Religion*. School for Advanced Research Press, Santa Fe, NM.
Franco C., José Luis
1961 Representaciones de la Mariposa en Mesoamerica. *El Mexico Antiguo* 9: 195–244.
Furst, Peter T.
1974 Ethnographic Analogy in the Interpretation of West Mexican Tomb Art. In *The Archaeology of West Mexico*, edited by Betty Bell, pp. 132–146. West Mexican Society for Advanced Study, Ajijic, Mexico.
Furst, Peter T., and Stuart Scott D.
1975 La Escalera del Padre Sol: Un Paralelo Etnográfico-Arqueológico Desde el Occidente de Mexico. *Boletín* 12: 13–20.
Ganot R., Jaime, and Alejandro A. Peschard F.
1995 The Archaeological Site of Cañón del Molino, Durango, México. In *The Gran Chichimeca: Essays on the Archaeology and Ethnohistory of Northern Mesoamerica*, edited by Jonathan E. Reyman, pp. 146–178. Avebury, Aldershot, UK.
1997 *Aztatlán: Apuntes para la historia y arqueología de Durango*. Gobierno del Estado de Durango, Secretaria de Educación, Cultura y Deporte, Durango.
García, Johan, and Catherine Liot
2006 Presentación de los Objetos de Metal. In *Transformaciones socioculturales y technológicas en el sitio de La Peña, Cuenca de Sayula, Jalisco*, edited by Catherine Liot, Susana Ramírez, Javier Reveles, and Otto Schöndube, pp. 399–406. Universidad de Guadalajara, México.
Garduño Ambriz, Mauricio
2007 Arqueología de rescate en la cuenca inferior del río Acaponeta. *Diario de Campo* 92: 36–52.

2011 *Proyecto: Investigaciones arqueológicas en el Cerro de Coamiles, Nayarit. Tomo I. Reporte técnico final/Temporada de campo 2009.* Archivo Técnico de la Sección de Arqueología del Centro INAH, Nayarit.

2012 *Proyecto: Investigaciones arqueológicas en el Cerro de Coamiles, Nayarit. Tomo I. Reporte técnico final/Temporada de campo 2010.* Archivo Técnico de la Sección de Arqueología del Centro INAH, Nayarit.

2013 *Análisis de Materiales de Excavación del Proyecto Arqueológico Coamiles.* Reporte Técnico de Análisis Cerámico/Temporadas 2007 y 2008. Report on file at Centro INAH, Nayarit.

Gifford, Edward W.

1950 *Surface Archaeology of Ixtlán del Río, Nayarit.* University of California Press, Berkeley.

Gill, George W.

1974 Toltec-Period Burial Customs within the Marismas Nacionales of Western Mexico. In *The Archaeology of West Mexico*, edited by Betty Bell, pp. 83–105. Sociedad de Estudios Avanzados del Occidente de Mexico, Ajijic.

Grimes, Joseph E.

1980 Huichol Life Form Classification II: Plants. *Anthropological Linguistics* 22(6): 264–274.

Hall, Robert L.

1997 *An Archaeology of the Soul: North American Indian Belief and Ritual.* University of Illinois Press, Urbana.

Hamann, Byron

1997 Weaving and the Iconography of Prestige: The Royal Gender Symbolism of Lord 5 Flower's/Lady 4 Rabbit's Family. In *Women in Prehistory: North America and Mesoamerica*, edited by Cheryl Claassen and Rosemary A. Joyce, pp. 153–172. University of Pennsylvania Press, Philadelphia.

Hendon, Julia A.

2006 Textile Production as Craft in Mesoamerica: Time, Labor and Knowledge. *Journal of Social Archaeology* 6(3): 354–378.

Hernández, Christine L., and Dan M. Healan

2008 The Role of Late Pre-Contact Colonial Enclaves in the Development of the Postclassic Ucareo Valley, Michoacan, Mexico. *Ancient Mesoamerica* 19(2): 265–282.

Hodder, Ian (editor)

1987a *The Archaeology of Contextual Meanings.* Cambridge University Press, Cambridge, UK.

1987b *Archaeology as Long-Term History.* Cambridge University Press, Cambridge, UK.

1989 *The Meanings of Things: Material Culture and Symbolic Expression.* Unwin Hyman, London.

2010 *Religion in the Emergence of Civilization.* Cambridge University Press, Cambridge, UK.

Hodder, Ian, and Scott Hutson

2003 *Reading the Past: Current Approaches to Interpretation in Archaeology*, 3rd ed. Cambridge University Press, Cambridge, UK.

Inomata, Takeshi, and Lawrence S. Coben (editors)
2006 *Archaeology of Performance: Theaters of Power, Community, and Politics*. AltaMira Press, Lanham, MD.
Insoll, Timothy (editor)
2004 *Archaeology, Ritual and Religion*. Routledge, London.
Jadot, Elsa
2016 *Productions céramiques et mobilités dans la région tarasque de Zacapu (Michoacán, Mexique): Continuités et ruptures techniques entre 850 et 1450 apr. J.-C.* Unpublished PhD dissertation, Université Paris 1 Panthéon-Sorbonne.
Jaúregui, Jesus, and Johannes Neurath (editors)
1998 *Fiesta, literatura, y magia en el Nayarit: Ensayos sobre coras, huicholes y mexicaneros de Konrad Theodor Preuss*. Centro de Estudios Mexicanos y Centroamericanos, Instituto Nacional Indigenista, Mexico City.
Jiménez Betts, Peter F.
2017 *Orienting West Mexico: The Mesoamerican World System, 200–1200 CE*. Unpublished PhD dissertation, Department of Historical Studies, University of Gothenburg, Sweden.
Kelley, J. Charles
1971 Archaeology of the Northern Frontier: Zacatecas and Durango. In *Archaeology of Northern Mesoamerica: Part II*, edited by Gordan F. Ekholm and Ignacio Bernal, pp. 768–801. *Handbook of Middle American Indians*, Vol. 11, Robert Wauchope, general editor. University of Texas Press, Austin.
1986 The Mobile Merchants of Molino. In *Ripples in the Chichimec Sea: New Considerations of Southwestern-Mesoamerican Interactions*, edited by Frances Joan Mathien and Randall H. McGuire, pp. 81–104. Southern Illinois University Press, Carbondale.
1990 The Early Post-Classic in Northern Zacatecas and Durango. In *Mesoamérica y Norte de México, Siglo IX–XII*, Vol. 2., edited by F. S. Miranda, pp. 487–520. Museo Nacional de Antropología, Mexico City.
1995 Trade Goods, Traders and Status in Northwestern Greater Mesoamerica. In *The Gran Chichimeca: Essays on the Archaeology and Ethnohistory of Northern Mesoamerica*, edited by Jonathan E. Reyman, pp. 102–145. Avebury, Aldershot, UK.
2000 The Aztatlán Mercantile System: Mobile Traders and the Northwestward Expansion of Mesoamerican Civilization. In *Greater Mesoamerica: The Archaeology of West and Northwest Mexico*, edited by Michael S. Foster and Shirley Gorenstein, pp. 137–154. University of Utah Press, Salt Lake City.
Kelley, J. Charles, and Howard D. Winters
1960 Revision of the Archaeological Sequence in Sinaloa, Mexico. *American Antiquity* 25(4): 547–561.
Kelly, Isabel T.
1938 *Excavations at Chametla, Sinaloa*. Ibero-Americana 14. University of California Press, Berkeley.
1945 *Excavations at Culiacán, Sinaloa*. Ibero-Americana 25. University of California Press, Berkeley.

Kent, Kate P.

1957 The Cultivation and Weaving of Cotton in the Prehistoric Southwestern United States. *Transactions of the American Philosophical Society*, n.s. 47(3): 457–732.

King, Stacie M.

2011 Thread Production in Early Postclassic Coastal Oaxaca, Mexico: Technology, Intensity, and Gender. *Ancient Mesoamerica* 22(2): 323–343.

Lekson, Stephen H.

2008 *A History of the Ancient Southwest*. School for Advanced Research Press, Santa Fe, NM.

Lister, Robert H.

1949 *Excavations at Cojumatlán, Michoacán, Mexico*. University of New Mexico Publications in Anthropology, No. 5. University of New Mexico, Albuquerque.

Long, Stanley V.

1966 *Archaeology of the Municipio of Etzatlán, Jalisco*. Unpublished PhD dissertation, Department of Anthropology, University of California, Los Angeles.

Mathiowetz, Michael D.

2011 *The Diurnal Path of the Sun: Ideology and Interregional Interaction in Ancient Northwest Mesoamerica and the American Southwest*. Unpublished PhD dissertation, Department of Anthropology, University of California, Riverside.

2018a The Sun Youth of the Casas Grandes Culture, Chihuahua, Mexico (A.D. 1200–1450). *Kiva: Journal of Southwestern Anthropology and History* 84(3):367–390.

2018b From this Day Forward I am your Way of Life: The *Capitán* Icon on Rio Grande Glaze Wares in the Pueblo IV-period American Southwest. *Journal of the Southwest* 60(3):699-752.

2019a A History of Cacao in West Mexico: Implications for Mesoamerica and U.S. Southwest Connections. *Journal of Archaeological Research* 27(3): 287–333. DOI: 10.1007/s10814-018-9125-7.

2019b El hijo de Dios que está en el Sol: Autoridad política y personificación del Dios Sol en el antiguo Noroccidente de México. In *Memoria del Primer Seminario-Taller Regional de la Arqueología Aztatlán*, edited by Laura Solar and Ben Nelson. Arizona State University and Instituto Nacional de Antropología e Historia, Mexico.

Mathiowetz, Michael, Polly Schaafsma, Jeremy Coltman, and Karl Taube

2015 The Darts of Dawn: The Tlahuizcalpantecuhtli Venus Complex in the Iconography of Mesoamerica and the American Southwest. *Journal of the Southwest* 57(1): 1–102.

McCafferty, Geoffrey G., and Sharisse D. McCafferty

2015 Mythstory and Archaeology: Of Earth Goddesses, Weaving Tools, and Buccal Masks. In *Bridging the Gaps: Integrating Archaeology and History in Oaxaca, Mexico*, edited by Danny Zborover and Peter C. Kroefges, pp. 97–111. University Press of Colorado, Boulder.

1991 Spinning and Weaving as Female Gender Identity in Post-Classic Central Mexico. In *Textile Traditions of Mesoamerica and the Andes: An Anthology*, edited by Margot B. Schevill, Janet C. Berlo, and Edward B. Dwyer, pp. 19–44. Garland, New York.

1994 Engendering Tomb 7 at Monte Albán: Respinning an Old Yarn. *Current Anthropology* 35(2): 143–166.

McCafferty, Sharisse D., and Geoffrey G. McCafferty

2008 Spinning and Weaving Tools from Santa Isabel, Nicaragua. *Ancient Mesoamerica* 19: 143–156.

McCarty, Kieran, and Dan S. Matson

1975 Franciscan Report on the Indians of Nayarit, 1673. *Ethnohistory* 22(3): 193–221.

Meighan, Clement

1976 *The Archaeology of Amapa, Nayarit*. Monumenta Archaeologica 2. Institute of Archaeology, University of California, Los Angeles.

Meighan, Clement W., and Leonard J. Foote

1968 *Excavations at Tizapan el Alto, Jalisco*. Latin American Studies, Vol. 11. University of California, Los Angeles.

Milbrath, Susan

2000 Xochiquetzal and the Lunar Cult of Central Mexico. In *Precious Greenstone, Precious Quetzal Feather. In Chalchihuitl in Quetzalli: Essays on Ancient Mexico in Honor of Doris Heyden*, edited by Eloise Quiñones Keber, pp. 31–55. Labyrinthos, Lancaster, CA.

Mountjoy, Joseph B.

1982 *El Proyecto Tomatlán Salvamento Arqueológico: Fondo Ethnohistórico y Arqueológico, Desarollo del Proyecto, Estudios de la Superficie*. Instituto Nacional de Antropología e Historia, Departamento de Monumentos Prehispánicos, Mexico City.

1990 El Desarollo de la Cultura Aztatlán Visto desde su Frontera Suroeste. In *Mesoamerica y Norte de Mexico, Siglo IX–XII*, Vol. 2., edited by F. S. Miranda, pp. 541–564. Museo Nacional de Antropología, Mexico City.

1995 Some Important Resources for Prehispanic Cultures on the Coast of West Mexico. In *The Gran Chichimeca: Essays on the Archaeology and Ethnohistory of Northern Mesoamerica*, edited by Jonathan E. Reyman, pp. 61–87. Avebury, Aldershot, UK.

2000 Prehispanic Cultural Development along the Southern Coast of West Mexico. In *Greater Mesoamerica: The Archaeology of West and Northwest Mexico*, edited by Michael S. Foster and Shirley Gorenstein, pp. 81–106. University of Utah Press, Salt Lake City.

Neff, Hector

1989a Origins of Plumbate Pottery Production. In *Ancient Trade and Tribute: Economies of the Soconusco Region of Mesoamerica*, edited by Barbara Voorhies, pp. 175–193. University of Utah Press, Salt Lake City.

1989b The Effect of Interregional Distribution on Plumbate Pottery Production. In *Ancient Trade and Tribute: Economies of the Soconusco Region of Mesoamerica*, edited by Barbara Voorhies, pp. 249–267. University of Utah Press, Salt Lake City.

Neff, Hector, and Ronald L. Bishop

1988 Plumbate Origins and Development. *American Antiquity* 53(3): 505–522.

Nicholson, Henry B.

1971 Religion in Prehispanic Central Mexico. In *Archaeology of Northern Mesoamerica, Part 1*, edited by Gordon F. Ekholm and Ignacio Bernal, pp. 395–446. *Handbook of Middle American Indians*, vol. 10, Robert Wauchope, general editor. University of Texas Press, Austin.

Obregón, Baltasár de

1924 Historia de los descubrimientos antiguos y modernos de la Nueva España, escrita por el conquistador en el año de 1584. Departamento Editorial de la Secretaría de Educación Pública, Mexico City.

Ohnersorgen, Michael A.

2007 *La Organización Socio-económica y la Interacción Regional de un Centro Aztatlán: Investigaciones Arqueológicas en Chacalilla, Nayarit.* Informe submitted to the Consejo de Arqueología, Instituto Nacional de Antropología e Historia.

Ortega, José

1754 [1944] Maravillosa reducción y conquista de la Provincia de S. Joseph del Gran Nayar, nuevo reino de Toledo. In *Apostólicos afanes de la Compañía de Jesús.* Editorial Layac, Mexico City.

Ortman, Scott

1998 Corn Grinding and Community Organization in the Pueblo Southwest, A.D. 1150–1550. In *Migration and Reorganization: The Pueblo IV Period in the American Southwest*, edited by Katherine A. Spielmann, pp. 165–192. Anthropological Research Papers No. 51, Arizona State University, Tempe.

2000 Conceptual Metaphor in the Archaeological Record: Methods and an Example from the American Southwest. *American Antiquity* 65(4): 613–645.

Pang, Hilda

1971 Archaeological Textiles from Chametla: Sinaloa. *Actas del XLI Congreso Internacional de Americanistas* I: 301–331.

Parsons, Mary

1972 Spindle Whorls from the Teotihuacan Valley, Mexico. In *Miscellaneous Studies in Mexican Prehistory*, edited by Michael W. Spence, Jeffrey Parsons, and Mary H. Parsons, pp. 45–80. Anthropological Papers No. 45. University of Michigan Museum of Anthropology, Ann Arbor.

Pauketat, Timothy R.

2013 *An Archaeology of the Cosmos: Rethinking Agency and Religion in Ancient America.* Routledge, London.

Pickering, Robert B., and Christopher S. Beekman

2016 Introduction: An Historical Overview of Shaft Tomb Archaeology in Western Mexico. In *Shaft Tombs and Figures in West Mexican Society: A Reassessment*, edited by Christopher S. Beekman and Robert B. Pickering, pp. 1–21. Thomas Gilcrease Institute of American History and Art, Tulsa, OK.

Pohl, John M.D.

1994 Weaving and Gift Exchange in the Mixtec Codices. In *Cloth & Curing: Community and Change in Oaxaca*, edited by Grace Johnson and Douglas Sharon, pp. 3–14. San Diego Museum Papers No. 32. San Diego Museum of Man, San Diego, CA.

2003a Ritual Ideology and Commerce in the Southern Mexican Highlands. In *The Postclassic Mesoamerican World*, edited by Michael E. Smith and Francis F. Berdan, pp. 172–177. University of Utah Press, Salt Lake City.

2003b Ritual and Iconographic Variability in Mixteca-Puebla Polychrome Pottery. In *The*

Postclassic Mesoamerican World, edited by Michael E. Smith and Francis F. Berdan, pp. 201–206. University of Utah Press, Salt Lake City.

2012 La tradición Aztatlán de Nayarit-Jalisco y el estilo Nahua-Mixteca de Cholula. *Arqueología Mexicana* 20: 60–65

2016 Transnational Tales: A Millennium of Indigenous Cultural Interaction between the United States and Mexico. In *The Forked Juniper: Critical Perspectives on Rudolfo Anaya*, edited by Roberto Cantú, pp. 253–288. University of Oklahoma Press, Norman.

Pollard, Helen

2015 The Prehispanic Heritage of the Tarascans (Purépecha). In *From Tribute to Communal Sovereignty: The Tarascan and Caxcan Territories in Transition*, edited by Andrew Roth-Seneff, Robert V. Kemper, and Julie Adkins, pp. 92–110. University of Arizona Press, Tucson.

Publ, Helmut

1986 *Prehispanic Exchange Networks and the Development of Social Complexity in Western Mexico: The Aztatlán Interaction Sphere*. PhD dissertation, Southern Illinois University, Carbondale.

1990 Interaction Spheres, Merchants, and Trade in Prehispanic West Mexico. *Research in Economic Anthropology* 12: 201–242.

Ramírez, Susana, and Cinthya Cárdenas

2006 Análisis de la Cerámica del Postclásico. In *Transformaciones Socioculturales y Tecnológicas en el Sitio de La Peña, Cuenca de Sayula, Jalisco*, edited by Catherine Liot, Susana Rámirez, Javier Reveles, and Otto Schöndube, pp. 307–372. Universidad de Guadalajara/INAH, Guadalajara.

Ramírez de Swartz, Susana, Catherine Liot, Javier Reveles, Otto Schöndube, Cinthya Cárdenas, Franco Mata, Carmen Melgarejo, Leonardo Santoyo, and Victoria B. Diego

2005 La Peña: Un Sitio Transición Entre el Epiclásico y el Postclásico Temprano en La Cuenca de Sayula, Jalisco. In *El Antiguo Occidente de México: Nuevas Perspectivas sobre el Pasado Prehispánico*, edited by Eduardo Williams, Phil C. Weigand, Lorenza López Mestas, and David C. Grove, pp. 305–329. El Colegio de Michoacán, Zamora.

Renteria, Jose A.

2018 *Xochipilli: Unveiling the Spiritual Concepts of a Nahua Deity*. Unpublished MA thesis. Department of Latin American Studies, San Diego State University, San Diego, CA.

Sahagún, Fray Bernardino de

1950–1982 *Florentine Codex: General History of the Things of New Spain*, translated by Arthur J. O. Anderson and Charles E. Dibble. Monographs of the School of American Research 14. School of American Research, Santa Fe, NM.

Sauer, Carl

1934 *The Distribution of Aboriginal Tribes and Languages in Northwestern Mexico*. Ibero-Americana 5. University of California Press, Berkeley.

1935 *Aboriginal Population of Northwestern Mexico*. Ibero Americana 10. University of California Press, Berkeley.

Sauer, Carl O., and Donald Brand

1932 *Aztatlán: Prehistoric Mexican Frontier on the Pacific Coast*. Ibero-Americana 5. University of California Press, Berkeley.

Schaafsma, Polly (editor)

2000 *Kachinas in the Pueblo World*. University of Utah Press, Salt Lake City.

Schaafsma, Polly

2013 Petitions for Rain: Textile and Pottery Designs in Rock Art. *International Newsletter on Rock Art* 66: 17–27.

Schaefer, Stacy B.

1989 The Loom and Time in the Huichol World. *Journal of Latin American Lore* 15(2): 179–194.

1993 The Loom as a Sacred Power Object in Huichol Culture. In *Art in Small-Scale Societies: Contemporary Readings*, edited by Richard L. Anderson and Karen L. Field, pp. 118–130. Prentice Hall, Englewood Cliffs, NJ.

1996 The Cosmos Contained: The Temple Where Sun and Moon Meet. In *People of the Peyote: Huichol Indian History, Religion, & Survival*, edited by Stacy B. Schaefer and Peter T. Furst, pp. 332–373. University of New Mexico Press, Albuquerque.

2002 *To Think with a Good Heart: Wixárika Women, Weavers, and Shamans*. University of Utah Press, Salt Lake City.

2015 *Huichol Women, Weavers, and Shamans*. University of New Mexico Press, Albuquerque.

Schöndube Baumbach, Otto

1974 Deidades Prehispánicas en el área de Tamazula-Tuxpán-Zapotlán en el Estado de Jalisco. In *The Archaeology of West Mexico*, edited by Betty Bell, pp. 168–181. West Mexican Society for Advanced Study, Ajijic, Mexico.

Sekaquaptewa, Emory, and Dorothy Washburn

2004 They Go along Singing: Reconstructing the Hopi Past from Ritual Metaphors in Song and Dance. *American Antiquity* 69(3): 457–486.

Smith, Michael E.

2003 Small Polities in Postclassic Mesoamerica. In *The Postclassic Mesoamerican World*, edited by Michael E. Smith and Francis F. Berdan, pp. 35–39. University of Utah Press, Salt Lake City.

Smith, Michael E., and Frances F. Berdan (editors)

2003 *The Postclassic Mesoamerican World*. University of Utah Press, Salt Lake City.

Smith, Michael E., and Frances F. Berdan

2003 Postclassic Mesoamerica. In *The Postclassic Mesoamerican World*, edited by Michael E. Smith and Francis F. Berdan, pp. 3–13. University of Utah Press, Salt Lake City.

Smith, Michael E., and Kenneth G. Hirth

1988 The Development of Prehispanic Cotton-Spinning Technology in Western Morelos, Mexico. *Journal of Field Archaeology* 15(3): 349–358.

Stark, Barbara L., Lynette Heller, and Michael A. Ohnersorgen

1998 People with Cloth: Mesoamerican Economic Change from the Perspective of Cotton in South-Central Veracruz. *Latin American Antiquity* 9(1): 7–36.

Sullivan, Thelma
1982 Tlazolteotl-Ixcuina: The Great Spinner and Weaver. In *The Art and Iconography of Late Post-Classic Central Mexico*, edited by Elizabeth Hill Boone, pp. 7–36. Dumbarton Oaks, Washington, DC.

Sweetman, Rosemary
1974 Prehistoric Pottery from Coastal Sinaloa and Nayarit. In *The Archaeology of West Mexico*, edited by Betty Bell, pp. 68–82. Sociedad de Estudios Avanzados del Occidente de Mexico, Ajijic.

Taube, Karl A.
1983 The Teotihuacan Spider Woman. *Journal of Latin American Lore* 9(2): 107–189.
2004 Aztec Religion: Creation, Sacrifice, and Renewal. In *The Aztec Empire*, edited by Felipe Solís, pp. 168–177. Guggenheim Museum, New York.
2005 Representaciones del Paraíso en el Arte Cerámico del Clásico Temprano de Escuintla, Guatemala. In *Iconografía y escritura teotihuacana en la costa sur de Guatemala y Chiapas*, edited by Oswaldo Chinchilla y Barbara Arroyo, pp. 35–54. Utz'ib: Serie Reportes 1(5). Asociación Tikal, Guatemala.
2006 Climbing Flower Mountain: Concepts of Resurrection and the Afterlife at Teotihuacan. In *Arqueología e Historia del Centro de México: Homenaje a Eduardo Matos Moctezuma*, edited by Leonardo López Luján, Davíd Carrasco, and Lourdes Cué, pp. 153–170. INAH, Mexico City.
2010 At Dawn's Edge: Tulúm, Santa Rita, and Floral Symbolism in the International Style of Late Postclassic Mesoamerica. In *Astronomers, Scribes, and Priests: Intellectual Interchange between the Northern Maya Lowlands and Highland Mexico in the Late Postclassic Period*, edited by Gabrielle Vail and Christine Hernández, pp. 145–191. Dumbarton Oaks, Washington, DC.

Taylor, R.E.
1970 The Shaft Tombs of Western Mexico: Problems in the Interpretation of Religious Function in Nonhistoric Archaeological Contexts. *American Antiquity* 35(2): 160–169.

Teague, Lynn
1998 *Textiles in Southwestern Prehistory*. University of New Mexico Press, Albuquerque.

Tello, Antonio
1858–1866 Fragmentos de una Historia de la Nueva Galicia. In *Colección de Documentos para la Historia de México*, Vol. 2, edited by Joaquín García Icazbalceta, pp. 343–438. Antigua Librería, Mexico City.

Tilley, Christopher
1999 *Metaphor and Material Culture*. Blackwell, Oxford.

Tilley, Christopher (editor)
1990 *Reading Material Culture*. Blackwell, Oxford.
1993 *Interpretative Archaeology*. Berg, Providence.

Turner, Andrew D.
2016 *Cultures at the Crossroads: Art, Religion, and Interregional Interaction in Central Mexico, AD 600–900*. Unpublished PhD dissertation, Department of Anthropology, University of California, Riverside.

Valadez, Susana Eger

2010 How Huichol Ceremonies Balance the World. In *Huichol Art and Culture: Balancing the World*, edited by Melissa S. Powell and C. Jill Grady, pp. 79–87. Museum of New Mexico Press, Santa Fe, NM.

VanPool, Christine S., Todd L. VanPool, and David A. Phillips Jr. (editors)

2006 *Religion in the Prehispanic Southwest.* AltaMira Press, Lanham, MD.

Vidal Aldana, Cinthya I., and Emmanuel Alejandro Gómez Ambríz

2017 Siguiendo el Camino del Sol: Pensamientos Cosmogónicos Compartidos entre la Costa Sinaloense y el Valle de Guadiana. *Arqueología* 54: 28–42.

von Winning, Hasso

1974 *The Shaft Tomb Figures of West Mexico.* Southwest Museum Papers, no. 24. Southwest Museum, Highland Park, Los Angeles.

Watson, Adam S., Stephen Plog, B. J. Culleton, P. A. Gilman, S. A. LeBlanc, P. M. Whiteley, S. Claramunt, and Douglas J. Kennett

2015 Early Procurement of Scarlet Macaws and the Emergence of Social Complexity in Chaco Canyon, NM. *Proceedings of the National Academy of Sciences of the United States of America* 112(27): 8238–8243.

Webster, Laurie

2007 Ritual Costuming at Pottery Mound: The Pottery Mound Textiles in Regional Perspective. In *New Perspectives on Pottery Mound*, edited by Polly Schaafsma, pp. 167–206. University of New Mexico Press, Albuquerque.

Webster, Laurie D., Kelley A. Hays-Gilpin, and Polly Schaafsma

2006 A New Look at Tie-Dye and the Dot-in-a-Square Motif in the Prehispanic Southwest. *Kiva* 71(3): 317–348.

Weigand, Phil C.

1985 Considerations on the Archaeology and Ethnohistory of the Mexicaneros, Tequales, Coras, Huicholes, and Caxcanes of Nayarit, Jalisco, and Zacatecas. In *Contributions to the Archaeology and Ethnohistory of Greater Mesoamerica*, edited by William J. Folan, pp. 126–187. Southern Illinois University Press, Carbondale.

Wells, E. Christian, and Karla L. Davis-Salazar (editors)

2007 *Mesoamerican Ritual Economy: Archaeological and Ethnological Perspectives.* University Press of Colorado, Boulder.

Whiteley, Peter M.

2004 Why Anthropology Needs More History. *Journal of Anthropological Research* 60(4): 487–514.

Whitley, David S., and Kelley Hays-Gilpin (editors)

2008 *Belief in the Past: Theoretical Approaches to the Archaeology of Religion.* Left Coast Press, Walnut Creek, CA.

Williams, Eduardo

1990 Huichol Ethnography and Archaeological Interpretation. *Papers from the Institute of Archaeology* 1: 6–15.

1991 The Stone Sculpture of Ancient West Mexico: Description and Interpretation. *Ancient Mesoamerica* 2(2): 181–192.

10

Conclusions

STEPHEN A. KOWALEWSKI

For those who know a little about West Mexico's archaeology (would anyone say they know a lot about it?), what might come to mind first could be the region's distinctive artifacts: its wonderful ceramic figurines, abundant obsidian, metallurgy, circular platform architecture, and shaft tombs. This is not a bad start. These things are neither irrelevant nor superfluous, as they each are commonly repeated expressions revealing a great deal about behavior and institutions at various times and places.

How to describe the archaeological record of West Mexico to someone who knows little about it? What words convey its dimensions? How large, how much? Concepts of physics and mathematics such as volume, amount, size, quantity, scale, bulk, weight, or density? We have to keep in mind that West Mexico is hardly a small or restricted place. In fact the West Mexicos represented by the places studied in this book are spread out over a distance of 1,000 kilometers.

Muñiz García and Sumano Ortega say that their recent archaeological project in the Santiago Bayacora Basin increased the total of known sites there to 100. In the Middle Balsas Basin, the survey by Punzo Díaz, Rangel, Ibarra, Zarco, and Castañón recorded 59 sites just in the limited area affected by the construction of a new dam. Heredia Espinoza's survey in the Tequila region (2017) found over 500 sites, an average of 1 site per square kilometer. Projecting from the known to the unknown, for all of West (or western) Mexico as usually defined (Beekman 2010), I would think the real number of sites must be at least in the hundreds of thousands. Fewer than 10 percent have been recorded in any way. The archaeological sites are not just pinpoints on a map, many are in the 10 to 100–hectare range, and a few are even larger.

How would one convince a nonspecialist that this archaeology demands attention? How does one make a compelling argument to someone not from Chalchihuites, but from Chile, Chad, or China, that one should spend time and money on this problem in this area rather than some other? What is the significance of this great mass of archaeological material and information, or some small part of the great mass? What might it mean, and for whom—individuals, teachers, scholars, employment, local communities, cities, states—for Mexico and for humanity in general? The magnitude of this archaeological record and how we can know more about it is a real and practical problem that I will return to at the end of this chapter.

The Third Radiocarbon Revolution

We know that the volume or mass of the archaeological record is greater and has more sites, more artifacts, and more information than we can easily comprehend today. As the recent research reported in this book shows, this mass of archaeology is not homogeneous or uniform. Confining ourselves for the moment just to its spatiotemporal dimensions, it is clear that the sites and the material are uneven, chunky, or granular, with more material pertaining to certain episodes or spans of time and less to others, which of course depends on the place of reference, just where one is talking about. Nowhere is the archaeological accumulation continuous, neither a straight line nor a smooth exponential curve. It is in this sense unpredictable. There was little settlement in the Middle Balsas region between AD 900 and 1250, according to Punzo Díaz, Rangel, Ibarra, Zarco, and Castañón. But in the north-central interior, growth and expansion took place earlier, roughly AD 600–900, as described by Caretta and Dueñas García in their study. It was the Postclassic after AD 900 that was the major time of growth and expansion on the coast, as Mountjoy, Cupul-Magaña, García de Quevedo-Machain, and López Mestas Camberos show in their chapter. I will return to this spatiotemporal unevenness later in the chapter, but now I want to focus on the chronologies themselves.

What archaeologists know about these periods of expansion and relative decline or absence is established through the usual combination of ceramic style and radiocarbon dating. When we have difficulty finding sites and materials dating to a particular period, we have three alternative hypotheses to check: (1) we have not looked hard enough or in the right places; (2) we do not know the ceramics or lithics well enough to recognize the artifacts associated with the period in question; and (3) there are few sites and materials because the area was not occupied very much during the time in question.

Archaeologists are aware that none of these three alternatives—failure to look, ignorance of the artifactual chronology, or abandonment—should be assumed, that all need to be investigated. Perhaps in 1980 or 1990, the first two hypotheses could still be entertained as reasonably possible for regions like the Middle Balsas, the northern interior, and others in West Mexico, because relatively few surveys, geomorphological studies, and detailed analyses of artifact styles had been carried out. But by 2010, with salvage work in a wide variety of settings, more archaeologists working in each state, better knowledge of artifact sequences, and sustained efforts by long-term projects in key areas, it became more probable that the hiatuses that still remained were due in large part to the third alternative; in other words, that the problem was not observer's error but a behavioral phenomenon, that population did decline in some periods and grow in others.

Still, ceramic (and lithic) chronologies often can be very imprecise and not offer much help, particularly when there were few artifacts produced, or their attributes were plain and simple, or they truly did not change very much. Likewise, even in sequences based on abundant, highly decorated, or variable artifacts, we often lack sufficiently large numbers in good contexts to make the chronological assignments that we can with the right samples. Also, we may want to date events for which time-sensitive artifacts are simply not available. To date events in those times of plain pottery and no projectile points, and even in other times of index-fossil points or elaborately painted pots, archaeologists turn to absolute techniques, the most important being radiocarbon.

But radiocarbon dating is a quantitative technique requiring largish numbers and a great deal of attention to well-documented context. For example, the lifespan of the Aztatlán style on the Pacific coast is bracketed from AD 900 to 1400 on the basis of 59 radiocarbon dates (Mountjoy et al., this volume), which might sound like a lot of dates, and it is, compared to some other places in Mesoamerica. Yet that is 59 dates spread over 500 years and about 1,000 kilometers from northern Sinaloa to Colima, one RC date per thousand or two thousand square kilometers. This brings us right to Almendros López's lead paper, which says we simply do not have enough good radiocarbon dates. "We may well have come late to the application of such [recent radiocarbon] studies compared to areas where they are no longer deemed 'novel' but, rather, 'routine.'"

Absolutely. Archaeology has already had its third radiocarbon revolution.

Colin Renfrew, the leading British archaeologist, observed 50 years ago that our field had experienced the first radiocarbon revolution in the 1950s after Willard Libby's discovery had been used successfully on archaeological materials. One spectacular result was the placement of the northwestern European Early Bronze Age and its megalithic monuments 1,000 years earlier than the Egyptian pyramids, from which the European developments had been supposed to diffuse. Renfrew's second revolution was the realization that radiocarbon dates had to be calibrated using the year-by-year tree-ring record, which pushed the start of the Bronze Age even farther back in time (Renfrew 1970).

Almendros López urges us to join the third radiocarbon revolution. The first was the new technique, the second was the calibration refinement, and the third is more dates, often more by an order of magnitude. The result has been a dramatic improvement in precision, roughly from 100-year to decadal uncertainties. This advancement came about because of two technologies, accelerator mass spectrometry (AMS) and Bayesian statistics.

Archaeologists gained more confidence in radiocarbon dates with AMS. A practicality is the fall in the real (adjusted for inflation) cost of a radiocarbon date. The costs of the other major items in an archaeological project's budget—wages, vehicles, fuel, supplies, overhead—have increased, but radiocarbon dates are cheaper now than they were in 1970.

The number of radiocarbon dates increased especially after the turn of the new millennium. An example of the rapid advances made in archaeology because of more and better radiocarbon dates is from Cerro Juanaqueña in Chihuahua. Physically, this was a hill covered with terraces and it looked like a site in the Trincheras tradition, which would have placed it a few hundred years before European contact—except there were no ceramics. Initially four radiocarbon dates placed Cerro Juanaqueña in the Late Archaic or early agricultural period, around 1000 BC (Hard and Roney 1998). Then 13 more samples were dated and the placement was refined to a brief but intensive and massive occupation between 1100 and 1000 BC, with a brief and lesser occupation between 300 and 200 BC (Hard and Roney 2005). Today we might quadruple those 13 dates.

As Almendros López remarks for West Mexico in particular, Mesoamericanists were not in the vanguard of the third radiocarbon revolution. Why? I will speak from my own perspective, as I was not an early adopter either. In the 1970s and 1980s, many of us thought that, with decorated ceramics, seriation provided finer chronological distinction than was possible with a few, often unreliable, conventional radiocarbon dates. Also, standards for reporting were lax—for example, we sometimes did not learn more about the specific context of a sample beyond "charcoal associated with X-phase material." That a good seriation provided more precision may have been true for a while, but not with the new radiocarbon techniques and lower radiocarbon lab costs, and it was never true for those periods dominated by plain, little-changing pottery. My work beginning in the 1970s focused on the regional settlement pattern survey in Oaxaca, and in my research proposals I always stated that we would use the current chronologies and provide some help to chronology building by improving the understanding of regional variation. Unfortunately, it was not until quite recently that my colleagues and I began to see significant numbers of radiocarbon dates from solid, representative contexts.

Imagine the following scene: Some of the archaeologists in this book, say, Christopher Beekman, José Luis Punzo Díaz and his coauthors, and Laura Almendros López, are eating *tacos de birría de chivo* in Tequila, Jalisco, as they finish writing their chapters for this volume. "We have managed to refine the Teuchitlán sequence with a good ceramic seriation," says Chris. "Now we know the order in which the Los Guachimontones platforms and ballcourts were constructed. Kimberly and Joshua say that helped them make sense of the way the builders created spaces and how the buildings related to one another. But it is frustrating, I am tearing my hair out, wishing our great colleague Phil Weigand had taken more carbon samples and written down exactly what they were supposed to date. I think we've done all we can without new excavations."

"You think your sequence is still rough," replied José Luis, "for Huetamo we only have 13 radiocarbon samples spread over 1,400 years! We've all been tearing our hair out, still dealing with periods, not even phases. We would like to get individual settlement histories, to see both local and region-wide events."

Laura laughs, "Not to be too personal or overcritical, but I did notice among us some odd coiffures, a lot of hair missing. No seriously, we have to be budgeting more money for radiocarbon dates. Solving problems like you are working on would be a significant advance for Mesoamerican archaeology, and I know we can do it with more dates—I mean hundreds—from contexts that we will document precisely."

Someone else says, "All well and good, but more money for carbon samples, does that mean less money for salaries? How are people who make their living in the field and in the lab going to support themselves if all the cash goes to radiocarbon labs? By the way, I am a little light right now, can someone pay for my taco?"

Now of course this scene never took place and none of the contributors said what I just imagined. But when you read their chapters you could agree that it might have gone something as I have told it. The archaeologists here generally know what needs to be done, but a big question is the means, the funding.

Cities

The papers in this book illustrate how much variety there is in West Mexico's archaeology. The distinctive variation is especially clear in its cities and towns. The most famous early cities are Los Guachimontones and its neighbors in the Volcán de Tequila region. This regional urban system flourished between 300 BC and AD 200. The city layouts are like nothing else in Mesoamerica. From their analysis, Kimberly Sumano Ortega and Joshua Englehardt conclude that the civic-ceremonial architecture and spatial plan does not conform to any existing, simple, or obvious model of social structure. They suggest that aspects of these urban plans reflect more corporate or collective, rather than individual aggrandizer or network, strategies. Verenice Heredia Espinoza reviews artifactual, architectural, settlement pattern, and funerary data pertaining to the Volcán de Tequila region in the Terminal Formative and Early Classic. She reaches a similar conclusion, that this system was complicated and does not conform to type.

To me, Teuchitlán resembles, in certain cultural evolutionary ways, the Indus Valley, Djenné-Djenno, Cahokia, and Chaco Canyon. All were large-scale, central-place systems that organized broad regions. They developed rather early in the historical trajectories of their areas. They lasted for a relatively brief period of time. Then they collapsed, leaving abandoned, monumental ruins not much used thereafter. While they all were spectacular developments, the way they were organized did not become the model for subsequent developments. And in all of these cases, many of the key institutions were organized using corporate or collective, rather than network or authoritarian, principles.

Very different from Teuchitlán are the hilltop towns of the interior. Nicolás Caretta and Manuel Dueñas García describe their recent studies of Cerro de Santiago, Aguascalientes. Cerro de Santiago is an almost 50 hectare settlement occupied between AD 600 and 900. It has a ceremonial zone with courtyards, mounds, and a ballcourt, plus the residential area on the slopes below. As a hilltop center, it might be compared with others of about the same time period such as El Encinal (discussed by Muñiz García and Sumano Ortega in their chapter) and the hilltop sites of Altavista and La Quemada, which have been known for a long time.

Archaeology and history have outgrown that stage when in the scholarly literature one would read arguments about whether a region's major sites, even places like Tikal or Tzintzuntzan, were really cities, or truly urban. Those essentialist or definitional debates have been overtaken by more analytical, processual concerns. The *guachimontón* centers of the Teuchitlán region in the Terminal Formative and Early Classic, the hilltop centers of the Epiclassic–Early Postclassic in the interior, the *malpais* city of Angamuco, and the Late Postclassic Purépecha capital of Tzintzuntzan are four distinct urban plans varying in population size, density, degree of centralization, internal divisions, and heterogeneity.

Elsewhere in Mesoamerica we also see variation in urban outcomes, between regions and over time. Classic-period centers in Veracruz are distinct from their contemporaries in the Maya area, Oaxaca, and Central Mexico. Thus we do not have a single model for "the Mesoamerican city," even if we restricted ourselves to one time period. The thesis explaining this urban variation has yet to be written.

Episodes of Growth and Decline

West Mexico's archaeological record—its materials and sites—is spatiotemporally uneven, chunky. The new archaeological work reported here strikingly illustrates this patchy pattern. Here we have case studies from six widely spaced regions that display strong growth and expansion in one or more periods, and equally marked decline in others. To review briefly, I will use the approximate dates as given, with the caveat that more work needs to be done in each region to describe more closely the sequence of events. Other regions in West Mexico such as the Bajío have their own histories (reviewed in Beekman 2010), but here I limit the discussion to the studies accessible in this book.

In the Volcán de Tequila region, the period between 300 BC and AD 200 saw great growth in population, numbers of sites, and material production (as detailed in Beekman's analysis of the chronology). Before that time, the size of the population was much lower. According to present evidence, demographic growth leveled off until sometime during the period AD 900–1500, when there was some increase (Heredia 2017).

The Middle Balsas (the chapter by Punzo Díaz, Rangel, Ibarra, Zarco, and Castañón), the Santiago Bayacora Basin in Durango (Muñiz García and Sumano Ortega), and the region of Cerro de Santiago in Aguascalientes (Caretta and Dueñas García) all had a pronounced apogee about AD 600–900, contrasting with less occupation before and after. The timing of the decline varied, with the Huetamo Middle Balsas having a seeming hiatus after AD 900, while there is evidence of some continued (but lesser) occupation in the Santiago Bayacora case.

The Pacific coastal plain and the region of Lake Pátzcuaro in the interior highlands grew and expanded after AD 1200 (Mountjoy et al., Mathiowetz, this volume; compare Punzo Díaz et al., this volume). To these regions of Late Postclassic expansion we could add the Volcán de Tequila region, too.

Spatiotemporal unevenness is hardly unique to West Mexico. Periods of growth and decline are a fact of the archaeological record across Mesoamerica from Honduras to Central Mexico. The spatial scale of these patterns may be quite broad (regional) or narrower (subregional). The times of growth and decline are not all coordinated or concordant. For example, the Late Classic was the big period of expansion in the southern Maya lowlands, while in the basin of Mexico and highland Oaxaca more growth occurred in the Early Classic. It may be that the majority of Mesoamerican regions experienced growth and expansion in the Late Postclassic, but even that generalization is not true everywhere, the southern Maya lowlands being a prominent example.

The broad outlines identifying periods of growth and their opposite, periods of decline, are accepted as a fact of the Mesoamerican archaeological record. This does not mean that archaeologists no longer need to address the three alternative hypotheses (discussed above) concerning spatiotemporal gaps (failure to look, lack of chronological control, real abandonment), for in practice archaeologists are always committed to looking, to discovery, and the work of chronological refinement is never done. If we are to understand and explain the behavioral processes leading to growth and decline, we will need many more systematic regional surveys to describe settlement change over time, and we will need much better chronological control. We need wider application of the tools of the third radiocarbon revolution, as Almendros López argues very effectively in her chapter. We also need to build theories that would have the power to guide us toward the crucial information. This point about theoretically designed research is the subject of the next section.

Explaining Growth and Decline

The most salient theme or lesson in this book has significance and relevance not just for West Mexico but for Mesoamerican archaeology and anthropology generally. It is a proposition that underlies the discussions in many of the chapters. This proposition is stated most directly by Caretta and Dueñas García: "waves of integration are a fundamental characteristic of world-systems. . . . These interaction networks . . . motivate massive and rapid changes." In other words, interaction across regions creates more or less regular or habitual social networks. These ties or pathways are channels that move material, people, and information. Peoples' behavior is motivated in part by the goods, people, and representations in the networks of which they are a part. The wave of integration may collapse, as for instance during the several centuries following AD 900 in southeastern Michoacán, where Punzo Díaz et al. (this volume) suggest "economic and social reorganization due to the breakdown of regional social networks that connected different areas in western Mesoamerica."

The proposition of waves of integration is a fairly complicated idea in that it has quite a few parts and implications. In the following paragraphs, I try to spell out how I understand it.

"Integration" is system-like, people are differentiated and interdependent, and they depend on each other through exchange. "Waves of integration" implies that sometimes people may be interdependent and involved in a wide network and at other times they are not and their networks are more restricted. We can think of these social networks forming or falling apart at varying spatial scales, from families to local communities to regions and macroregions or world-systems. Caretta and Dueñas García are saying that the Cerro Santiago region and other places in the north were part of a network of exchange that spread and became more and more involving between AD 600 and 900. This is what was happening in the Postclassic along the Pacific coast. It happened in the highlands, in the wide region centering on the Volcán de Tequila between 300 BC and AD 200, and perhaps at times in the Formative, as manifest in Colima with what archaeologists call Capacha.

Caretta and Dueñas García say that world-systems are known to be subject to waves of integration (see also Jiménez Betts 2017). The converse is that the big networks and interdependencies fall apart. For example, the Mediterranean world developed a wide and deep network of integration beginning in Archaic and Classical times; this was transformed in Roman times, and it broke up with the fall of the empire.

Networks wide and narrow in scope are important because they motivate and direct peoples' behavior. Why? How can an abstraction cause someone to do one thing rather than another? This question could take us into psychological theory, but for practical archaeological purposes it is still worth considering. An individual actor might wish to make a *huipil*, for example, because she wants to consume another good, perhaps to pay for an addition to the house or for the copal and charcoal for her contribution to a community ritual. Or she may gain some return by embroidering the hem and neck with just the right yellow-and-red design recalling the hummingbird, which she knows is what people will want these days. For the Maya area, Halperin (2017) has written from a feminist perspective about cosmopolitanism, in which people are active participants in all senses: technological, social, and communicative. An actor could be motivated by a negative sanction (pay your taxes), status competition, conformity to canon, rebellion against the canon, and so on. The point is that these material or ideological motivations are delivered over the network, so the wider the network, the wider the common understandings of value, significance, and what is good. I think something like these ideas underlies Mathiowetz's descriptions (this volume) of a symbolically and ritually infused labor process in the making of cotton textiles in Postclassic times along the Pacific coastal plain and beyond.

"Integration" sounds abstract, which it is, but our challenge is to get at the behavioral process, the actions of actors, that makes for different degrees of integration. Actors' strategies are what Heredia Espinoza, and Sumano Ortega and Englehardt (this volume), intend in their analyses of architecture and artifacts in the Volcán de Tequila region. They suggest that some of these archaeologically identifiable patterns are consistent with corporate strategic behavior. "Integrated social networks" do not imply that all the participating individuals and groups use individualizing strategies, since in many contexts people form corporate groups and pursue objectives using collective strategies. Thus, within wider social networks, some of the constituent subnetworks, such as communities or states, may be relatively closed. Those subnetworks can behave as actors. How institutions map onto social networks is a significant new research question.

Ceramic style zones have long been an indication of interaction for archaeologists. Their interpretation—specifying the behavior that made the pattern—is anything but straightforward. As Beekman, Mountjoy et al., and Almendros López (this volume) illustrate, chronological refinement is a crucial step forward. Caretta and Dueñas García's Table 9.1 and Figure 9.8 suggest that pottery styles had a very wide, overlapping, open-network distribution, and that people in any given region had access to a variety of styles. We might envision the sophisticated network analysis that could be done with systematic collections from hundreds of sites located by regional survey.

As several of the studies in this book say, the exchange spheres created in periods of integration were not just "trade" networks for goods—they were economic, but also involved other cultural institutions, beliefs, and practices. When economists talk about "market integration," they are usually referring to the transferability of goods and factors of production between economic sectors. Integration also has a spatial dimension. So if we say "integrated world market for wheat," we mean that the price of wheat and its production factors are roughly the same in Australia as they are in Hungary; therefore, among other things, farmers in those places are competing against each other. Archaeologists as anthropologists have a broader perspective on the meaning of integration. Mathiowetz, Caretta and Dueñas García, and Mountjoy et al. show that in ancient Mesoamerica, when waves of integration formed strong ties at the world-system scale, the networks were not only for the movement of goods, but also the means by which other aspects of culture were channeled, including symbolic representations, style and fashion, and understandings of worth and value. We are in a position to pose questions that are at once economic and more broadly cultural. If households in different places were producing cotton garments, we could ask: Were they in competition with each other? Did supply and demand result in roughly similar prices? Were the factors of production roughly similar in their cost? Are the producers remunerated at about the same rate? And are producers in different places able to obtain similar goods, services, and information?

A crucial element of this explanatory proposition about waves of integration is that it generates implications we can identify and measure archaeologically. In the words of Caretta and Dueñas García, "interaction networks are empirically determinable connections." For example, the chapters by Mathiowetz and Mountjoy et al. conclude on the basis of strong evidence that, in the Late Postclassic, people along the coast were producing cotton and textiles for external consumption. What are the behavioral chains of cotton growing, thread making, dyeing, and weaving? What do producers get in return for the products of their labor? Are producer households able to consume things that they might not if they were not so involved? Do we see similarities in household strategies and outcomes throughout the area, from Colima to Sinaloa? Household excavations could document the production and consumption processes and measure the degree of involvement and commitment in this exchange system.

What were the institutions that stimulated and enabled regional and long-distance integration? International trade theorists and institutional economists emphasize that actors at a distance recognize the mutual benefits of exchanging goods and will engage in trade if the transaction costs are low enough to allow a profit. The key factor is the transaction cost, which has two components, one being transportation, and the other the entire bag of human-to-human relations. This includes security, spoilage, fees, bribes, time, disputes, salaries, licenses, credit, expertise, and so on, that might be involved in exchanging one good for another at a distance. Technological investments (canoes, roads, bridges, ferries) can reduce the friction of physical distance. People come up with particular institutions (merchant guilds, letters of credit, truces and safe-conducts, standards of count and quality, monies, *khipu*, and so on) to reduce the social transaction costs (for a recent review of the theoretical literature in a practical archaeological context, see Bresson 2016).

During the times of West Mexico's waves of integration, what institutions did people establish to allow these archaeologically proven levels of exchange to take place during the several centuries that they did? Today this is a research-frontier question.

Several of the authors here suggest that "leaders" or elite (meaning political office holders, nobles?) were the creators and key nodes of the network. Cross-culturally, the direct exchange of goods among royalty and nobles plays a part in the ancient urban economy, for example in the endowment of temples and monasteries, dowries, feasts, funerary gifts, tribute gifts, and so on. Such appropriations and expenditures can have structural effects, for instance in stimulating or retarding production in other economic sectors. It should of course not be assumed that exchange in the royal-noble networks structures the other 90 percent of society, which is a question for empirical evaluation.

The integrated networks of large geographical scale in the periods AD 600–900 and the Late Postclassic extended across many small states (more neutrally, polities). Neither of these situations was an empire under a single political sovereignty. Likewise, these networks must have involved many different languages and language families, although the linguistic map is not well known. If it was not common polity or common language, what did bind the participants together, what understandings of worth and value, what material currency of cost and reward? What institutions lubricated the movement of goods, people, and information?

To my mind, in these West Mexican and broader Mesoamerican contexts, the political elite (nobles, royalty), in their capacity as leaders of states or statelets, were unlikely to have been the principal network builders and maintainers. To me, it is difficult to see how the symbols of any of the petty states would become by consent or force a set of international conventions for exchange. The primary interests of the elite, as I understand the way our Mesoamericanist colleagues employ this term, would have been to keep and expand local power and authority. To also imagine them adept at penetrating all borders, and transforming wealth here into more wealth there, would be to create in them insoluble conflicts of interest, or else give them truly superhuman superhero powers. One possibility would be that the usual Mesoamericanist assumptions about autocratic polity rulership are incorrect. If polities had more distributed leadership (such as in the Athens or Venetian Republic), then conceivably merchants could be rulers for a time and vice versa. Perhaps the governance of polities, say, at Cinaloa, Culiacán, Piaxtla, Chametla, Aztatlán, or Sentispac, is still an open question?

Alternatively, other institutions apart from the state structured international exchange. What might those have been? Here I consider four broad categories: individual artists, cults, merchant guilds, and markets. These are not mutually exclusive, and several were probably operating.

Individually identifiable artists or brands were known in early modern Europe and Asia, but not in Mesoamerica. Works were not signed, as far as I know. There were artisans of the highest order in all kinds of fine work such as gold and feathers (for example, Berdan 2016), but in Mesoamerica we do not have evidence of named great artists, their schools, or individually led movements in style or fashion. We have not detected individuals who might have had the political-economic placement and trend-defining role of a Frida Kahlo. Contrast with the semimythical Quetzalcoatl, the closest I can think of to an individualized Mesoamerican style setter, a person dressed with symbols to represent or personify ideas or principles, a state officeholder.

The notion of "cult" seems to be a closer fit to the Mesoamerican cultural context. We must be careful, however, about what we mean by cult.

If Mesoamericanists neglect to specify what they mean by "elite," they are even more negligent about "cult." Fortunately, there has been some critical consideration of the concept of cult as it might be applied by archaeologists. Knight (1986) used the cultural anthropologist Wallace's (1966) concept of cult institution, which consists of common rituals and beliefs supported by an organized group of people. Archaeologically, we are first presented with *sacra*, meaning the material symbols of belief and ritual, but locating and describing the organization supporting them is an important but often demanding task.

A cult institution would have personnel, material resources, beliefs, rituals and other activities, objectives, membership, definable limits or boundaries, norms, and means of promulgating and enforcing norms. In Mesoamerica we know of oracles, temples, shrines, and pilgrimage centers with their priests, rituals, feasts, and followers (Ix Chel in Yucatán, and centers like Achiutla and Mitla in Oaxaca). These institutions also had their commercial functions. They were localized in that they were anchored to specific places, but they had wider reputations and they drew followers from afar. This place-specific institution is probably not the notion of cult that Mathiowetz has in mind for the geographically very extended occurrence of material symbols along the Pacific coast in the Late Postclassic.

Perhaps instead, the cult or complex suggested by Mathiowetz was not an identifiable organization with members, facilities, rules, finances, and so on, but was a cloud of *sacra*, symbols, and performance that could be widely shared and modified (compare Heredia and Englehardt 2015). It would be more like a fashion, wave, or trend that appealed to participants (especially women?) in part because it was new, cosmopolitan, engaging, different, or antipatriarchal? The ideas and practices would have their promoters but would not be the property or estate of a church. This notion of a cloud of symbols is not equivalent to what the concept of cult means in the comparative study of religions, but it does seem in keeping with Mesoamerican religion in which animistic spirits or forces are called to mind by overlapping, fuzzy sets of conventional representations, not fixed gods but transformable clouds of life forces.

Groups of merchants who operate together at home and abroad are well known in Mesoamerica. These have been referred to as *pochteca* in the literature (especially Sahagún 1959: Book 9). Recent studies supply more clarification and detail (for example, Hirth and Pillsbury 2013; Hirth 2016). All regions had their merchants, including provincial regions from Guatemala to perhaps Sinaloa, so the *pochteca* described by Sahagún were not the only ones. Merchant groups provided organization and institutionalized ways of lowering transactions costs, for example: by organizing transportation, by handling credit, by offering incentives for successful trade, by gathering expertise and information, by their ability to cross political boundaries, and by articulating local and regional markets.

The ethnohistoric literature makes it clear that merchants operated not independently but in the context of regional marketplace systems. As merchants were the channels of exchange, marketplaces were the nodes. Marketplaces have been documented by historical sources and archaeology at so many places that I think we can assume that all towns and cities had designated, supervised, public places for exchange. Where were the marketplaces in West Mexico cities, from the Late Formative to the Postclassic? How large were they? Did marketplaces have special, architectural features? Did some craft production take place near marketplaces? Were they closely associated with civil or religious buildings?

Marketplaces were institutions that had physical places, regular schedules, rules and regulations, fees or taxes, means of resolving disputes, and the personnel to carry out these functions. Markets were frequented by buyers from all sectors of society, sellers who were often professionals and, if recent experience is any indication, venders ambulatory or on the margins.

Households were market-dependent, producing for and consuming from the market, for most goods in daily use. Distributional studies in Classic and Postclassic cities and towns in the Maya area, Veracruz, Oaxaca, and central Mexico consistently show that households obtained most of the pottery, chipped stone, and other archaeologically identifiable goods from a common pool, rather than each household from separate sources as was sometimes the case in the earlier Formative. Household self-sufficiency is highly unlikely.

Market systems could be integrated at different spatial scales. During periods of weak integration, local markets were tied to each other in a small region, but beyond that, exchange might be irregular. These would be times of somewhat greater regional or even local self-sufficiency, but less consumption, less commerce, and less growth. But in a period of a "wave of integration," the network of regularly, predictably exchanging markets could extend unbroken for many hundreds of kilometers (more on market dependence and market integration can be found in Kowalewski, in press).

All the institutions discussed here—states and their leaders, individual artists, cults, merchant guilds, and markets—were involved in maintaining these world-systems, which do tend to be total cultural phenomena. But to me it is the market economy (marketplaces and associated ways of buying and selling) that is essential. Waves of integration imply dependable articulation between local, regional, and macroregional market systems. Market integration requires a degree of peace and security for travel between polities, one way in which political actors play a role. Mesoamericans did not acclaim individual artists, instead they knew works of style by the places in which they were made. Integration requires institutions that facilitate interregional exchange, which probably means organizations of professional merchants. Integration also means that producers and consumers have adequate information. I had previously thought of this information as being about supply and demand, but after reading the studies in this book, I see that "information" should be broadened to include the beliefs, values, ideas, and symbols that motivate people to make and use things, analogous but not identical to the information that in our times we call advertising, style, and fashion.

Conclusions

The studies here in *Ancient West Mexicos: Time, Space, and Diversity* are a small sampling of the rich, fascinating, diverse and important archaeology in this part of the world. Orders of magnitude more things remain to be discovered than we have in hand, or know about at present. We can expect more fascinating sites from all time periods, greater diversity, and more surprises. If the archaeological record of West Mexico were a thick, five-volume book, then we have read only the first page of each volume. There will be many surprises and new lessons for all to learn.

Archaeologists are now able to understand a lot more about the past than we could 30 years ago. We might have known that two sites dated to somewhere between 1500 BC and 1 AD, but that is about as accurate as saying that Julius Caesar could invite Hernando Cortés to his house for a feast. The third radiocarbon revolution and other archaeological techniques give us the capacity to sort out events into their correct decades and moments. Now we know a lot more about the history of building at Los Guachimontones, we know that for most of the time when it was a major center, it never looked the way it appears to us now.

West Mexico experienced almost 2,000 years of urbanism, the same length of time as the rest of Mesoamerica. Its cities were diverse in form and organization, and the same can be said for cities in different regions in the rest of Mesoamerica. We can expect that archaeologists in West Mexico will soon have more to say about comparative urbanism, considering how many new discoveries they have made in the last decade.

Urban systems have cycles of growth and collapse. The studies in this book describe growth and collapse at different times and places in West Mexico. The authors not only describe these cycles, they provide the tools and directions for explaining what happened. Why such cycles of growth and decline? One factor seems to be that growth depends on having geographically wide connections, and decline happens as these interregional connections are for one reason or another severed. Given that what we know comes only from a tiny sampling of what is in the ground, no specialist is ready to give a definitive explanation.

Since the material remains of the past are so vast and the archaeologists so few, how can we provide the information for answering significant questions such as the rise and fall of societies? Because it is the archaeologists who confront this problem every day, I think we are the ones who know what it would take to make some real advances. What is needed? Personnel, infrastructure including laboratories and curation facilities, equipment, big projects—in other words, substantial and steady funding, and good ways of directing limited resources to the most significant objectives.

References

Beekman, Christopher S.

2010 Recent Research in Western Mexican Archaeology. *Journal of Archaeological Research* 18(1): 41–109.

Berdan, Frances F.

2016 Featherwork as a Commodity Complex in the Late Postclassic Mesoamerican World System. In *Alternative Pathways to Complexity: A Collection of Essays on Architecture, Economics, Power, and Cross-Cultural Analysis in Honor of Richard E. Blanton,* edited by Lane F. Fargher and Verenice Y. Heredia Espinoza, pp. 131–154. University Press of Colorado, Boulder.

Bresson, Alain

2016 *The Making of the Ancient Greek Economy: Institutions, Markets, and Growth in the City-States.* Translated by Steven Rendall. Princeton University Press, Princeton, NJ.

Halperin, Christina T.

2017 Ancient Cosmopolitanism: Feminism and the Rethinking of Maya Inter-regional Interactions during the Late Classic to Postclassic Periods (ca. 600–1521 CE). *Journal of Social Archaeology* 17(3): 349–375.

Hard, Robert J., and John R. Roney
1998 A Massive Terraced Village Complex in Chihuahua, Mexico, 3000 Years before Present. *Science* 279(5357): 1661–1664.
2005 The Transition to Farming on the Río Casas Grandes and the Southern Jornada Mogollon Region. In *The Late Archaic across the Borderlands: From Foraging to Farming,* edited by Bradley J. Vierra, pp. 141–186. University of Texas Press, Austin.
Heredia Espinoza, Verenice Y.
2017 Long-Term Regional Landscape Change in the Northern Tequila Region of Jalisco, Mexico. *Journal of Field Archaeology* 42(4): 298–311.
Heredia Espinoza, Verenice Y., and Joshua D. Englehardt
2015 Simbolismo Pan-Mesoamericano en la Iconografía Cerámica de la Tradición Teuchitlán. *Travaux et Recherches dans les Amériques du Centre* 68: 9–34.
Hirth, Kenneth G.
2016 *The Aztec Economic World: Merchants and Markets in Ancient Mesoamerica.* Cambridge University Press, New York.
Hirth, Kenneth G., and Joanne Pillsbury (editors)
2013 *Merchants, Markets, and Exchange in the Pre-Columbian World.* Dumbarton Oaks Research Library and Collections, Washington, DC.
Jiménez Betts, Peter F.
2017 *Orienting West Mexico: The Mesoamerican World System, 200–1200 CE.* PhD dissertation, Department of Historical Studies, University of Gothenburg, Sweden.
Knight, Vernon James, Jr.
1986 The Institutional Organization of Mississippian Religion. *American Antiquity* 51(4): 675–687.
Kowalewski, Stephen A.
In press Finding the Marketplaces in Prehispanic Mesoamerica: A Review. In *Weights and Market Places: The Phenomenology of Places of Exchange within a Diachronic and Multi-Cultural Perspective*, edited by Lorenz Rahmstorf and Edward Stratford. Wachholtz Murmann, Neumünster, Germany.
Renfrew, Colin
1970 The Tree-Ring Calibration of Radiocarbon: An Archaeological Evaluation. *Proceedings of the Prehistoric Society* 36: 280–311.
de Sahagún, Bernardino
1959 *General History of the Things of New Spain: Book 9: The Merchants*, edited by Arthur J. O. Anderson and Charles E. Dibble, Vol. 10. School of American Research, University of Utah Press, Salt Lake City.
Wallace, Anthony F. C.
1966 *Religion: An Anthropological View.* Random House, New York.

Contributors

Laura Almendros López is a native of Barcelona, Spain, and received her undergraduate degree in archaeology from the University of Barcelona. She has worked professionally in Mexico since the late 1990s on various archaeological projects in the Bajío region (Guanajuato and Querétaro), as well as in central-northern Mesoamerica (San Luis Potosí). Since 2004, she has been a research professor at the Centro INAH-Colima, specializing in the Formative period contexts of far western Pacific Mexico. She is currently pursuing her doctoral degree at the Autonomous University of Barcelona, where her research focuses on the chronological study of early groups in the western region of Mexico.

Christopher S. Beekman is associate professor of anthropology at the University of Colorado–Denver. His research focuses on sociopolitical organization in ancient western Mexico and that region's interaction with its neighbors. Dr. Beekman received his PhD from Vanderbilt University in 1996. He has held fellowships at Dumbarton Oaks and the Sainsbury Research Unit at the University of East Anglia, and has been an invited professor at the Institut d'Art et Archéologie, Université Paris I–Panthéon Sorbonne. He is a coauthor on the first volume of the *Historia de Jalisco*, edited by José M. Muriá. He has coedited several books, including *Nonlinear Models for Archaeology and Anthropology* (with William Baden), *La Tradición Teuchitlán* (with Phil Weigand and Rodrigo Esparza), and *Shaft Tombs and Figures in West Mexican Society: A Reassessment* (with Robert Pickering).

Mijaely Castañón received her undergraduate degree in archaeology from the Escuela Nacional de Antropología e Historia (ENAH), and earned a master's degree in anthropology at the Universidad Nacional Autónoma de México. She specializes in materials analysis and archaeometry. She is currently a professor at ENAH and collaborates on the Proyecto Arqueología y Paisaje del Área Centro Sur de Michoacán, under the direction of Dr. José Luis Punzo.

Fabio Germán Cupul-Magaña has been a research professor at the Centro Universitario de la Costa, University of Guadalajara, in Puerto Vallarta, Jalisco, since 1992. He obtained his PhD (in sustainable development) at the University of Guadalajara. During the past seven years he has studied, in collaboration with Dr. Joseph B. Mountjoy and Rafael García de Quevedo Machain, faunal and osteological remains from archeological sites in west Jalisco, Mexico.

Manuel Dueñas García is a Mexican archaeologist with bachelor's and master of arts degrees in archaeology and Latin American studies from the Autonomous University of San Luis Potosí. Currently, he is a graduate student in the Interdisciplinary Humanities program at the University of California–Merced. His primary research foci include the archaeology of northern Mesoamerica, violence in prehistory, World Systems Theory, and 3D records of archaeological materials.

Joshua D. Englehardt (PhD in anthropology, Florida State University, 2011) is Profesor-Investigador at the Centro de Estudios Arqueológicos of El Colegio de Michoacán, Mexico, and a Level I National Investigator of the Mexican National Council on Science and Technology (CONACYT). He specializes in Mesoamerican archaeology and epigraphy, with a research focus on the development of Mesoamerican writing systems in the Formative period. He is codirector of the Mesoamerican Corpus of Formative Period Art and Writing, an NEH-funded interdisciplinary project that is developing an online database and digital tools for the investigation and presentation of early Mesoamerican visual cultures. He is also currently a member of a multidisciplinary team currently exploring the cultural uses and significance of cycads as part of the CONACYT-funded project *Cícadas y la Domesticación de Maíz en el Paisaje Mesoamericano*. Recent publications include *Archaeological Paleography*, the edited volumes *Agency in Ancient Writing* and *Interregional Interaction in Ancient Mesoamerica*, as well as over 20 articles and book chapters that stem from his fieldwork throughout Mexico and 20 years of living, working, or studying in the Global South.

Rafael García de Quevedo-Machain has been a research professor at the Centro Universitario de la Costa, University of Guadalajara, in Puerto Vallarta, Jalisco, since 1990. He holds an MSc in aquatic and marine livestock sciences from the University of Colima. During the past seven years he has studied, in collaboration with Dr. Joseph B. Mountjoy and Dr. Fabio Germán Cupul-Magaña, the remains of mollusks, crustaceans, and bone fish at archaeological sites in west Jalisco, Mexico.

Verenice Y. Heredia Espinoza received a PhD in anthropological archaeology from Purdue University in 2005. She is professor at El Colegio de Michoacán in the Centro de Estudios Arqueológicos, and served as chair of her department from 2010 to 2014. Her primary interests include social complexity, cross-cultural analysis, archaeological method and theory, economic anthropology, political economy, and regional analysis. Dr. Heredia has worked in several areas of Mesoamerica, including Oaxaca, Tlaxcala, and Jalisco. Her publications include edited books, articles, and book chapters that deal mainly with alternative pathways to complexity. She currently directs the Teuchitlán Archaeological Project at the Los Guachimontones site in Teuchitlán, Jalisco, Mexico.

Erika Ibarra is an archaeologist with wide-ranging interests. She has worked on several archaeological projects in Durango (PIACOD and PROCUMA, in collaboration with José Luis Punzo, 2007–2010), Oaxaca (Monumental Center of Atzompa Archaeological Project, 2011–2012), and Santa Ana Tlacotenco and Milpa Alta (2012), where she participated in excavations of Pleistocene megafauna. She has also carried out excavations and research within the Migdal archaeological project in Magdala, Israel (2011–2012). Most recently, she worked as field coordinator for the archaeological project at the Chigüero Dam site, in the municipality of Huetamo, Michoacán. She has taught several courses at the National School of Anthropology and History and currently serves as the coordinator of the Laboratorio de Materiales Arqueológicos Provenientes de Rescates Arqueológicos in Bogotá, Colombia.

Stephen A. Kowalewski received his PhD in anthropology from the University of Arizona in 1976. He has taught at Lehman College and Hunter College, CUNY, in New York City, and since 1978 at the University of Georgia, where he is professor emeritus. He has done archaeological fieldwork in Arizona and Georgia and carried out systematic settlement pattern surveys covering the Valles Centrales, Peñoles, central Mixteca Alta, and Coixtlahuaca valley regions in Oaxaca, Mexico.

Martha Lorenza López Mestas Camberos received her undergraduate degree in archaeology from the Universidad Autónoma de Guadalajara, a master's degree in regional history from the Universidad de Colima, and a doctoral degree in social sciences, with a specialization in social anthropology, from the Centro de Investigaciones y Estudios Superiores en Antropología Social. Her research focuses on the archaeology of western Mexico.

Michael Mathiowetz earned a PhD from the University of California–Riverside in 2011. Over the past two decades he has participated in a number of archaeological projects in West and northwest Mexico and the southwestern United States. His work focuses on the history and legacy of pre-Hispanic cultural connections between societies in Mesoamerica and the U.S. Southwest.

Joseph B. Mountjoy received a BA degree in anthropology from the University of Illinois in 1963 and earned a PhD in anthropology, with a specialization in archaeology, from Southern Illinois University in 1970. His dissertation centered on the archaeology of the municipality of San Blas, Nayarit, focusing on the local pre-Hispanic cultural sequence and coastal contacts. Subsequently, he undertook investigations at Cholula, Puebla, before returning to West Mexico with projects in the Jalisco municipalities of Teuchitlán, Tomatlán, Puerto Vallarta, San Sebastián del Oeste, Mascota, Purificación, Ayutla, and Cabo Corrientes. His research interests include cultural ecology, rock art, metallurgy, ceramics, stone sculptures, stelas, shaft and chamber tomb use, coastal contacts with Central and South America, the relationship between ethnographic and archaeological data, the Middle Formative colonization of far western Mexico, and the southward coastal expansion of the Aztatlán culture. Dr. Mountjoy is currently a research professor at the University of Guadalajara Coastal Center in Puerto Vallarta, Mexico.

David Arturo Muñiz García earned his BA in archaeology at the Escuela Nacional de Antropología e Historia, where he has also taught courses in archaeological survey and colonial Mexican history. He also received a BA in history at the Universidad Nacional Autónoma de México, and has an MA in archaeology from El Colegio de Michoacán. He has worked on a number of archeological projects, as well as historic and museum exhibits, in Mexico City, the State of Mexico, Michoacán, Jalisco, and Durango. His main research interests focus on spatial distribution and the construction of knowledge in Pre-Hispanic and Contact periods, topics on which he has published articles and book chapters. He currently works as a professor-researcher in the History Department at the Universidad Autónoma de Ciudad Juarez in Chihuahua, Mexico, where he teaches courses on geography and the archaeology of ancient and colonial period Mexico.

M. Nicolás Caretta studied archaeology at the National School of Anthropology and History in Mexico City. Later he carried out his doctoral studies in the University of Leiden, The Netherlands. He has participated as Head Archaeologist or Collaborator in several archaeological projects in Mexico and Europe. At the Universidad Autónoma de San Luis Potosí, Dr. Nicolás Caretta was the Founding Member and Head of the archaeology program and a faculty member of the Faculty of Social Science and Humanities. He is director of the Santiago Archaeological Project in Aguascalientes, Mexico, Coresearcher of the Vasagård Archaeological Project, Denmark, and coordinator of the ArchaeoBalt–Laying fixed foundations for innovative Archeotourism, a new "green" Archaeoroute in the Southern Baltic Sea Region, a project funded by the Interregional South

Baltic and European Unions. He is currently a researcher and director of the International Office at the Bornholm Archaeological Research Center of the Bornholm Museum in Denmark.

José Luis Punzo Díaz obtained his doctoral degree in archaeology at the Escuela Nacional de Antropología e Historia in Mexico City. Since 2004, he has been a Researcher with the the Instituto Nacional de Antropología e Historia and is currently a Member of the National Research System (SNI) level 1. Previously, he served as director of the Museo de las Culturas del Norte in Paquimé, Chihuahua. His principal research interests are the relations between the U.S. Southwest, Northern Mexico, and Mesoamerica, archeometallurgy, the Chalchihuites culture of northern Mexico, and the archaeology of Michoacán. His extensive archaeological fieldwork has focused on Durango and Michoacán, and he has been director of the Ferrería archeological zone in Durango, as well as Tzintzuntzan and Tingambato in Michoacán. He has published more than 30 archeological research papers and five books.

Diego Rangel received his undergraduate degree in archaeology from the Escuela Nacional de Antropología e Historia. He has worked for the Instituto Nacional de Antropología e Historia on a variety of projects, focusing on a range of issues that include prehistoric megafauna, cliff dwellings in the Sierra de Durango, rock art in Guanajuato and Querétaro, and archaeological salvage projects in the Tierra Caliente region of Michoacán. He has also collaborated on cultural projects in Mexico City in conjunction with UNESCO, as well as museum displays and exhibitions for the Ministry of Culture.

Kimberly Sumano Ortega was born and raised in Mexico. She earned a BA in history from the Universidad Nacional Autónoma de México and an MA in archaeology from El Colegio de Michoacán. Kimberly is currently studying her PhD in borderlands history at the University of Texas at El Paso. Her research focuses on both space and discourse analysis at the intersection of archaeology and history. She has published articles related to the contact period in Central México, rock art from Michoacán, and spatial analysis in Jalisco and Durango. Her current research focuses on the phenomenological analysis of indigenous and Spanish perceptions of the Camino Real de Tierra Adentro, a space discursively created during the sixteenth century and experienced in many ways on a daily basis.

Jesús Zarco received his BA in archaeology form the Escuela Nacional de Antropología e Historia (ENAH). He has worked on various archaeological projects, principally in the states of Michoacán and Baja California. He has also participated in excavations at Teotihuacan and the Templo Mayor in Mexico City. His research focuses primarily on the development and evolution of lithic industries in these contexts. Currently he is a professor in the undergraduate program in archaeology at ENAH. He is also a member of the graduate program in anthropology at the Universidad Nacional Autónoma de México, where he studies the hunter-gatherer groups of northern Mexico.

Index

www.ingramcontent.com/pod-product-compliance
Lightning Source LLC
Chambersburg PA
CBHW071945260626
47231CB00052B/180